Milestones in Drug Therapy

More information about this series at
http://www.springer.com/series/4991

Antonio Guglietta
Editor

Drug Treatment of Sleep Disorders

Springer

Editor
Antonio Guglietta
Independent Pharmaceutical Consultant
Barcelona
Spain

ISSN 2296-6056 ISSN 2296-6064 (electronic)
ISBN 978-3-319-11513-9 ISBN 978-3-319-11514-6 (eBook)
DOI 10.1007/978-3-319-11514-6
Springer Cham Heidelberg New York Dordrecht London

Library of Congress Control Number: 2014957395

Printed on acid-free paper

Springer is part of Springer Science+Business Media (www.springer.com)

Preface

Sleep disorders are very common in the modern society. In the United States it is calculated that between 50 and 70 million of people suffer from some kind of sleep disorder while it is estimated that in the developing countries across the world these conditions affect approx. 150 million people. Disorders of the sleep are on the rise worldwide and affect people of all ages, gender, and ethnicity. An inadequate sleep poor in quantity and quality and an excessive daytime sleepiness negatively affect daily activities causing, for example, poor concentration, memory difficulties, and impaired driving ability.

Given the magnitude and impact on the society, it is not surprising that there is an increasing interest by governments, universities, and media on sleep and sleep-related disorders. Several scientific societies such as The American Academy of Sleep Medicine, The National Sleep Foundation, and the Sleep Research Society have been established and disorders of sleep are now recognized as a separate medical subspecialty with specific training courses being offered in medical schools worldwide to prepare doctors to properly diagnose and treat these disorders.

The past decade has witnessed major advances in the understanding of sleep physiology and pharmacology which have provided a better understanding of the mechanisms that underlie sleep and have prompted promising research in this field which in turn have led to the development of new and better drugs to treat these conditions. The FDA has recently approved new drugs to treat disorders of sleep and other molecules are in advanced phase of clinical development or have just completed the development process. Furthermore, new regulatory and clinical guidelines for the development of drugs and treatment of these conditions have been issued or are in preparation.

The idea behind this book is to review some of the recent major breakthroughs in the drug treatment of sleep disorders. The drugs reviewed in the book, whether recently approved drugs (i.e., Doxepin), variations of previously approved molecules (i.e., Zolpidem sublingual preparation), or new chemical entities in late stage of clinical development (i.e., Lorediplon), have significantly changed or are expected to change the drug treatment of these disorders. Each chapter of the book was written by an expert in the field and is structured in such way that can be read as stand-alone chapter or as part of the whole book. The final result is a comprehensive yet practical book that will bring all the scientists, clinicians, and drug developers up to date in this area[1].

Barcelona, Spain
August 2014

Antonio Guglietta

[1] After this book went to press, on Aug 13th, 2014, the US Food and Drug Administration approved Suvorexant tablets (Belsomra) to treat difficulty in falling and staying asleep (insomnia). Suvorexant is the first approved drug of the orexin antagonists class.

Contents

Part I General Concepts

An Overview of Sleep Physiology and Sleep Regulation 3
Chiara Berteotti, Matteo Cerri, Marco Luppi, Alessandro Silvani, and Roberto Amici

Neuronal Networks Regulating Sleep and Arousal: Effect of Drugs 25
Elemer Szabadi

Classification of Sleep Disorders 71
Michael Thorpy

Development of New Therapeutical Agents for Treatment of Insomnia and Other Sleeps Disorders 85
Gloria Martin and Antonio Guglietta

Part II Drugs to Treat Insomnia

Low-Dose Doxepin for Insomnia 111
Dimitri Markov and Karl Doghramji

Lorediplon: A New GABA$_A$ Modulator Drug for Treatment of Insomnia 121
Fabiana d' Aniello, Benjamin Santos, and Antonio Guglietta

Zolpidem Sublingual Formulations 147
Luc Staner

Eszopiclone: Review and Clinical Applications in Chronic and Comorbid Insomnia 159
Jadwiga S. Najib

Orexin/Hypocretin Antagonists in Insomnia: From Bench to Clinic 191
Alberto K. De la Herrán-Arita, Ana C. Equihua-Benítez, and René Drucker-Colín

Part III Drugs to Treat Excessive Daytime Sleepiness

Modafinil and Armodafinil 211
Karel Šonka, Peter Šóš, and Marek Susta

Sodium Oxybate (Xyrem®): A New and Effective Treatment for Narcolepsy with Cataplexy 231
Ahmed S. BaHammam, David Nelson Neubauer, and Seithikurippu R. Pandi-Perumal

Cell Therapy to Treat Narcolepsy 249
Oscar Arias-Carrión

Part IV Drugs to Treat Circadian Sleep Disorders

Tasimelteon 261
David N. Neubauer, Ahmed S. BaHammam, and Seithikurppu R. Pandi-Perumal

Part V Conclusive Remarks

The Past, Present, and Future of Drug Development and Treatment in Sleep Disorders 273
Antonio Guglietta

Index 289

Part I
General Concepts

An Overview of Sleep Physiology and Sleep Regulation

Chiara Berteotti, Matteo Cerri, Marco Luppi, Alessandro Silvani, and Roberto Amici

Abstract Sleep is a complex behavior that cyclically alternates with wakefulness and is fundamental for both mental and physical wellness. During normal sleep the body acquires a specific posture; any active behavior is absent and mental activity typically oscillates between a state of "loss of consciousness" and the experience of dreaming.

Sleep is, by itself, a cyclical process characterized by the alternation of two phases, non-REM (NREM) sleep and "rapid eye movement" (REM) sleep, during which the typical synchronized electroencephalographic (EEG) pattern of NREM sleep is substituted by a "paradoxical" EEG desynchronization.

NREM sleep is a state of minimal energy expenditure and motor activity, during which cardiorespiratory and thermoregulatory variables are driven by the autonomic nervous system at a lower and more stable level compared to wakefulness. During REM sleep, beyond the occurrence of REMs, posture control is lost and dreams become more vivid. Furthermore, autonomic activity is highly unstable leading to centrally driven surges in heart rate and blood pressure, breathing becomes irregular, and thermoregulation is suspended or depressed, suggesting a derangement of the integrative regulation of physiological functions.

Overall sleep quality is one of the most important parameters considered for the subjective assessment of quality of life and general health status. However, although sleep function(s) is intuitively associated with some kind of rest that seems to be mostly required by the central nervous system, and many theories on sleep function have been proposed, a full comprehension of sleep function has not yet been achieved and is probably not imminent.

C. Berteotti • M. Cerri • M. Luppi • A. Silvani • R. Amici (✉)
Department of Biomedical and Neuromotor Sciences—Physiology, Alma Mater Studiorum-University of Bologna, Piazza P.ta S. Donato, 2, 40126 Bologna, Italy
e-mail: chiara.berteotti@unibo.it; matteo.cerri@unibo.it; marco.luppi@unibo.it; alessandro.silvani3@unibo.it; roberto.amici@unibo.it; http://www.unibo.it/Faculty/default.aspx?UPN=roberto.amici%40unibo.it

A. Guglietta (ed.), *Drug Treatment of Sleep Disorders*, Milestones in Drug Therapy,
DOI 10.1007/978-3-319-11514-6_1

1 Neurophysiology and Neurobiology of Sleep

1.1 What Is Sleep?

Sleep is a specific behavior, common to humans and other animals, that cyclically alternates with wakefulness and is undoubtedly fundamental for both mental and physical wellness. Everyone is aware of the importance of having a good sleep and a proper daily sleep time. Sleep quality is among the most important parameters considered for the subjective assessment of quality of life and general health status.

During sleep the body needs to acquire a typical posture; any active behavior is absent and, following a normal sleep time, there is a clear perception of bodily restoration as well as that of the mind. In fact, the strong "pressure" to fall asleep after either a long period of waking or a very tiring activity is a very common subjective experience. The mental activity during sleep is also quite peculiar and after the subjective experience of the "loss of consciousness" when falling asleep, the experience of dreaming is characteristic: a mental phenomenon that, historically, represents one of the main topics in cultures and arts.

Falling asleep is a kind of progressive reduction of the awareness of self and environment; hence, in the past, sleep was considered to be a mere passive process, in which brain activity was reduced, and even absent, compared to wakefulness (Dement 2011). Some clinical observations by the neurologist Constantin von Economo in 1916, as well as experimental data obtained in the 1930s by Frédéric Bremer with his milestone experimental preparations on cats named "cerveau isolé" and "encéphale isolé", apparently supported this view. Von Economo observed that patients affected by "encephalitis lethargica" typically presented lesions localized at the boundary between the mesencephalon and the diencephalon, while Bremer, on the basis of his experimental results, postulated that sleep was the result of the reduction of afferent sensory stimuli to the forebrain.

In 1929, Hans Berger introduced electroencephalography (EEG) as a tool to record the cortical electrical activity from the scalp. The EEG was an amazing technical advancement in neuroscience and soon it was considered of primary importance for the investigation of the sleeping brain. In support of Bremer's theory, Giuseppe Moruzzi and Horace Magoun found, in the late 1940s, that the stimulation of the pontine reticular formation induced a waking EEG in the anesthetized cat. Therefore, they suggested that the "deafferentation" of the sensory signals to the telencephalon was reasonably due to a transient change in the functional activity of the reticular formation (Kumar 2010).

Although the "passive" theory of sleep was the most widely accepted, some experimental results suggested that sleep was actively induced by some neural structure. Some of the patients studied by Von Economo presented an insomnia that was associated with a lesion of the anterior hypothalamus. Moreover, in the 1940s, Rudolph Hess showed that a low frequency stimulation of the thalamus could induce sleep in the cat, while Walle Nauta proposed, on the basis of brain-lesion studies in the rat, that the rostral part of the hypothalamus was the site for the "capacity of sleep" (Kumar 2010).

However, the definitive turning point leading to the concept that sleep is a state that is actively promoted by the brain was the discovery of an "active" sleep phase; it was in 1953 that Eugene Aserinsky and Nathaniel Kleitman described the existence of a sleep phase which cyclically interrupted the typical synchronized EEG pattern of sleep. During this phase it was possible to observe frequent eye movements under the closed eyelids and the EEG signal was desynchronized and very similar to that observed in wakefulness.

It took several years before this functional state was recognized as a distinct sleep phase. It was Michel Jouvet who, in 1959, defined this particular state as "paradoxical sleep" (Kumar 2010). Soon, sleep would be divided into two main phases: the "rapid eye movement" (REM) phase and the nonREM (NREM) phase (Dement 2011). Also, William Dement, while working with Kleitman, associated REM sleep with dream occurrence (Dement 2011). REM sleep was soon considered a very peculiar sleep phase, due to the paradoxical EEG signal and the occurrence of REMs, as well as several other physiological features, such as the paralysis of anti-gravitational muscles, the occurrence of ponto-geniculo-occipital (PGO) waves, and, as firstly assessed by Pier Luigi Parmeggiani, a dramatic functional change in physiological regulation, with a large autonomic instability.

Therefore, sleep is now considered to be a complex and active behavioral state, during which the brain works differently from normal waking. Sleep cyclically alternates with wakefulness every single day and is, by itself, a cyclical process characterized by the alternation of two phases, NREM sleep and REM sleep.

1.2 Sleep Architecture

In every species studied, sleep propensity is closely related to the time of day and consequently to the circadian rest-activity cycle, and in humans, the majority of sleep time is concomitant with the nocturnal hours (Achermann and Borbély 2011).

During NREM sleep, the EEG becomes typically synchronized, i.e., rich in slow-frequency and high-voltage waves, in contrast to the fast-frequency and low-voltage waves of the typical waking EEG (Brown et al. 2012). In humans the structure of NREM sleep is further subdivided into four stages (S1–S4, Rechtschaffen and Kales classification, R&K) or three stages (N1–N3, American Academy of Sleep Medicine classification, AASM) (Carskadon and Dement 2011) based on the degree (intensity) of the EEG synchronization. During sleep occurrence, the NREM sleep stages and REM sleep appear in a precise order:

1. NREM sleep S1 (N1) represents the transition from wakefulness to sleep and is considered to be the first step of EEG synchronization; theta (4.0–7.0 Hz) activity is predominant.

2. The beginning of S2 (N2) is characterized by the appearance in the EEG of two typical signs, the "K-complex" (a single high voltage negative wave, followed by a single slow positive wave) and the "spindle" (a 12–14 Hz oscillation with an increasing and then a decreasing amplitude, lasting around 1–2 s).
3. The deeper stages of NREM sleep are S3 and S4 of the R&K classification, which essentially constitute together the N3 stage of the AASM classification. In these stages, the EEG is rich in (S3) or fully dominated by (S4) high-amplitude activity in the slow delta range (0.5–2.0 Hz). This slow-wave activity (SWA) is considered to be the macroscopic result of a cellular phenomenon, called "slow-oscillation," during which cortical neurons oscillate between two functional states, the hyperpolarized and silent "down-state" and the high-frequency (around 40 Hz) discharging "up-state" (Steriade et al. 2001). Actually, the "slow-oscillation" is a travelling wave, involving almost all the cortical areas, apparently starting from the frontal cortex and then propagating towards the posterior regions (Massimini et al. 2004).
4. During a normal NREM-REM sleep cycle, S4 is followed by a reverse progression, through S3, to light NREM sleep stages (S2–S1), until REM sleep occurs. REM sleep is followed by either an awakening or by S1–S2. In the latter case, the whole sequence to S3–S4 and back to S2–S1 and REM sleep usually starts again.

Typically, in humans, this cyclic process lasts about 90–100 min (Carskadon and Dement 2011) and during a standard night-time sleep period 4–5 cycles are completed. The deepest NREM sleep stage (S4) is usually reached only during the first part of the night. Also, late cycles are generally characterized by the occurrence of longer REM sleep phases, lasting up to 20–25 min in the very last cycle before waking (Carskadon and Dement 2011). Normally, morning awakening directly follows a completed REM sleep phase.

In the young adult the overall night sleep duration is between 7 and 8 h, and REM sleep represents 20–25 % of the total sleep amount. The daily amount of sleep and the quantity of SWA progressively decrease from childhood to old age, while the proportion of REM sleep, which is apparently almost double in the newborn compared to the young child, is then maintained at a stable level throughout the life span.

1.3 Sleep Mechanisms

In recent years, great steps forward have been made in the understanding of the brain mechanisms underlying sleep and wakefulness control and regulation (cf. Brown et al. 2012). Schematically, the cyclical alternation of wakefulness and sleep is considered to be mainly due to the cyclical activation and inhibition, respectively, of a key neural structure defined as the ascending reticular activating system. This system consists of a network of fibers projecting from the brainstem to the forebrain, via two main pathways: the dorsal and the ventral pathways. The dorsal pathway originates from the cholinergic and glutamatergic neurons within the pontine and midbrain reticular formation and within the cholinergic laterodorsal and pedunculopontine tegmental nuclei. This pathway projects to the intralaminar and midline thalamic nuclei and, consequently, widely innervates the cerebral cortex. The ventral pathway originates from the same ponto-mesencefalic structures of the dorsal pathway, but, in this case, projections are directed towards the lateral hypothalamus and the tuberomamillary nucleus, as well as to the basal forebrain. The basal forebrain also receives afferent projections from both the orexinergic neurons in the lateral hypothalamus and the histaminergic neurons in the tuberomamillary nucleus, and widely projects to the cerebral cortex. Both pathways receive important contributions from the noradrenergic locus coeruleus and the serotonergic dorsal raphe (Brown et al. 2012). Sleep neurochemistry will be addressed in more detail in Chap. "Neuronal Networks Regulating Sleep and Arousal: Effect of Drugs."

In this conceptual framework, NREM sleep is apparently induced by the inhibition of the wake-promoting neurons of the ascending reticular activating system. The mechanism that triggers this inhibitory activity is not completely known. It is possible that the inhibition of the ascending reticular activating system mainly begins with the increased GABAergic activity of the median preoptic nucleus; then, once started, sleep is maintained by the subsequently increased activity of other GABAergic neurons, in particular those located within the ventrolateral preoptic nucleus (Brown et al. 2012). At a cellular level, the increasing inhibitory activity gradually leads to the hyperpolarization of the thalamic and cortical neurons: this is the functional condition of the neural membrane that is conducive to rhythmic bursting, as described by Mircea Steriade (Steriade et al. 2001).

As stated before, NREM sleep episodes are cyclically interrupted by REM sleep. The neurophysiology of REM sleep is considered to be based on the reciprocal interaction between two main systems: the "REM-on" and the "REM-off" neurons (Saper et al. 2005). In particular, the cholinergic "REM-on" system of the laterodorsal tegmentum and pedunculo-pontine tegmentum excites the glutamatergic neurons in the sublaterodorsal nucleus, promoting REM sleep onset. These glutamatergic neurons also excite GABAergic neurons of the dorsal paragigantocellular reticular nucleus, that, in turn, inhibit adjacent REM-off neurons of the locus coeruleus (noradrenergic) and the dorsal raphe (serotonergic). In support of this mechanism is another one, based on the inhibition of the "REM-off" GABAergic activity of the ventrolateral periaqueductal gray and of the deep mesencephalic reticular nucleus; also in this case, the disinhibition of these structures activates other GABAergic "REM-on" neurons leading to the inhibition of the locus coeruleus and the dorsal raphe. In these structures, REM-off (i.e., noradrenergic and serotonergic) neurons normally inhibit the REM-on neurons in the laterodorsal tegmentum and the pedunculo-pontine tegmentum, and they can also inhibit themselves in support of the activation of other afferent inhibitory projections, thus promoting REM sleep onset (Brown et al. 2012; Luppi et al. 2013).

During REM sleep, the activation of cortical activity is mainly due to two systems: (1) the glutamatergic projection from the sublaterodorsal nucleus to the intralaminar thalamic nuclei and (2) the rostral cholinergic system in the basal forebrain (Luppi et al. 2013). From the sublaterodorsal nucleus the glutamatergic REM-on neurons also project caudally and have a key role in the activation of GABAergic and glycinergic neurons located in the alpha and ventral gigantocellular and in the lateral paragigantocellular reticular nuclei, as well as in the raphe magnus, all projecting to the spinal cord; it has been proposed that these pathways are responsible for the induction of muscle atonia during REM sleep (Luppi et al. 2013).

Recently the lateral hypothalamus has gained major attention for its involvement in both REM sleep induction and regulation. Orexinergic neurons are peculiar to this area, and are functionally involved in the maintenance of the stability of wakefulness; the lack of these neurons is associated with narcoleptic/cataleptic symptoms. Within the lateral hypothalamus, intermingled with orexinergic neurons, are also melanin concentrating hormone (MCH) REM-on neurons and GABAergic nonorexinergic/non-MCH neurons, that seem to importantly contribute to the inhibition of the REM-off neurons within the ventrolateral periaqueductal gray and the histaminergic tuberomamillary nucleus (Luppi et al. 2013), promoting REM sleep.

The lateral hypothalamus may have a pivotal role in REM sleep regulation and the REM-on cells within this area may start to inhibit REM-off areas; subsequently, other REM-on areas may be facilitated in their activity and a REM sleep episode can start and be maintained. When these REM-on areas decrease their activity, the REM-off areas may start up their activity again, allowing the reactivation of the arousal system. In fact, REM sleep episodes are very frequently terminated by an arousal (Luppi et al. 2013).

2 Sleep Regulation

2.1 *Sleep Chronobiology*

In physiological conditions and in a natural environment, the distribution of sleep and wakefulness depends on the time of day, following a 24-h (nicthemeral) rhythm. In the absence of external time cues such as ambient light, this rhythm persists and becomes free running, i.e., its period deviates from its original 24 h value by 10–20 min. Synchronization (entrainment) of a near-24 h endogenous rhythm to environmental time cues is the hallmark of a circadian rhythm.

The circadian sleep rhythm is generated in the hypothalamic suprachiasmatic nucleus (SCN), which projects to the adjacent subparaventricular zone (SPZ), and thence to the dorsomedial nucleus of the hypothalamus (DMH). The DMH targets sleep-promoting neurons in the ventrolateral preoptic nucleus of the hypothalamus

(VLPO) and wake-promoting neurons in the lateral hypothalamus, such as orexinergic neurons (Chou et al. 2003). The DMH also projects to the paraventricular nucleus of the hypothalamus (PVH), which modulates the anterior hypophysis to drive the circadian rhythm of adrenal cortisol release. On the other hand, the circadian rhythms of melatonin and body temperature are mediated by neural pathways that originate upstream of the DMH. The SCN directly projects to the PVH, thereby modulating in sequence the sympathetic preganglionic neurons in the thoracic spinal cord, the sympathetic neurons in the superior cervical ganglion, and the pineal gland cells that release melatonin.

The body temperature rhythm is largely mediated by projections from the SPZ to the medial preoptic area, which is part of a thermoregulatory pathway involving the DMH and the medullary raphe pallidus. These parallel pathways, that irradiate from the SCN, ensure phase locking between the sleep rhythm and the circadian rhythms of disparate physiological variables such as cortisol, melatonin, and body temperature. These multisynaptic pathways are also flexible, allowing for differences in diurnal preference between species, the critical switch probably being located in the SPZ. Accordingly, the peak in SCN neural activity and the minimum value (nadir) of melatonin secretion invariably occur during the subjective light period, thus being associated with reduced sleep propensity in humans and with increased sleep propensity in nocturnal laboratory rodents. Both in diurnal humans and in nocturnal laboratory rodents, however, the maximal sleep propensity is associated with the nadir of body temperature, which physiologically occurs just before habitual wake time and is followed by the peak in cortisol secretion.

The circadian SCN rhythm is shifted in phase by environmental time cues (Zeitgeber) until its period achieves a precise 24-h value, as if the hands of the SCN clock were manually turned until synchronization with the environmental clock is achieved. Light is usually the most powerful Zeitgeber. The phase of the SCN circadian rhythm is delayed or anticipated by exposure to light early or late in the subjective night, respectively. Light increases the activity of SCN neurons by stimulating excitatory neurotransmitter release from terminals of the retinohypothalamic tract, which connects specialized retinal ganglion cells to the SCN. In addition to receiving information from light-sensitive rods and cones, these specialized ganglion cells express the photopigment melanopsin, which makes them intrinsically sensitive to light, particularly in the blue-green (450–500 nm) portion of the spectrum.

2.2 *Sleep Homeostasis*

The term homeostasis was coined at the beginning of the last century by the American physiologist Walter B. Cannon to name the peculiar steady states of living organisms, characterized by slight and controlled instability rather than by fixed and rigid constancy. Some 50 years later, the term was borrowed by the sleep research community to describe rebounds in sleep duration and intensity after sleep deprivation or reductions in sleep propensity after sleep episodes (Achermann and Borbély 2011). In this particular sense, the concept of homeostasis should be interpreted loosely, as it is still unclear which sleep-related physiological variable, if any, is kept at a steady state.

The amount of NREM sleep time gained during sleep rebound is usually lower than the loss occurred during previous sleep deprivation. However, spectral power of the EEG in the delta band also increases during rebound NREM sleep as a function of the duration of prior wakefulness, and these changes roughly parallel the increases in the arousal threshold. On these bases, it is thought that recovery of NREM sleep mainly occurs by increasing NREM sleep intensity or depth, as indexed by increased SWA. Accordingly, SWA during NREM sleep is the highest at the beginning of each sleep cycle and progressively decreases in the course of each NREM sleep episode and throughout the whole sleep period. Interestingly, NREM sleep homeostasis in terms of SWA manifests at both a local and a global brain scale. Thus, SWA during NREM sleep is higher in frontal than in posterior EEG derivations, especially at the beginning of the sleep period or after sleep deprivation, and brain regions experiencing increases in activity during wakefulness develop increases in SWA during subsequent NREM sleep. The mechanism underlying local and global NREM sleep homeostasis in terms of SWA may be that increases in cortical synaptic strength during wakefulness increase neuronal synchrony during subsequent NREM sleep, leading to more frequent and ample EEG slow waves. In turn, EEG slow waves may then contribute to downscale the strength of cortical synapses during NREM sleep (Vyazovskiy et al. 2011).

At variance with NREM sleep homeostasis, the homeostatic response to REM sleep loss is thought to occur mainly, if not solely, in terms of REM sleep amount. Selective REM sleep deprivation by cold exposure in rats is followed by an immediate and sustained rebound of REM sleep amount, which allows the animals to fully restore the previous REM sleep loss within 3 days of recovery. In humans, however, REM sleep rebound after total sleep deprivation typically occurs only after the SWA rebound in NREM sleep, if it does at all. In part, this discrepancy is because the homeostatic pressure to increase NREM sleep SWA conflicts with and prevails on REM sleep homeostasis. Moreover, allometric scaling of REM sleep rebound with body mass in different species suggests that small animals such as rats have a lower tolerance to REM sleep loss than large animals such as humans (Amici et al. 2008).

2.3 Interaction Between Circadian and Homeostatic Sleep Control

Sleep and wakefulness may be viewed as the key intermediate mechanisms, through which the circadian rhythm modulates physiological functions. In this respect, an important layer of sophistication and complexity is added by the simultaneity of circadian and homeostatic sleep control. As a result, in humans, sleep propensity is sustained by high homeostatic pressure because of previous wakefulness at the beginning of the night and by high circadian pressure at the end of the night. To some extent, circadian and homeostatic sleep control may be disentangled experimentally with the aid of appropriate study designs. Thus, continual sleep deprivation and constancy of posture, feeding, and environment in constant routine protocols allow circadian control to be singled out at least partially from the effects of sleep homeostasis. Forced desynchrony protocols force subjects to follow wake–sleep cycles with extreme periods (typically 20 h or 28 h) that make entrainment of the endogenous circadian rhythm impossible, causing sleep to occur at different circadian phases over several days. These experimental efforts have been combined with advanced mathematical modeling techniques, starting from the seminal two-process (i.e., circadian and homeostatic) model of sleep regulation in the 1980s (Achermann and Borbély 2011). Altogether, it has emerged that the interaction between circadian and homeostatic sleep control is often nonadditive and complex, and basic research has started to suggest potential mechanisms of this complexity. A simple example is that sleep is associated with eyelid closure, which decreases retinal light exposure, thus feeding back to the SCN via the retinohypothalamic tract. Furthermore, at a cellular level, sleep decreases the activity of SCN neurons, which in turn determine circadian sleep propensity via the SPZ-DMH pathway. Most importantly, circadian and homeostatic sleep controls interact at a genetic level, which will be addressed in the next section.

2.4 Genetics of Circadian and Homeostatic Sleep Control

The endogenous circadian rhythm in SCN neuron activity results from sophisticated interplay between transcription, translation, and posttranslational modifications involving a relatively small set of core clock genes. At the basis of this interplay lie two linked and delayed negative feedback loops (Feng and Lazar 2012). In the first loop, homo- and heterodimers formed by CRY proteins (coded by two Cryptochrome genes, Cry1-2) and PER proteins (coded by three Period genes, Per1-3) inhibit transactivation of their own genes, which is exerted by another heterodimer between the proteins CLOCK (coded by the Clock gene, Circadian Locomotor Output Cycle Kaput) and BMAL1 (coded by the Bmal1 gene, brain and muscle ARNT-like protein). This inhibition occurs after a delay resulting from Per and Cry gene transcription, translation into proteins, and protein migration to the nucleus. Continuous and regulated degradation of PER and CRY proteins eventually disinhibits CLOCK-BMAL1 transactivation of Per and Cry genes and starts a new cycle of the molecular clock. In the second loop, CLOCK-BMAL1 heterodimers transactivate the orphan nuclear receptor gene Rev-erbα, which codes for a protein that inhibits transcription of the Bmal1 and Cry genes, thus explaining the circadian fluctuations in their mRNA. Notably, the Clock gene transcript is not in itself rhythmical, whereas the nuclear vs. cytoplasmic localization of its protein product is.

These basic negative feedback loops are connected with other loops of gene transcription and translation and expressed ubiquitously in cells, eventually driving circadian expression of large fractions (3–20 %) of the transcriptome in different tissues (Feng and Lazar 2012). The SCN neurons collectively constitute the master circadian clock of the organism, which entrains the myriad molecular clocks in peripheral cells by means of neural, hormonal, and metabolic signals.

Recent and exciting data are unraveling links between circadian rhythms, sleep, metabolism, and the epigenome. Thus, the wake-sleep circadian rhythm causes feeding and activity cycles, which change energy balance and redox state of peripheral cells. This is reflected in changes in ATP/AMP and NADH/NAD^+ ratios. These ratios modulate the clock gene feedback loops either directly or indirectly through the enzymes AMP Kinase and Sirtuin (SIRT) 1, which is a NAD^+-dependent histone deacetylase. The CLOCK-BMAL1 heterodimers recruit other histone deacetylases as well as histone acyl-transferases. Histone acetylation remodels chromatin, leading to rhythmic epigenomic programming of gene expression that integrates information from circadian clocks, wake-sleep behavior, and energy balance (Feng and Lazar 2012).

The tight link between sleep and circadian rhythms at the genetic level is supported by the finding that mutations of core clock genes, such as Clock, Bmal1, Per, and Cry, cause not only impairments in the circadian sleep rhythm but also important alterations in sleep duration and architecture. Recent gene expression studies suggest that the Per2 gene plays a prominent role in the cross talk between circadian rhythms and sleep homeostasis. Sleep homeostasis itself is under strong genetic control. In particular, data on mice point to the short splice variant of the Homer1 gene, which codes for a protein involved in glutamatergic signaling and synaptic plasticity, as a key regulator of the rate of accumulation of NREM sleep SWA during extended wakefulness (Maret et al. 2007).

Studies performed on monozygotic and dizygotic human twins have revealed a striking heritability of some sleep traits, which approaches 100 % for REM density and EEG spectral power during NREM sleep at 8–16 Hz (De Gennaro et al. 2008). Linkage analysis in rodents has identified key genes involved in EEG delta power during NREM sleep (*Rarb1*, retinoid acid receptor beta) and EEG theta frequency in REM sleep (*Acads*, short-chain acylcoenzyme A dehydrogenase). On the other hand, genome-wide association studies (GWAS) in humans are progressively uncovering the loci that increase susceptibility to sleep disorders, such as narcolepsy and restless leg syndrome (Sehgal and Mignot 2011). However, the genetic landscape of physiological sleep traits in humans remains largely uncharted by these studies because of issues related to replicability and statistical power. Finally, microarray studies performed on rats have demonstrated that the link between sleep and gene transcription is bidirectional. In the cerebral cortex, in particular, sleep upregulates transcription of several genes coding for proteins involved in depression of synaptic plasticity, membrane trafficking, GABAergic neurotransmission, and membrane hyperpolarization. Conversely, wakefulness upregulates gene transcripts related to acquisition of synaptic plasticity, energy metabolism, stress and unfolded protein responses, and glutamatergic neurotransmission (Cirelli et al. 2004).

3 Physiological Functions During Sleep

3.1 A Physiological View of Sleep

The three main wake-sleep states, i.e., wakefulness, NREM sleep, and REM sleep are usually recognized on the basis of EEG rhythms and levels of muscle tone, and the presence of eye movements associated with PGO waves. However, the physiological definition and understanding of these states cannot disregard the assessment of the respiratory, cardiovascular, and metabolic parameters, which are under the integrated control of the autonomic and the endocrine system (Parmeggiani 2005). This definition surpasses the standard one, largely based on the level of brain cortical and somatomotor activity. The hypothalamus plays a key role in this complex integrative activity (Thompson and Swanson 2003), which is critical for the maintenance of body homeostasis, for body survival (fight or flight response), and for reproduction. This activity leads to the shaping of bodily functions, largely on the basis of external and internal sensory information, in accordance with the physiological meaning and goals of the different behavioral states.

During NREM sleep, physiological regulation clearly operates favoring the maintenance of body homeostasis. NREM sleep is a state of minimal energy expenditure and motor activity, during which cardiovascular, respiratory, and thermoregulatory variables are driven by the autonomic nervous system (ANS) at a lower level compared to wakefulness and are kept stable by the autonomic reflexes.

On the contrary, during REM sleep posture control is lost, ANS is highly unstable, centrally driven surges in heart rate and blood pressure occur, breathing becomes irregular, and thermoregulation is suspended or depressed. The integrative function of the hypothalamus becomes imbalanced and this modality of physiological regulation has been described as “poikilostatic,” ποικιλο meaning “diverse” in Greek, (Parmeggiani 2005), because it is not apparently aimed at the maintenance of the broad stability (homeostasis) of the physicochemical properties of the extracellular compartments that underlie cellular survival.

3.2 Sensory and Motor Functions

The transmission of sensory information to the central nervous system (CNS) is attenuated during sleep. Brain processing of sensory information continues across the different sleep stages, but a thalamic gating system operates to modulate the access of sensory information to central nervous areas, even if to a different extent, during both NREM sleep and REM sleep, favoring sleep continuity (Peever and Sessler 2011). It follows that a stimulus can interrupt sleep only if it is strong enough, or coincident with sufficient levels of arousal to allow a full cortical processing of the information. Somatosensory processing, including that related to pain, is also reduced during sleep (Peever and Sessler 2011).

Somatomotor control is largely influenced by the different wake-sleep states (Chase 2013): muscle activity is high during active wakefulness, progressively decreases during quiet wakefulness and NREM sleep, and disappears during tonic REM sleep, when muscle atonia occurs due to a deep inhibition of spinal and brainstem motoneurons. During phasic REM sleep, atonia is interrupted by brief twitches and jerks of limb and eye muscles (REMs). The REM sleep pattern of somatomotor control affects all somatic muscles with the exception of pure respiratory muscles, such as the diaphragm, which is spared by REM sleep motor inhibition, and the middle ear musculature, which is activated in order to depress auditory inputs. While the progressive reduction of muscle tone from wakefulness to NREM sleep represents a continuum within a common mechanism, REM sleep atonia depends on a change to a different central control mechanism.

The degree of muscle activity during the different wake-sleep states apparently depends on changes in the balance between excitatory (glutamatergic/monoaminergic) and inhibitory (GABAergic/glycinergic) inputs to the motoneurons. As shown by experiments on intracellular recording at the level of trigeminal and spinal motoneurons, REM sleep atonia is due to a very large inhibitory input to motoneurons that overcomes any possible excitatory drive. However, during phasic REM sleep, this inhibition, which is even larger than that occurring during tonic REM sleep, can be overcome by intermittent excitatory inputs leading to the generation of twitches. The inhibitory drive to motoneurons during both tonic and phasic REM sleep seems to be glycinergic (Chase 2013).

3.3 Respiratory Functions

During sleep, the regular breathing pattern of quiet wakefulness is replaced by a large respiratory variability. This is the result of the loss of voluntary control of respiratory muscles and changes in both ventilatory control and resistance of the upper airways.

At sleep onset, ventilation regularly decreases, and respiratory instability may emerge, characterized by increases and decreases in breathing amplitude; these changes may be either aperiodic or regular and periodic (Krieger 2005). Periodic breathing has been described in 40–70 % of normal subjects; its frequency increases with age and it can comprise brief apneas (central or obstructive). At sleep onset, when sleep is not stabilized, changes in the amplitude of breathing parallel changes in the level of vigilance, with an increase in ventilation occurring during spontaneous brief arousals. Periodic breathing disappears when stable deep sleep is settled.

During stable NREM sleep, ventilation is very regular. In humans, minute ventilation and tidal volume progressively drop from wakefulness to NREM sleep stages S3 and S4. Data regarding changes in minute ventilation, tidal volume, and respiratory frequency on passing from NREM sleep to REM sleep do not follow any evident trend nor show any significant difference. During REM sleep, breathing becomes irregular, displaying phasic changes in respiratory amplitude and frequency that follow the occurrence of REMs; respiratory irregularities may comprise central apneas (Krieger 2005). Minute ventilation decreases compared to wakefulness, and it is lower in phasic REM sleep than in tonic REM sleep (Douglas 2005).

During wakefulness upper airway muscle activity is maintained by reflex-activation of dilator muscles, but at sleep onset and, principally, during REM sleep, this reflex activity is reduced. Thus, even in normal subjects, upper airway resistance increases during sleep, producing an increment in inspiratory resistance. During sleep, the cough response is suppressed, occurring only after arousal. During REM sleep, hypotonia or atonia of the upper airway muscles are accompanied by intercostal muscle atonia, greatly reducing costal breathing (Chase 2013).

3.4 Endocrine Functions

The circadian system and sleep occurrence largely influence the daily profile of the basal secretion of the principal components of the endocrine system (Van Cauter and Tasali 2011). However, 24-h sleep deprivation studies or experiments in which sleep occurrence has been shifted to an unfavorable circadian phase have highlighted that the weight of the circadian and the wake-sleep ultradian modulation on hormone secretion varies for different hormones (Van Cauter and Tasali 2011).

Modulatory effects of sleep on endocrine release are not limited to the hormones belonging to the hypothalamic-pituitary axes, but also apply to hormones controlling carbohydrate metabolism, appetite regulation, and body-fluid balance. The molecules involved in energy homeostasis (glucose plasma levels) are maintained at a constant level during the night sleep period in humans despite prolonged fasting, thanks to the decrease in brain and body energy consumption during the first half of the night.

The overall framework is very complex, mostly due to the differences observed, according to gender, to age, and to the effect of lifestyle and sleep profile on basal hormonal secretion. In addition, the slow time constant of the endocrine regulation often masks the NREM sleep-REM sleep cycling ultradian influences on hormone secretion, except for cases in which secretion is strongly influenced by the activity of the ANS.

As far as the growth hormone axis is concerned, the most reproducible pulse of growth hormone occurs shortly after sleep onset of normal nighttime sleep, concomitantly with the peak in SWA. A different regulatory modality has been shown for the hypothalamo-pituitary-adrenocortical system, in which the circadian component is largely prevalent over the sleep-related component, leading to a peak of cortisol plasma levels during the early morning hours. Daytime levels of plasma thyroid-stimulating hormone are low and relatively stable and are followed by a rapid elevation starting in the early evening and culminating in a nocturnal maximum occurring around the beginning of the sleep period. Under normal conditions, prolactin plasma levels undergo a major nocturnal elevation starting shortly after sleep onset and culminating around mid-sleep, and are maintained at lower levels during daytime (Van Cauter and Tasali 2011).

Finally, the study of mean daily profiles of both glucose levels and insulin-secretion rates (ISR) in healthy subjects, in which the normal meal schedule was replaced by intravenous glucose infusion at a constant rate, showed that sleep influence on these parameters was stronger than that of the circadian system.

3.5 Autonomic and Cardiovascular Functions

The activity of the ANS greatly changes during the wake-sleep cycle (Khairandish and Shapiro 2013). In quiet wakefulness, the activity of the parasympathetic and the sympathetic branches of the ANS controls cardiovascular, respiratory, thermoregulatory, gastrointestinal, and endocrine functions, in order to maintain body homeostasis. On passing from wakefulness to NREM sleep, the contribution of the parasympathetic branch increases compared to that of the sympathetic section, according to the reduced metabolic and somatic activity of this sleep stage. During wakefulness and NREM sleep, ANS activity still works to preserve body homeostasis. During REM sleep, sympathetic activity shows a considerable variability, which is accompanied by phasic changes in parasympathetic discharge. In this state, ANS apparently works according to a "poikilostatic" modality (Parmeggiani 2005).

The regulation of cardiac and circulatory functions deeply changes across the wake-sleep cycle as a consequence of the variations in physiological regulation and autonomic outflow (Franzini 2005). The blood supply to organs must match specific metabolic requirements and it responds to the animal's behavior, through changes in blood pressure and vascular resistance. The regulation of these two variables is critically modified on the basis of autonomic activity directed to the heart and blood vessels. In turn, autonomic output to heart and blood vessels varies according to the hypothalamic integrative activity, coordinating somatic, autonomic, and endocrine functions (Silvani 2008). In NREM sleep, body temperature, metabolism, and muscle tone decrease, and both heart rate and blood pressure decrease compared to wakefulness (Silvani 2008).

The main autonomic features of REM sleep are phasic fluctuations in sympathetic and parasympathetic activity, with instability of cardiovascular and respiratory variables. In particular, it is possible to notice surges in cardiac sympathetic and parasympathetic activity associated with bursts of phasic REMs, PGO waves, myoclonic twitches, and breathing irregularities. Phasic surges in heart rate (approximately a 35 % increase) are followed by bradycardia due to the baroreceptor response to the increase in blood pressure (Khairandish and Shapiro 2013).

In most human subjects, blood pressure shows spontaneous diurnal changes, with a decrease to its lowest levels during nighttime sleep ("dipping"), primarily related to sleep-dependent blood pressure changes, rather than to the endogenous circadian rhythm. A blunted sleep-related blood pressure reduction (non-dipping status, defined as <10 % decrease in blood pressure during sleep) is considered to be one of the most sensitive predictors of cardiovascular mortality.

3.6 *Thermoregulation and Metabolism*

The wake-sleep cycle is tightly coupled to the regulation of body temperature and metabolism (Krauchi and de Boer 2011). In fact, the most opportune moment of the day for sleep occurrence is the rest period, when the circadian system drives a decrease in body temperature and energy expenditure and the probability of an active interaction with the external environment is largely reduced.

Although in the human adult the decrease in energy expenditure during sleep is apparently moderate compared to quiet wakefulness, sleep may play a more relevant role in energy conservation in animals with a less favorable surface-to-volume ratio (infants or small mammals), in which energy conservation is more pressing (Krauchi and de Boer 2011). In general, body energy expenditure decreases during sleep.

In different species, brain energy metabolism largely decreases during NREM sleep and increases during REM sleep to levels similar to, or even slightly larger than, those of wakefulness. The relationship between the different wake-sleep states and the thermoregulatory process has been widely investigated. Thermoregulatory responses may be experimentally elicited by the delivery of either external or internal thermal loads. Such responses are present in both quiet wakefulness and NREM sleep, but absent (in small mammals) or depressed (in humans) in REM sleep (Parmeggiani 2005). While the thermoregulatory differences between wakefulness and NREM sleep only depend on the different levels of energy metabolism in the two states, a deep functional change occurs during REM sleep. This change shifts the normal closed-loop homeostatic regulatory modality to an open-loop one (Parmeggiani 2005). In particular, during REM sleep episodes occurring under a positive (warm) thermal load, peripheral vasodilation and tachypnea are suppressed in animals, while sweating is abolished at the beginning of the episode and subsequently depressed in humans. Also, under a negative (cold) thermal load shivering is suppressed, and heat exchanger vasoconstriction is reduced in animals.

The involvement of central nervous structures in these processes has been confirmed by studies showing that the capacity of the preoptic-hypothalamic thermosensitive neurons to respond to a direct thermal stimulation is impaired during REM sleep (Parmeggiani 2005). Interestingly, recent experiments in the rat have shown that this is not the case for the hypothalamic osmosensitive neurons, since the degree of the release of antidiuretic hormone (ADH) following the intracerebroventricular administration of an hyperosmotic solution does not differ across the different wake-sleep states (Amici et al. 2013).

4 Sleep Evolution and Sleep Functions

4.1 Sleep Phylogenesis

Sleep is a kind of behavior that can be considered to be universally distributed across mammals and birds. In both classes, sleep can be unambiguously identified through both behavioral (reduced responsiveness to stimuli, homeostatic regulation) and neurophysiological (EEG activity) parameters. The commonness of sleep among the different species of the two phylogenetically most recent classes suggests that such a behavior was positively selected by natural selection and that primordial signs of it should be found in less recent classes.

Unfortunately, few studies have satisfactorily investigated the comparative physiology of sleep. Aside from behavioral traits, the neurophysiological signature of sleep appears to be drastically different in classes other than mammals and birds. In reptiles, for instance, a high-amplitude EEG is observed during periods of activity, while a low-amplitude EEG characterizes periods of rest. Indeed, the anatomical difference in the structure of the CNS throughout the animal kingdom prevents the use of the commonly used mammalian-like EEG activity as an identifier for sleep.

In general, a succession of periods of activity and rest has been described in almost every species, from insects to mammals, but whether the period of rest may be labeled as sleep is still controversial. To compensate for the unavailability of mammalian-like EEG in other species (like the Drosophila Melanogaster), the search for molecular markers of sleep have been used as a possible alternative (Zimmerman et al. 2008). This kind of approach is based on the underlying idea that sleep serves a very basic cellular function common to most of the species. While this may be true, it is worth considering that evolution may hijack older regulations to fulfill new objectives, and that, therefore, the function of sleep may differ from class to class. For instance, speculatively speaking, homeothermy made its appearance on the evolutionary scene with mammals and birds (the only two classes to present both NREM and REM sleep) and the brain of these two classes had to spend its resting period at quite a high temperature, possibly forcing a drastic change in the function and regulation of sleep. In conclusion, comparative physiology could be used as a powerful tool to investigate a possible ancestral function of sleep, while not forgetting that evolution may have redirected it towards newer needs.

4.2 *Effects of Sleep Deprivation*

The logic that stands behind the use of sleep deprivation in sleep research is very intuitive: in order to understand what a body function is for, such a function must be impeded. The consequences of deprivation speak about the physiological role of the function itself in two ways: (1) if a rebound in the time/intensity of the deprived function is observed during the recovery following deprivation, the conclusion is that such a function plays a critical/vital role for the organism, and (2) the adverse consequences of the deprivation on the systems physiology of the organism may suggest what that function is vital for. Sleep deprivation is therefore used as a means to try to understand the yet-to-be-understood functions of sleep.

When compared to other forms of function deprivation, like, for instance, food or water deprivation, sleep deprivation presents some peculiarities. In particular, while food and water deprivation may occur in the natural world, sleep deprivation is quite uncommon, although it has been hypothesized that migrating birds do not sleep during long flights, and also the torpor bouts in hibernators have been considered to count as a very peculiar kind of sleep deprivation. This implies that animals in a laboratory setup have to be sleep deprived by some kind of environmental interference that removes the possibility to fall asleep for the animal. Several methods of sleep deprivation have been developed in past years, among which the most relevant are: (1) forced locomotion; (2) disk over water; (3) gentle handling, while more specific for REM sleep deprivation are (4) flower pot; (5) cold exposure (see Revel et al. 2009 for review).

Independently from the methods used, a very consistent observation following sleep deprivation experiments has been that sleep loss generates a drive for a sleep rebound during the following recovery period. In other words, sleep presents the typical homeostatic kind of regulation that has been described for other behaviors such as feeding or drinking (Amici et al. 2008). As previously discussed in Sect. 2.2, the rebound response for NREM sleep occurs mostly in terms of its "intensity" (increase in SWA, Achermann and Borbély 2011), while that for REM sleep occurs mostly in terms of its "duration" (Amici et al. 2008).

The functionality of the cerebral cortex seems to be very sensitive to sleep deprivation, and such sensibility may account for the degradation of cognitive performance induced by sleep deprivation. The general excitability of cortical neurons is increased during sleep deprivation, while neighborhood neurons tend to fire synchronously, impairing the computational ability of the cortex and, therefore, the quality of tasks depending on such functionality (Vyazovskiy et al. 2013).

However, following prolonged sleep deprivation, more severe general consequences arise. In rodents, 4 weeks of sleep deprivation lead to death by a unique syndrome characterized by a set of incoherent symptoms, such as weight loss, a massive increase in food intake, and hypothermia (Rechtschaffen and Bergmann 2002). These symptoms seem to suggest that prolonged sleep deprivation causes a lethal disruption to the central control of metabolism. A disruption in metabolic regulation was also shown to be evoked in humans by 48 h of sleep deprivation (Van Cauter et al. 2008). Such metabolic dysregulation may very well be the result of the effects of sleep deprivation on the central nervous areas controlling metabolism. It is worth noting that some neuronal populations placed within these areas (for instance, the orexinergic neurons of the lateral hypothalamus) can also modulate the activity of the cerebral cortex, therefore, possibly accounting for the effects of sleep deprivation on the cortex itself. In conclusion, the sleep rebound induced by sleep deprivation suggests that the function of sleep is very relevant; moreover, while the most visible consequences of sleep deprivation are related to an impairment in cortical functions, the central network controlling body metabolism may be a biological target of sleep deprivation.

4.3 Sleep and Memory

A growing corpus of evidence has investigated the effects that sleep has over memory traces, following the hypothesis that memory traces can be strengthened or weakened by the neurophysiological events taking place in the brain during sleep. While the process of memory consolidation is not completely known in detail, it is intuitive that it requires changes in neuronal connection and metabolism, processes that, ultimately, can be identified with the expression "neuronal plasticity."

In general, the three main hypotheses on how memory may be modulated by sleep have been proposed (Peigneux and Smith 2011): (1) the first hypothesis suggests that a continuous replay of recently created engrams during sleep (both REM and NREM) promotes their consolidation; (2) an alternative proposal suggests that during NREM sleep recently acquired memories are transferred from the hippocampus to the cerebral cortex, where new engrams are stored as consolidated memories; and (3) finally, the synaptic homeostasis theory (Tononi and Cirelli 2003) suggests that synapses on local cortical circuits are downscaled in a use-dependent mode during NREM sleep, enhancing the signal-to-noise ratio of relevant memory traces compared to nonrelevant ones. Interestingly, with memory consolidation as a goal, NREM sleep seems to be more critical than REM sleep. In particular, at least according to the synaptic homeostasis theory, the electrophysiological event that signals the cortical neurons to initiate the synaptic downscaling is the slow wave. Of course, the view that NREM sleep is the only determinant of memory consolidation is an oversimplification, since several data show an effective role in memory consolidation for REM sleep as well, suggesting that the clear-cut association of memory reconsolidation processes with precise sleep states may not be so strict.

4.4 *Theories of Sleep Functions*

Sleep function is intuitively associated with some kind of rest that seems to be mostly required by the CNS. Many theories on the function of sleep have been proposed, but a theory of sleep would not just have to interpret the existing evidence but also to provide testable predictions.

One legitimate hypothesis on the function of sleep is that during sleep some sort of toxic, wake-related substance is mobilized or metabolized, especially in the brain. Such a long-existing theory was recently revived by the finding that during NREM sleep, a drastic change in the clearance of the cerebrospinal fluid takes place (Xie et al. 2013). This finding was paralleled by the observation that sleep deprivation favors the accumulation of β-amyloid. This theory provides a theoretical frame for the consideration of sleep as a general functional process, but still leaves major questions open. For instance, no hypothesis on the function of REM sleep is provided and it is unclear why a change in brain fluid clearance should require cortical synchronization and loss of consciousness.

Another claim on the function of sleep was recently suggested within the synaptic homeostasis theory frame. The synaptic homeostasis theory correlates the synaptic plasticity of the cerebral cortex with the high degree of synchronized neuronal activity that occurs during NREM sleep (Tononi and Cirelli 2003). The theory is supported by experimental data, even if alternative explanations of these findings have recently been proposed (Frank 2012). The link between SWA in the cortex and neuronal plasticity was experimentally confirmed and used as a measure of the importance of sleep for all cognitive functions that depend on a well-functioning cortex, such as memory consolidation. But, slowly, the maintenance of cerebral cortex plasticity, from being the target of the positive effects of sleep, became the main determinant of sleep (Tononi and Cirelli 2014). This subtly modified interpretation of the theory overlooks the relevance of underlying autonomic effects as the basis of sleep generation and regulation. It is hard to interpret all the autonomic changes required for sleep to occur merely as the result of a feedback signal coming from the cortex and directed to the sleep central network, which carries information regarding the state of synaptic plasticity in the cortex. In other words, if the only determinant of sleep was the state of neuronal plasticity in the cortex, sleep-related autonomic changes would appear to be meaningless. They would be an evolutionary accident conserved for no apparent reason, and cortical synchronization could occur without autonomic changes. Moreover, the synaptic homeostasis theory does not provide a clear explanation of the REM sleep function. Finally, it is interesting to underline that the synaptic homeostasis theory, as well as the "brain clearance" one, are apparently unable to predict the effects of prolonged sleep deprivation or to provide an explanation on why such a procedure is lethal.

A full comprehension of sleep functions is probably not imminent.

References

Achermann P, Borbély AA (2011) Sleep homeostasis and models of sleep regulation. In: Kryger MH, Roth T, Dement WC (eds) Principles and practice of sleep medicine, 5th edn. Elsevier Saunders, Philadelphia, pp 431–444

Amici R, Cerri M, Ocampo-Garces A et al (2008) Cold exposure and sleep in the rat: REM sleep homeostasis and body size. Sleep 31:708–715

Amici R, Cerri M, Parmeggiani PL (2013) Overview of physiological processes during sleep. In: Kushida CA (ed) The encyclopedia of sleep. Academic, Waltham, pp 385–389

Brown RE, Basheer R, McKenna JT et al (2012) Control of sleep and wakefulness. Physiol Rev 92:1087–1187

Carskadon MA, Dement EC (2011) Normal human sleep: an overview. In: Kryger MH, Roth T, Dement WC (eds) Principles and practice of sleep medicine, 5th edn. Elsevier Saunders, Philadelphia, pp 16–26

Chase MH (2013) Motor control during sleep and wakefulness: clarifying controversies and resolving paradoxes. Sleep Med Rev 17:299–312

Chou TC, Scammell TE, Gooley JJ et al (2003) Critical role of dorsomedial hypothalamic nucleus in a wide range of behavioral circadian rhythms. J Neurosci 23:10691–10702

Cirelli C, Gutierrez CM, Tononi G (2004) Extensive and divergent effects of sleep and wakefulness on brain gene expression. Neuron 41:35–43

De Gennaro L, Marzano C, Fratello F et al (2008) The electroencephalographic fingerprint of sleep is genetically determined: a twin study. Ann Neurol 64:455–460

Dement WC (2011) History of sleep physiology and medicine. In: Kryger MH, Roth T, Dement WC (eds) Principles and practice of sleep medicine, 5th edn. Elsevier Saunders, Philadelphia, pp 3–15

Douglas NJ (2005) Respiratory physiology: control of ventilation. In: Kryger MH, Roth T, Dement WC (eds) Principles and practice of sleep medicine, 4th edn. Elsevier Saunders, Philadelphia, pp 224–231

Feng D, Lazar MA (2012) Clocks, metabolism, and the epigenome. Mol Cell 47:158–167

Frank MG (2012) Erasing synapses in sleep: is it time to be SHY? Neural Plast 2012:264378

Franzini C (2005) Cardiovascular physiology: the peripheral circulation. In: Kryger MH, Roth T, Dement WC (eds) Principles and practice of sleep medicine, 4th edn. Elsevier Saunders, Philadelphia, pp 203–212

Khairandish A, Shapiro CM (2013) Peripheral nervous system and sleep. In: Kushida CA (ed) The encyclopedia of sleep. Academic, Waltham, pp 494–502

Krauchi K, de Boer T (2011) Body temperature, sleep, and hibernation. In: Kryger MH, Roth T, Dement WC (eds) Principles and practice of sleep medicine, 5th edn. Elsevier Saunders, Philadelphia, pp 323–334

Krieger J (2005) Respiratory physiology: breathing in normal subjects. In: Kryger MH, Roth T, Dement WC (eds) Principles and practice of sleep medicine, 4th edn. Elsevier Saunders, Philadelphia, pp 232–244

Kumar VM (2010) Sleep is neither a passive nor an active phenomenon. Sleep Biol Rhythms 8:163–169

Luppi PH, Clément O, Fort P (2013) Paradoxical (REM) sleep genesis by brainstem is under hypothalamic control. Curr Opin Neurobiol 23:786–792

Maret S, Dorsaz S, Gurcel L et al (2007) Homer1a is a core brain molecular correlate of sleep loss. Proc Natl Acad Sci USA 104:20090–20095

Massimini M, Huber R, Ferrarelli F et al (2004) The sleep slow oscillation as a traveling wave. J Neurosci 24:6862–6870

Parmeggiani PL (2005) Physiologic regulation in sleep. In: Kryger MH, Roth T, Dement WC (eds) Principles and practice of sleep medicine, 4th edn. Elsevier Saunders, Philadelphia, pp 185–191

Peever JH, Sessler BJ (2011) Sensory and motor processing during sleep and wakefulness. In: Kryger MH, Roth T, Dement WC (eds) Principles and practice of sleep medicine, 5th edn. Elsevier Saunders, Philadelphia, pp 348–359

Peigneux P, Smith C (2011) Memory processing in relation to sleep. In: Kryger MH, Roth T, Dement WC (eds) Principles and practice of sleep medicine, 4th edn. Elsevier Saunders, Philadelphia, pp 335–347

Rechtschaffen A, Bergmann BM (2002) Sleep deprivation in the rat: an update of the 1989 paper. Sleep 25:18–24

Revel FG, Gottowik J, Gatti S et al (2009) Rodent models of insomnia: a review of experimental procedures that induce sleep disturbances. Neurosci Biobehav Rev 33:874–899

Saper CB, Scammell TE, Lu J (2005) Hypothalamic regulation of sleep and circadian rhythms. Nature 437:1257–1263

Sehgal A, Mignot E (2011) Genetics of sleep and sleep disorders. Cell 146:194–207

Silvani A (2008) Physiological sleep-dependent changes in arterial blood pressure: central autonomic commands and baroreflex control. Clin Exp Pharmacol Physiol 35:987–994

Steriade M, Timofeev I, Grenier F (2001) Natural waking and sleep states: a view from inside neocortical neurons. J Neurophysiol 85:1969–1985

Thompson RH, Swanson LW (2003) Structural characterization of a hypothalamic visceromotor pattern generator network. Brain Res Brain Res Rev 41:153–202

Tononi G, Cirelli C (2003) Sleep and synaptic homeostasis: a hypothesis. Brain Res Bull 62:143–150

Tononi G, Cirelli C (2014) Sleep and the price of plasticity: from synaptic and cellular homeostasis to memory consolidation and integration. Neuron 81:12–34

Van Cauter E, Tasali E (2011) Endocrine physiology in relation to sleep and sleep disturbances. In: Kryger MH, Roth T, Dement WC (eds) Principles and practice of sleep medicine, 5th edn. Elsevier Saunders, Philadelphia, pp 291–311

Van Cauter E, Spiegel K, Tasali E et al (2008) Metabolic consequences of sleep and sleep loss. Sleep Med 9(Suppl 1):S23–S28

Vyazovskiy VV, Cirelli C, Tononi G (2011) Electrophysiological correlates of sleep homeostasis in freely behaving rats. Prog Brain Res 193:17–38

Vyazovskiy VV, Olcese U, Cirelli C et al (2013) Prolonged wakefulness alters neuronal responsiveness to local electrical stimulation of the neocortex in awake rats. J Sleep Res 22:239–250

Xie L, Kang H, Xu Q et al (2013) Sleep drives metabolite clearance from the adult brain. Science 342:373–377

Zimmerman JE, Naidoo N, Raizen DM et al (2008) Conservation of sleep: insights from non-mammalian model systems. Trends Neurosci 31:371–376

Neuronal Networks Regulating Sleep and Arousal: Effect of Drugs

Elemer Szabadi

Abstract The three vigilance states (wakefulness [W], slow wave sleep [SWS], rapid eye movement sleep [REMS]) are controlled by distinct, but interconnected, networks of neurons. The sleep/arousal network consists of separate systems of W-promoting and SWS-promoting neurons, located in nuclei in the basal forebrain, diencephalon and brainstem. Each neuronal system operates via a distinct neurotransmitter, providing its unique "neurochemical signature". W-promoting neurons are active during W and quiescent during SWS, whereas SWS-promoting neurons are active during SWS and cease to fire during W. The level of arousal at any one time reflects the intricate balance between W-promoting and SWS-promoting systems. W is the result of cortical activation by W-promoting neurons; sleep ensues when SWS-promoting neurons switch off the W-promoting systems. REMS is regulated by a network of REMS-promoting and REMS-inhibiting neurons located in the brainstem and hypothalamus. A third network is responsible for the regulation of the circadian rhythmicity of the wakefulness/sleep cycle. The neurochemical signatures of W-promoting and SWS-promoting neurons make it possible to develop drugs that, by targeting specific neuroreceptors and synaptic mechanisms, have predictable effects on sleep and arousal. Arousal-modifying drugs act by tipping the balance between W-promoting and SWS-promoting neuronal activity. Thus a sedative drug, useful for the treatment of insomnia, may act by activating a SWS-promoting system (e.g. benzodiazepines, melatonin receptor agonists) or inhibiting a W-promoting system (e.g. H1-antihistamines, orexin receptor antagonists). Conversely, an alerting drug, useful for the treatment of excessive daytime sleepiness, may inhibit a SWS-promoting system or activate a W-promoting system (e.g. psychostimulants, H3 histamine receptor antagonists).

Abbreviations

BF Basal forebrain
DMH Dorsomedial hypothalamus

E. Szabadi (✉)
Developmental Psychiatry, Queen's Medical Centre, E Floor, South Block, Nottingham NG7 2U, UK
e-mail: elemer.szabadi@nottingham.ac.uk

A. Guglietta (ed.), *Drug Treatment of Sleep Disorders*, Milestones in Drug Therapy,
DOI 10.1007/978-3-319-11514-6_2

DR	Dorsal raphe nucleus
LC	Locus coeruleus
LDT	Laterodorsal tegmental nucleus
LGN	Lateral geniculate nucleus
LH	Lateral hypothalamic area
MCH	Melanin concentrating hormone
PBN	Parabrachial nucleus
PC	Precoeruleus nucleus
PF	Perifornical area
PPT	Pedunculopontine tegmental nucleus
PVN	Paraventricular nucleus
SCN	Suprachiasmatic nucleus
SLD	Sublaterodorsal nucleus
SPZ	Subparaventricular zone
Th	Thalamus
TMN	Tuberomamillary nucleus
TPR	Tegmentopontine reticular nucleus
vGC	Ventral gigantocellular nucleus
vlPAG	Ventrolateral periaquductal grey (matter)
VLPO	Ventrolateral preopic nucleus
VPAG	Ventral periaquductal grey (matter)
VTA	Ventral tegmental area

1 Introduction

The mode of action of drugs used to treat the two major categories of sleep disorder (insomnia and hypersomnia) used to be interpreted on the basis that the level of arousal reflected the general excitability of the brain (Esplin 1970). Each level of general neuronal activity is reflected in the dominant frequency in the EEG and corresponds to a well-defined state of arousal along the "maximum depression (coma)" ↔ "maximum excitation (convulsions)" continuum. The drugs to treat insomnia (hypnotics) used to be regarded as general (nonselective) CNS depressants that depressed the activity of all neurons and shifted the levels of neuronal excitability and arousal downwards (towards sedation), whereas drugs to treat hypersomnia were classified as general (nonselective) CNS stimulants that increased the activity of all neurons and shifted the levels of excitability and arousal upwards (towards alertness). The effect of a given dose of a sedative or stimulant drug was dependent on the baseline (pre-treatment) level of activity of the CNS, a sedative drug causing a greater degree of sedation if the CNS was already depressed (Esplin 1970).

The lack of selectivity of sedative drugs (e.g. barbiturates, benzodiazepines) was supported by observations that, apart from hypnotic effects, they also possess some other actions (e.g. anaesthetic, anxiolytic, anticonvulsant) that can be related to general CNS depression. However, on the other hand, it was also realized that even the nonselective drugs show some selectivity (e.g. different barbiturates have variable potencies as hypnotics, anaesthetics, anxiolytics and anticonvulsant). Furthermore, different groups of neurons show differential sensitivities: inhibitory interneurons are especially sensitive, leading to a paradoxical excitatory effect after the administration of relatively small doses of a sedative drug, whereas some neurons controlling vital functions (e.g. respiration) are relatively resistant (Esplin 1970).

It has been discovered over the past 20 years or so that arousal-modifying drugs are not nonselective, as used to be believed, but rather show a great degree of selectivity. This selectivity is both anatomical and neurochemical: the drugs act on specific neurochemically identified sites ("targets") in the brain that are critical for the regulation of sleep and wakefulness (Szabadi 2014). The functional significance of each target depends on its location within a well-defined arousal-modifying network (see Sect. 2).

Unravelling the sleep/arousal networks creates an opportunity for the classification of arousal-modifying drugs according to mechanism of action. The basis for this classification is the understanding that the level of arousal at any one time reflects the intricate balance between sleep-promoting and wake-promoting neuronal systems. Thus the level of arousal will be decreased by a drug that either activates a sleep-promoting system (e.g. benzodiazepines) or inhibits a wake-promoting system (e.g. antihistamines), whereas it will be increased by a drug that either activates a wake-promoting system (e.g. amphetamine, modafinil) or inhibits a sleep-promoting system. There are no therapeutically useful examples of alerting (stimulant) drugs that act by inhibiting a sleep-promoting system (e.g. the GABA receptor antagonists are also convulsants) (Szabadi 2014).

Targets of arousal-modifying drugs are specified both at the level of the arousal system and the cellular mechanism within that system. For example, a drug may be aimed at blocking release-inhibiting H3 receptors in the histaminergic wake-promoting system (see Sect. 4.1.3), or at stimulating melatonin receptors in the sleep-promoting suprachiasmatic nucleus/melatonin system (see Sect. 3.4.2).

2 Neuronal Networks Regulating Sleep and Arousal

On the basis of the level of arousal, it is possible to distinguish between three vigilance states (Fort et al. 2009): wakefulness (W), slow wave sleep (SWS) and rapid eye movement (REM) sleep (REMS). Each vigilance state is defined by the EEG: W and REMS are characterized by desynchronized, low amplitude, high frequency (alpha and beta) waves, whereas in SWS the EEG is synchronized with high amplitude low frequency (delta and theta) waves (Lin et al. 2011). The three vigilance states are regulated by distinct neuronal networks, consisting of neuronal groups (nuclei) and their connections, located in the basal forebrain, diencephalon and brainstem. Each neuronal group within a network has its unique "neurochemical signature" derived from the neurotransmitter used. It is possible to distinguish between three separate, but interconnected, networks: the W/SWS network, the REMS network and the circadian network.

2.1 Sleep/Arousal Network

The level of arousal at any one time reflects the intricate interplay between distinct W-promoting and SWS-promoting nuclei. Each of these nuclei is defined both anatomically and neurochemically, i.e. by the neurotransmitter utilized. The W-promoting nuclei are maximally active during wakefulness, show reduced activity during SWS and are quiescent during REMS. On the other hand, the SWS-promoting nuclei are maximally active during sleep and quiescent during wakefulness. The W-promoting and SWS-promoting nuclei, together with their major connections, are shown in Fig. 1. The W-promoting neurons send excitatory projections to the cerebral cortex and other W-promoting nuclei, and inhibitory outputs to SWS-promoting nuclei. The SWS-promoting neurons exert inhibitory influence on the W-promoting nuclei (Szabadi 2014).

W-promoting neurons can be found in the basal forebrain (BF), the diencephalon (thalamus and hypothalamus) and the brainstem. The BF contains W-promoting cholinergic neurons. There are W-promoting glutamatergic neurons in the thalamus. Hypothalamic W-promoting nuclei include the histaminergic tuberomamillary nucleus (TMN) and the orexinergic neurons of the lateral hypothalamic/perifornical are (LH/PF). W-promoting nuclei in the brainstem are the noradrenergic locus coeruleus (LC), the dopaminergic ventral tegmental area (VTA), the serotonergic dorsal raphe nucleus (DR) and the cholinergic pedunulopontine tegmental/laterodrosal tegmental nuclei (PPT/LDT). The PPT/LDT also contain REMS-promoting cholinergic neurons that are inhibited by the LC (see Sect. 2.2) (Szabadi 2014).

The basal forebrain contains some SWS-promoting GABAergic inhibitory neurons projecting to the cerebral cortex: these neurons are inhibited by a noradrenergic input from the LC (Fig. 5) (Szabadi 2013). The major SWS-promoting nucleus is the ventrolateral preoptic nucleus (VLPO) of the hypothalamus. GABAergic inhibitory neurons from the VLPO project to the TMN and LC. During wakefulness VLPO activity is switched off by an inhibitory input from the LC. Neurons containing the neuropeptide melanin concentrating hormone (MCH), located in the lateral hypothalamus, intermingled with orexinergic neurons, exert a SWS-promoting action by sending inhibitory projections to all W-promoting nuclei. There are also short-axon GABAergic inhibitory interneurons in the brainstem that exert a SWS-promoting effect by inhibiting W-promoting noradrenergic and dopaminergic neurons (Szabadi 2014).

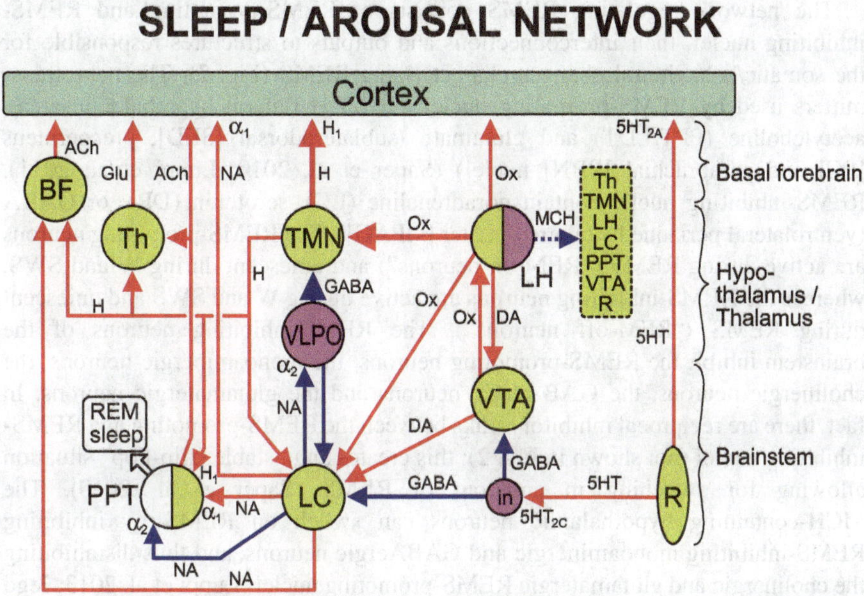

Fig. 1 Schematic diagram of the connections within the neuronal network regulating slow wave sleep and wakefulness ("sleep/arousal network"). *Wake-promoting nuclei* (*yellow*): BF, basal forebrain; TMN, tuberomamillary nucleus; LH, lateral hypothalamic area; Th, thalamus; LC, locus coeruleus; VTA, ventral tegmental area; PPT, pedunculopontine tegmental nucleus; R, raphe nuclei. *Sleep-promoting nuclei* (*purple*): VLPO, ventrolateral preoptic nucleus; in, GABAergic interneurons; GABAergic interneurons; LH, lateral hypothalamic area. *REM sleep-promoting nucleus* (*white*): PPT, pedunculopontine tegmental nucleus. *Connections*: *arrows* (*red*: excitatory; *blue*: inhibitory). *Lettering next to arrow*: neurotransmitter (ACh: acetylcholine; NA: noradrenaline; H: histamine; Ox: orexin; GABA: γ-aminobutyric acid; DA: dopamine; 5HT: 5-hydroxytryptamine; Glu: glutamate; MCH: melanin-concentrating hormone). *Receptors*: α_1, excitatory α_1-adrenoceptors; α_2, inhibitory α_2-adrenoceptors; H_1, excitatory H_1 histamine receptors; $5HT_{2A}$ and $5HT_{2C}$, excitatory 5HT receptors. See text (Sect. 2.1) for details. Reproduced, with permission, from Szabadi (2014)

2.2 *REM-Sleep Network*

REM sleep is also called paradoxical or active sleep since, although the level of consciousness is restricted, the EEG is almost indistinguishable from that associated with W (McCarley 2007). REMS is also characterized, by definition, by the presence of phasic conjugate saccadic eye movements ("rapid eye movements") (Peigneux et al. 2001). In association with rapid eye movements electrical field potentials can be recorded in the pons, lateral geniculate bodies and occipital cortex (ponto-geniculo-occipital [PGO] spikes). PGO spikes are best recorded in experimental animals using deep electrodes, but they are also likely to occur in humans (McCarley 2007). Other features associated with REMS are muscle atonia (Chase 2013), penile erections (Hirshkowitz and Schmidt 2005) and dreaming (Hobson 2009).

The network regulating REMS consists of REMS-promoting and REMS-inhibiting nuclei, their interconnections and outputs to structures responsible for the somatic/behavioural changes characterizing REMS (Fig. 2). The neurotransmitters used by REMS-promoting nuclei are MCH (lateral hypothalamic area), acetylcholine (PPT/LDT) and glutamate (sublaterodorsal [SLD], precoeruleus [PC] and parabrachial [PBN] nuclei) (Saper et al. 2010; Luppi et al. 2011). REMS-inhibiting nuclei contain noradrenaline (LC), serotonin (DR), or GABA (ventrolateral periaqueductal grey matter [vlPAG]). The REMS-promoting neurons are active during REMS ("REM-on neurons") and quiescent during W and SWS, whereas the REMS-inhibiting neurons are active during W and SWS and quiescent during REMS ("REM-off neurons"). The REMS-inhibiting neurons of the brainstem inhibit the REMS-promoting neurons: the monoaminergic neurons, the cholinergic neurons, the GABAergic neurons and the glutamatergic neurons. In fact, there are reciprocal inhibitory links between the REMS-promoting and REMS-inhibiting nuclei (not shown in Fig. 2): this creates an unstable "flip-flop" situation allowing for switching in and out of REMS (Saper et al. 2010). The MCH-containing hypothalamic neurons can switch on REMS by inhibiting REMS-inhibiting monoaminergic and GABAergic neurons, and thus disinhibiting the cholinergic and glutamatergic REMS-promoting nuclei (Luppi et al. 2013; Jego et al. 2013).

Cortical activation during REMS is brought about by direct excitatory outputs from the cholinergic and glutamatergic REMS-promoting nuclei. The activation of the visual cortex, via an excitatory output from the lateral geniculate nucleus (LGN), originating from the PPT/LDT, underlies the generation of PGO spikes and dreams during REMS. Impairment of the balance between excitatory cholinergic and inhibitory serotonergic inputs to the LGN has been implicated in the generation of complex visual hallucinations (Manford and Andermann 1998).

Horizontal saccadic eye movements are controlled by premotor neurons in the tegmentopontine reticular nucleus (TPR) (Büttner-Ennever and Horn 1997). The activation of the TPR by a glutamatergic input from the SLD may be responsible for the rapid eye movements in REMS (Sánchez-López and Escudero 2011).

The activity of striated muscles is stimulated by a noradrenergic facilitatory influence on motoneurons originating from the LC, and is inhibited by a glycinergic/GABAergic influence on motoneurons originating from the ventral gigantocellular nucleus (vGC) of the medulla and a network of medullary and spinal interneurons (Saper et al. 2010). The atonia of the skeletal muscles and the relaxation of the facial (e.g. chin) muscles are due partly to the withdrawal of the noradrenergic stimulation of motoneurons (McGregor and Siegel 2010; Peever 2011) and partly to the activation of the glycinergic/GABAergic neurons in the vGC (Luppi et al. 2011; Chase 2013). Trigeminal motoneurons also cease their activity during REMS leading to the relaxation of the masseter muscle (Peever 2011). Recently a descending glutamatergic pathway from the SLD to the spinal inhibitory interneuron pool has been implicated in the causation of REMS-related atonia (Krenzer et al. 2011). In rapid-eye-movement-sleep behaviour disorder (RBD), a parasomnia associated with REMS, skeletal muscle atonia is absent

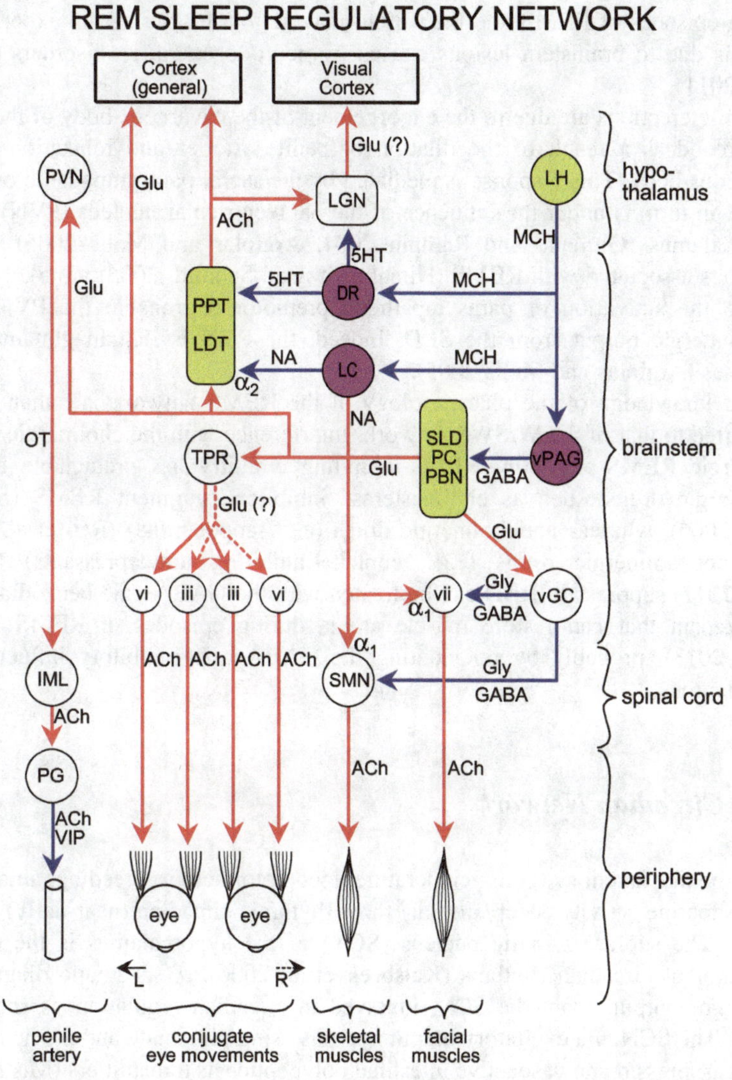

Fig. 2 Schematic diagram of the connections within the neuronal network regulating REM sleep ("REM sleep network"). *REMS-promoting nuclei* (*yellow*): LH, lateral hypothalamic area; PPT/LDT, pedunculopontine/laterodrosal tegmental nuclei; SLD, sublaterodorsal nucleus; PC, precoeruleus nucleus; PBN: parabrachial nuclei. *REMS-inhibiting nuclei* (*purple*): vPAG, ventral periaqueductal grey matter; LC, locus coeruleus; DR, dorsal raphe nucleus). *Target nuclei* (*white*): LGN, lateral geniculate nucleus; PVN, paraventricular nucleus; TPR, tegmentopontine reticular nucleus; vGC, ventral gigantocellular nucleus; III, VI, VII, cranial nerve motor nuclei (III: oculomotor; VI: abducens; VII: facial); SMN, spinal motoneurons; IML, intermediolateral column of spinal cord; PG, pelvic ganglion. *Connections*: *arrows* (*red*: excitatory; *blue*: inhibitory). *Lettering next to arrow*: neurotransmitter (MCH: melanin concentrating hormone; GABA: γ-aminobutyric acid; Gly: glycine; Glu: glutamate; NA: noradrenaline; 5HT: 5-hydroxytryptamine [serotonin]; ACh: acetylcholine; OT: oxytocin; VIP: vasoactive intestine polypeptide). *Receptors*: α_1, excitatory α_1-adrenoceptors; α_2, inhibitory α_2-adrenoceptors. See text (Sect. 2.2) for details

during episodes of REMS and the patient may act out his/her dream experiences. RBD is due to brainstem lesions caused by neurodegenerative disorders (Luppi et al. 2011).

Penile erections are due to the engorgement of the cavernous body of the penis with blood as a result of the dilation of penile arteries and relaxation of the cavernous body. This response is mediated by the sacral parasympathetic outflow, which, in turn, is under the influence of the paraventricular nucleus (PVN) of the hypothalamus (Giuliano and Rampin 2004; Argiolas and Melis 2005). Penile erections associated with REMS (Hirshkowitz and Schmidt 2005) are likely to be due to the activation of parasympathetic premotor neurons in the PVN by a glutamatergic output from the SLD. Indeed, the PVN is rich in glutamatergic synapses (Argiolas and Melis 2005).

Our knowledge of the pharmacology of the REMS network is rather patchy compared to that of the W/SWS network. Interference with the cholinergic/monoaminergic REMS-promoting/REMS-inhibiting circuitry has predictable effects: cholinergic drugs, such as cholinesterase inhibitors, augment REMS (Schredl et al. 2006), whereas anticholinergic drugs (e.g. scopolamine) (Rao et al. 2004) and monoaminergic drugs (e.g. reuptake-inhibiting antidepressants) (Göder et al. 2011) suppress it. RBD can be treated with the GABAergic benzodiazepine clonazepam that can restore muscle atonia during episodes of REMS (Luppi et al. 2013), probably by potentiating the GABAergic inhibitory influence on motoneurons.

2.3 Circadian Network

Many bodily functions (body temperature, locomotor activity, feeding, autonomic and endocrine activity, sleep/arousal) show rhythmic circadian (near-daily) fluctuations. The suprachiasmatic nucleus (SCN) of the hypothalamus is the central generator of circadian rhythms (Kalsbeek et al. 2006). A schematic diagram of the major outputs from the SCN, involved in circadian regulation, is shown in Fig. 3. The SCN, via excitatory output neurons using glutamate and the neuropeptides vasopressin and vasoactive intestinal polypeptide as transmitters (Mistlberger 2005), projects to two hypothalamic areas: the subparaventricular zone (SPZ) and the dorsomedial hypothalamus (DMH) (Fuller et al. 2006). The dorsal SPZ has been implicated in regulating the circadian rhythm of body temperature. The ventral SPZ projects to the DMH. The DMH is involved in the regulation of the rhythmicity of locomotor activity and sleep and arousal. Both glutamatergic excitatory and GABAergic inhibitory outputs from the DMH increase the level of arousal: the excitatory output by activating the LH and LC and the inhibitory output by inhibiting the VLPO. Light exerts its wake-promoting effect via this pathway: light stimulates the retina that sends a glutamatergic excitatory output to the SCN. This effect countermands the sleep-promoting effect of light arising from the direct stimulation of the VLPO (see also Sect. 4.5.1, Photomodulation).

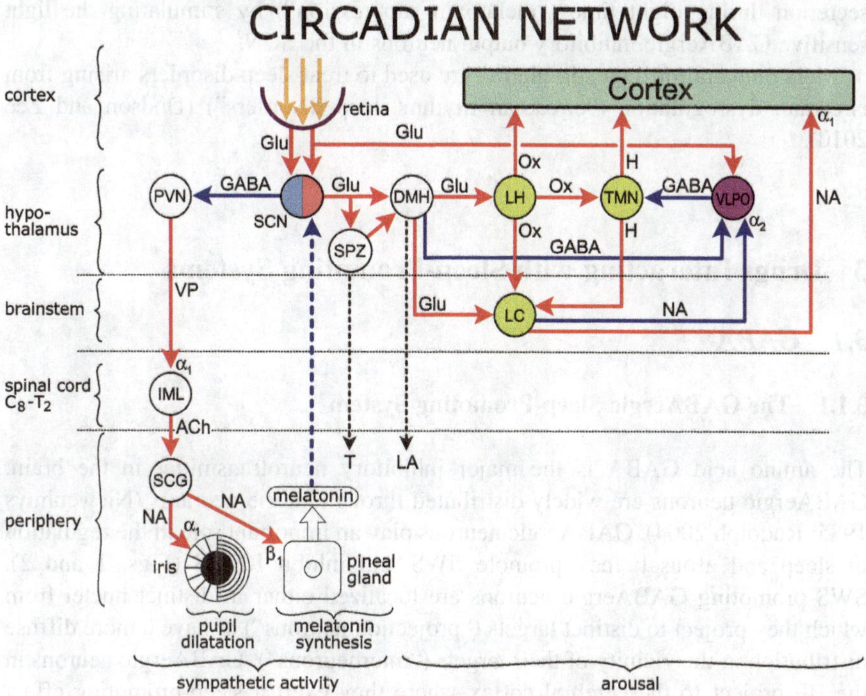

Fig. 3 Schematic diagram of the connections within the neuronal network controlling circadian regulation ("circadian network"). *Wake-promoting nuclei* (*yellow*): LH, lateral hypothalamic area; TMN, tuberomamillary nucleus; LC, locus coeruleus). *Sleep-promoting nucleus* (*purple*): VLPO, ventrolateral preoptic nucleus. *SCN: suprachiasmatic nucleus* (*blue*: inhibitory output neurons; *red*: excitatory output neurons). *Target and relay nuclei* (*white*): DMH, dorsomedial hypothalamus; SPZ, subparaventricular zone; PVN, paraventricular nucleus; IML, intermediolateral column of spinal cord; SCG, superior cervical ganglion. *Target functions*: T, body temperature; LA, locomotor activity. *Connections*: *arrows* (*red*: excitatory, *blue*: inhibitory, *broken blue*: inhibitory hormonal; *broken black*: functional). *Lettering next to arrow*: neurotransmitter (Glu: glutamate, GABA: γ-aminobutyric acid; NA: noradrenaline; Ox: orexin; H: histamine; VP: vasopressin; ACh: acetylcholine). *Receptors*: α_1, excitatory α_1-adrenoceptors; β_1, excitatory β_1-adrenoceptors. See text (Sect. 2.3) for details

There is a GABAergic inhibitory output from the SCN to the premotor sympathetic neurons of the PVN that control sympathetic outflow from the lower cervical/upper thoracic spinal cord (C_8-T_2) to the dilator muscle of the iris and the pineal gland (Kalsbeek et al. 2000). The pineal gland synthesizes and secretes the "sleep hormone" melatonin. Melatonin has sleep-promoting propensity due to the stimulation of inhibitory MT_1 receptors on wake-promoting SCN neurons (Szabadi 2014). The GABAergic output neurons in the SCN control the circadian activity of melatonin secretion: during day time, when these neurons are maximally active, no melatonin is secreted, while during night time, when these neurons are quiescent, melatonin is synthesized and secreted. Light can switch off melatonin secretion during night time (“melatonin suppression”) by stimulating the light sensitive GABAergic inhibitory output neurons in the SCN.

Melatonin and/or light stimulation are used to treat sleep disorders arising from circadian dysregulation (“circadian rhythm sleep disorders”) (Dodson and Zee 2010).

3 Drugs Interacting with Sleep-Promoting Systems

3.1 GABA

3.1.1 The GABAergic Sleep-Promoting System

The amino acid GABA is the major inhibitory neurotransmitter in the brain. GABAergic neurons are widely distributed throughout the neuraxis (Niewenhuys 1985; Rudolph 2004). GABAergic neurons play an important role in the regulation of sleep and arousal: they promote SWS and inhibit REMS (Figs. 1 and 2). SWS-promoting GABAergic neurons are localized either in distinct nuclei from which they project to distinct targets (“projection neurons”) or have a more diffuse distribution in the vicinity of their targets (“interneurons”). GABAergic neurons in the BF project to the cerebral cortex where they exert a sleep-promoting effect (Manns et al. 2003). These neurons, like the GABAergic neurons of the VLPO, are inhibited by the LC (Fig. 5). Paradoxically, the GABAergic neurons of the BF can also mediate a W-promoting effect by inhibiting inhibitory interneurons in the cerebral cortex, and thus disinhibiting cortical activity (Lin et al. 2011). GABAergic neurons in the VLPO, together with neurons containing the inhibitory neuropeptide galanin, project to the major W-promoting nuclei, such as the TMN and LC (España and Scammell 2011). There is a reciprocal inhibitory connection from the LC to the VLPO (Fig. 1). GABAergic interneurons in the brainstem, in the vicinity of the VTA and LC, exert an inhibitory influence on these catecholaminergic W-promoting nuclei. The activity of the GABAergic interneurons is facilitated by an excitatory serotonergic input from the DR stimulating $5HT_{2C}$ receptors (Gobert et al. 2000). Recently GABAergic neurons have been identified in the nucleus accumbens (ventral striatum): they project to W-promoting nuclei and thus may promote SWS (Lazarus et al. 2011).

3.1.2 GABA Receptors

GABA exerts its inhibitory effects by interacting with distinct receptors. Three GABA receptors have been identified: $GABA_A$, $GABA_B$ and $GABA_C$. Of these the $GABA_A$ receptor is the most important in the regulation of sleep and arousal, and most available GABAergic drugs act via $GABA_A$ receptors.

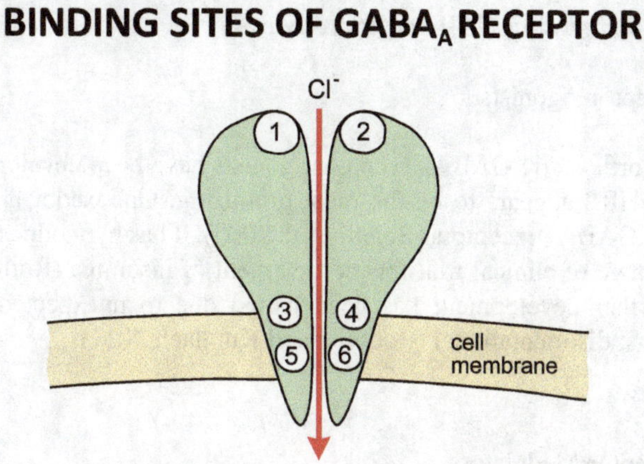

Fig. 4 Schematic diagram of the $GABA_A$ receptor. The receptor is a chloride-gated ion channel that is opened when the receptor is activated by agonists. Numbers indicate binding sites for agonists. *1* orthosteric binding site for the natural transmitter (GABA); *2* to *6* binding sites for allosteric modulators (*2* for benzodiazepines; *3* for anaesthetics; *4* for ethanol; *5* for barbiturates; *6*: for neurosteroids). Positive allosteric modulators enhance the effect of GABA. See text (Sect. 3.1.2) for details. Modified, with permission, from Rudolph (2004)

The $GABA_A$ receptor is an ionotropic receptor: it mediates the passage of chloride ions into the cell leading to membrane hyperpolarization. In fact, the $GABA_A$ receptor is a ligand-gated chloride channel (Fig. 4). It has a pentameric structure corresponding to five subunits. $GABA_A$ receptors containing the α_1 subunit are the most relevant for sedative drug action. The $GABA_A$ receptor, apart from a binding site for the natural neurotransmitter GABA ("orthosteric site"), also contains a number of binding sites for different drugs that modulate the activity of the receptor ("allosteric sites"). The positive allosteric $GABA_A$ receptor modulators, that include benzodiazepines, some anaesthetics, ethanol, barbiturates and neurosteroids, enhance the effect of endogenous GABA (Rudolph 2004; Sigel and Steinmann 2012). $GABA_A$ receptors occur both postsynaptically in close association with GABAergic nerve terminals and extrasynaptically on neurons located at some distance from the site of release. Extrasynaptic $GABA_A$ receptors mediate a slow tonic inhibitory response (Winsky-Sommerer 2009; Rudolph and Knoflach 2011).

The $GABA_B$ receptor is a metabotropic receptor that signals via G-proteins. These receptors also mediate neuronal inhibition (Pinard et al. 2010). $GABA_B$ receptors occur both presynaptically and postsynaptically. These receptors have been identified on W-promoting cholinergic neurons in the LDT and serotonergic neurons in the DR, mediating the sleep-promoting effect of GABA (Kohlmeier et al. 2013).

Relatively little is known about $GABA_C$ receptors. A number of ligands have been developed for these receptors with the aim of exploring their physiological role and the therapeutic potential of drugs interacting with them (Johnston et al. 2003).

3.1.3 Sleep-Promoting GABAergic Activators

GABA$_A$ Receptor Agonists

A number of orthosteric GABA$_A$ receptor agonists have been developed, of which gaboxadol (THIP) appears to be the most promising. Gaboxadol is selective for extrasynaptic GABA$_A$ receptors (Belelli et al. 2005). It has hypnotic effects and has reached the stage of clinical trial for the treatment of insomnia (Roth et al. 2010). However, further development has been halted due to unexpected side effects (hallucinations, disorientations) (Rudolph and Knoflach 2011).

GABA$_A$ Receptor Modulators

Drugs acting at the *benzodiazepine site* (also referred to as the "benzodiazepine receptor") include the benzodiazepines themselves (e.g. diazepam, temazepam) and the "Z" drugs (zolpidem, zopiclone/eszopiclone, zaleplon). These are the most commonly prescribed drugs for insomnia (Wilson et al. 2010; Proctor and Bianchi 2012). The adverse effects of the benzodiazepines (hangover effects causing daytime sedation, cognitive and psychomotor impairment, development of tolerance and dependence) are well recognized, and the Z drugs also share some of the disadvantages of benzodiazepines. However, it may be possible to improve the clinical usefulness of GABA$_A$ receptor modulators by fine-tuning the properties of new drugs according to the subtype (constellation of subunits) of GABA$_A$ receptor targeted (Winsky-Sommerer 2009; Nutt and Stahl 2010; Rudolph and Knoflach 2011).

The barbiturates act at the *barbiturate site*. These drugs, apart from acting as allosteric modulators of the GABA$_A$ receptor, also have some direct agonistic activity at the receptor. During the first half of the twentieth century, prior to the discovery of the benzodiazepines, these drugs were the most commonly used hypnotics. They are still used as anticonvulsants (phenobarbital) and anaesthetics (pentobarbital).

A number of intravenous (etomidate, propofol) and volatile (enflurane, isoflurane) anaesthetics produce sedation by enhancing GABAergic activity by binding to the *anaesthetic site* of the GABA$_A$ receptor. It should be noted, however, that the GABA$_A$ receptor may not be the only target of these drugs. It has been proposed that GABAergic anaesthetics, that also include the anaesthetic barbiturates (e.g. pentobarbital, thiopental), may act by stimulating GABA$_A$ receptors at the projection targets of VLPO neurons, such as the TMN. The increased inhibition of the wake-promoting histaminergic neurons of the TMN would lead to withdrawal of the histaminergic stimulation of the cerebral cortex, and this in turn would result in sedation (Nelson et al. 2002). It should be noted, however, that this model has been challenged recently (Zecharia et al. 2012).

There is an allosteric binding site on the $GABA_A$ receptor for *neurosteroids* (or neuroactive steroids) that can evoke rapid changes in neuronal excitability via the potentiation of GABAergic neurotransmission. Some steroids of the pregnane class, synthesized by neurons and glia cells, act as powerful endogenous regulators of $GABA_A$ receptor function (Lambert et al. 2009). There is evidence that neurosteroids (e.g. pregnalone, progesterone) have sleep-promoting effects both in experimental animals and humans (Steiger 2007). It is likely that synthetic steroid anaesthetics (e.g. alphaxolone) exert their sedative effects via this site (Lambert et al. 2009).

There is an allosteric binding site for *ethanol* that is different from the benzodiazepine and barbiturate sites. Ethanol has sleep-promoting effects (Dijk et al. 1992), and is often used as a hypnotic (Johnson et al. 1998).

$GABA_B$ Receptor Agonists

Baclofen is an orthosteric $GABA_B$ receptor agonist available for the treatment of spasticity. Drowsiness is one of its side effects, consistent with its sleep-promoting propensity (Bowery 2006).

γ-Hydroxybutyric acid (GHB, sodium oxybate) is an orthosteric partial agonist at $GABA_B$ receptors. In addition, it also interacts with some high-affinity binding sites ("GHB receptors") that have recently been identified as a subtype of the $GABA_A$ receptor (Bay et al. 2014). GHB is sleep promoting (Kohlmeier et al. 2013). It is licensed for the treatment of narcolepsy with cataplexy: there is evidence that GHB ameliorates all the cardinal symptoms of this disorder (i.e. it reduces cataplexy and EDS/daytime sleep attacks and improves sleep architecture) (Boscolo-Berto et al. 2012).

GABA Reuptake Inhibitors

Released GABA is removed from the synaptic cleft into GABAergic nerve terminals by an active transport mechanism operated by GABA transporter GAT-1 localized in the membrane of the nerve terminal. *Tiagabine* inhibits GAT-1 leading to the blockade of the reuptake of GABA: this leads to enhanced activation of $GABA_A$ and $GABA_B$ receptors (Winsky-Sommerer 2009). Tiagabine has well-documented sleep-promoting effects (Matthias et al. 2001).

3.1.4 Wake-Promoting GABAergic Inhibitors

Although inhibition of central GABAergic neurotransmission can increase the level of alertness, inhibition of the GABAergic system leads to a number of other changes (e.g. cognitive enhancement ["pro-cognitive effect"], anxiety, convulsions) that may mask a wake-promoting effect.

GABA$_A$ Receptor Antagonists

The GABA$_A$ receptor antagonists bicuculline and gabazine, when administered systemically, are potent convulsants (Johnston 2013). However the wake-promoting effect of bicuculline is revealed when it is microinjected into the cholinergic wake-promoting nucleus PPT (Torterolo et al. 2002).

Benzodiazepine Receptor Antagonists

Flumazenil is a *competitive antagonist* at the benzodiazepine binding site (Hoffman and Warren 1993). It has been shown to possess a procognitive effect in animal models of memory (Lal et al. 1988). It has no wake-promoting effect. Intravenously administered flumazenil is used as an antidote to benzodiazepine overdose and to reverse sedation after midazolam-induced anaesthesia (Hoffman and Warren 1993).

Inverse agonists of the benzodiazepine binding site (e.g. beta-carbolines) have procognitive, anxiogenic, proconvulsant (increased susceptibility to convulsions) and convulsant effects: these effects are observed as the dosage of the beta-carboline is increased from very low to high levels (Venault and Chapouthier 2007). The beta-carbolines also have wake-promoting effects, which, however, are superseded by convulsions (Massotti et al. 1985). Recently benzodiazepine receptor inverse agonists have been developed that are selective for the α5 subtype of the GABA$_A$ receptor. These drugs have marked procognitive effects without anxiogenic and (pro)convulsant propensities (Atack 2010), and thus may have a clinical potential for the treatment of cognitive disorders (Martinez-Cué et al. 2014).

GABA$_B$ and GABA$_C$ Receptor Antagonists

Antagonists of both GABA$_B$ and GABA$_C$ receptors have wake-promoting effects (Gottesmann 2002). GABA$_B$ receptor antagonists also possess procognitive propensity (Stäubli et al. 1999), and some of these drugs have reached the stage of clinical trial for the treatment of cognitive disorders (Froestl et al. 2004).

GABA Receptor Autoantibodies

In some autoimmune disorders neurotransmitter receptors can act as antigens leading to the generation of specific autoantibodies against the receptors (Bien et al. 2012). Autoantibodies have been identified against both GABA$_A$ (Petit-Pedrol et al. 2014) and GABA$_B$ (Lancaster et al. 2010) receptors in cases of autoimmune encephalitides. The autoantibodies close down the GABA receptors: this may explain the prevalence of seizures in these disorders. However, in a rare form of autoimmune disorder, total insomnia ("agrypnia") was reported in association with $GABA_B$ receptor autoantibodies (Frisullo et al. 2007).

3.2 Melanin Concentrating Hormone

3.2.1 The MCHergic Sleep-Promoting System

MCH is a cyclic neuropeptide localized mainly in neurons of the lateral hypothalamus. MCH-containing neurons are intermingled with orexin-containing neurons. There is a reciprocal discharge pattern of the two groups of neurons: orexinergic neurons are active during W and quiescent during SWS and REMS, whereas MCH neurons are quiescent during W, show some activity during SWS, and are maximally active during REMS (Hassani et al. 2009). MCH neurons, like the orexin-containing neurons, project diffusely to many structures of the brain. In particular, there are projections to the cholinergic and monoaminergic wake-promoting nuclei (Fig. 1) and the GABAergic REMS-inhibiting neurons of the midbrain (Fig. 2). There are two G-protein-coupled receptors (MCHR1 and MCHR2) stimulated by MCH; both receptors mediate neuronal inhibition. By inhibiting wake-promoting neurons, MCH promotes SWS, and by inhibiting REMS-inhibiting neurons, MCH promotes REMS (see Sect. 2; Monti et al. 2013; Tsunematsu et al. 2014). The main function of the MCH sleep-promoting system is likely to be to facilitate REMS (Jego et al. 2013). In fact, the optogenetic activation of MCH neurons can switch SWS into REMS (Tsunematsu et al. 2014).

3.2.2 Wake-Promoting MCH Inhibitors

MCHR1 receptor antagonists have been reported to increase wakefulness and suppress both SWS and REMS (Ahnaou et al. 2008), suggesting that these receptors may become a target for the development of wake-promoting drugs.

3.3 Galanin

3.3.1 The Galaninergic Sleep-Promoting System

Galanin is a neuropeptide functioning both as an inhibitory neurotransmitter and a co-transmitter/neuromodulator. Galaninergic neurons have been identified in the VLPO where they are intermingled with GABAergic neurons (Gaus et al. 2002). These galaninergic neurons have the same projection targets as the GABAergic neurons. Galanin also occurs in other neurons, such as the noradrenergic neurons of the LC and serotonergic neurons of the DR, where it functions as a co-transmitter, modulating (inhibiting) the release of the principal transmitter (Le Maître et al. 2013). Via synergism of the sleep-promoting GABAergic neurons of the VLPO and inhibition of the wake-promoting noradrenergic and serotonergic neurons, galanin exerts a sleep-promoting action.

Three galanin receptors (GalR1, GalR2, GalR3) have been identified, and agonists and antagonists have been developed for them (Webling et al. 2012).

3.3.2 Sleep-Promoting Galanin Receptor Agonists

According to one report, galanin injected intravenously alters the sleep EEG consistent with promotion of both SWS and REMS (Murck et al. 1999). Like the GABA receptor agonists and positive allosteric modulators, galanin receptor agonists also possess anxiolytic (Rajarao et al. 2007) and anticonvulsant (Lerner et al. 2008) propensities. Galanin receptors are a promising target for the development of anticonvulsant drugs (Hoyer 2010).

3.3.3 Wake-Promoting Galanin Receptor Antagonists

The development of these drugs is at an early stage. It is predicted that they would enhance the level of alertness. Furthermore, they may have antidepressant effects due to the disinhibition of central noradrenergic and serotonergic neurotransmission (Ögren et al. 2006).

3.4 *Melatonin*

3.4.1 The Melatoninergic Sleep-Promoting System

The hormone melatonin is secreted by the pineal gland. Its synthesis is under strict circadian control by the SCN, and is modulated by light (Sect. 2.3). Melatonin has a large number of targets where it stimulates MT_1 and MT_2 receptors (Pandi-Perumal et al. 2006). It has sleep-promoting and sleep-modulating (chronobiotic) effects by interacting with these receptors in the SCN. The sleep-promoting effect has been attributed to the stimulation of MT_1 receptors and the chronobiotic effect to the stimulation of MT_2 receptors (Hardeland 2012).

3.4.2 Sleep-Promoting Melatoninergic Activators

Melatonin itself has sleep-promoting effect (Pandi-Perumal et al. 2006). This is likely to be due to the stimulation of inhibitory MT_1 receptors on wake-promoting glutamatergic output neurons in the SCN. This in turn would lead to dampening of the activity of the wake-promoting "SCN → DMH → LC → cortex" circuit (Fig. 3), resulting in reduction in the level of arousal (Szabadi 2014).

Apart from a sustained release formulation of melatonin (Circadin®), a number of MT_1/MT_2 receptor agonists have been developed for the treatment of insomnia: ramelteon, tasimelteon and 6-chloromelatonin (LY156735) (Szabadi 2014).

3.4.3 Chronobiotic Melatoninergic Activators

$MT1/MT_2$ receptor agonists also exert chronobiotic effects that involve resetting the circadian clock and introducing phase shifts. These drugs are useful for the treatment of some circadian rhythm disorders, such as the initial insomnia of delayed sleep phase disorder and the free-running disorder of blind people (Dodson and Zee 2010).

3.4.4 Wake-Promoting Melatoninergic Inhibitors

The synthesis of melatonin is driven by the sympathetic outflow via the "PVN → spinal preganglionic neurons → postganglionic neurons in superior cervical ganglion" pathway. The postganglionic noradrenergic neurons innervate the pineal gland cells where they stimulate β_1-adrenoceptors (Fig. 3).

Inhibition of melatonin synthesis by β-adrenoceptor antagonists leads to wake-promoting and sleep-disrupting effects; this effect can be reversed by the administration of melatonin (Van den Heuvel et al. 1997). Melatonin has been recommended to treat insomnia associated with the use of beta-blockers (Fares 2011).

3.5 Adenosine

3.5.1 The Adenosine-Mediated Sleep-Promoting System

Adenosine, a breakdown product of adenine nucleotides, accumulates in the brain during prolonged wakefulness (Huang et al. 2011). It is an endogenous sleep-inducing agent ("somnogen"): by increasing the tendency to fall asleep ("sleep propensity" or "sleep pressure"), it acts as a homoeostatic regulator of sleep (Porkka-Heiskanen 2013). The sleep-promoting effects of adenosine are mediated via the inhibition of wake-promoting neurons and the stimulation of sleep-promoting neurons.

3.5.2 Adenosine Receptors

Adenosine, released from metabolically active cells, interacts with specific cell surface receptors. Of the four adenosine receptors identified, two (A_1 and A_{2A}) are important for the promotion of sleep (Landolt 2008). The A_1 receptors mediate an inhibitory response. They are localized on wake-promoting cholinergic neurons of the BF, noradrenergic neurons of the LC, serotonergic neurons of the DR, and orexinergic neurons of the LH/PF. These receptors are located partly on the cell bodies and partly on the nerve terminals of these neurons where they inhibit the release of the neurotransmitter. The A_{2A} receptors mediate an excitatory response, either directly (Gallopin et al. 2005) or indirectly by inhibiting inhibitory interneurons and thus disinhibiting the target neuron (Morairty et al. 2004). A_{2A} receptors have been identified on or near GABAergic sleep-promoting neurons in the VLPO (Scammell et al. 2001; Gallopin et al. 2005). Recently, A_{2A} receptors have been detected on putative GABAergic sleep-promoting neurons in the nucleus accumbens (ventral striatum): these neurons may inhibit the activity of wake-promoting neurons via their widespread projections (Zhang et al. 2013).

3.5.3 Sleep-Promoting Adenosine Receptor Agonists

Agonists of both A_1 (Benington et al. 1995) and A_{2A} (Scammell et al. 2001) receptors have sleep-inducing effects in experimental animals. However, these drugs have not been studied in humans.

3.5.4 Wake-Promoting Adenosine Receptor Antagonists

The psychostimulant *caffeine* exerts its wake-promoting effect via antagonizing adenosine receptors (Landolt 2008). Although caffeine has affinity for both A_1 and A_{2A} receptors, recent evidence indicates that its alerting effect is mediated primarily via the A_{2A} receptor (Huang et al. 2005). Furthermore, A_{2A} receptors in the nucleus accumbens have been implicated in this effect (Lazarus et al. 2011).

SYN115 is a synthetic A_{2A} adenosine receptor antagonist developed for the treatment of movement disorders resulting from dopaminergic deficiency (e.g. Parkinson's disease) (Kulisevsky and Poyurovsky 2012). The basis for this is that A_{2A} receptors are co-localized with D_2 dopamine receptors in the basal ganglia where they modulate (inhibit) dopamine receptor function. As expected, SYN115 possesses marked wake-promoting effects (Lane et al. 2012).

4 Drugs Interacting with Wake-Promoting Systems

4.1 *Histamine*

4.1.1 The Histaminergic Wake-Promoting System

Histaminergic neurons are localized in the TMN of the posterior hypothalamus and send diffusely arborizing projections to most areas of the brain and spinal cord. The histaminergic neurons, like most other wake-promoting neurons, are maximally active during wakefulness and quiescent during sleep (Ko et al. 2003). These neurons, via their precisely targeted connections within the W/SWS network (Fig. 1), play an essential role in the maintenance of wakefulness (Haas and Lin 2012; Szabadi 2014). The importance of the posterior hypothalamus, including the TMN, in the maintenance of wakefulness was first highlighted by von Economo who correlated lesions in this area with the hypersomnia in encephalitis lethargica (Triarhou 2006). Central histaminergic neurons are activated by an excitatory orexinergic input from the lateral hypothalamus, and inhibited by a GABAergic input from the VLPO. In fact, the sleep-promoting effect of the VLPO is mediated largely by the switching off of TMN activity. VLPO activity is kept in check during wakefulness by an inhibitory output from the LC. Histaminergic outputs from the TMN mediate an excitatory effect via stimulation of H1 histamine receptors at their target areas, including the cerebral cortex and a number of wake-promoting nuclei (LC, PPT/LDT, BF and thalamus). The activation of cortically projecting cholinergic neurons in the BF may play a major role in mediating the wake-promoting effect of histamine (Zant et al. 2012).

4.1.2 Histamine Receptors

There are four types of histamine receptor, of which three (H1, H2 and H3) occur in the central nervous system. H1 and H2 receptors are excitatory postsynaptic receptors, whereas H3 receptors are inhibitory autoreceptors located on the histaminergic neuron itself (Szabadi 2014). In a somatodendritic location, H3 receptors suppress neuronal firing, whereas located on presynaptic terminals they inhibit transmitter synthesis and release. H3 receptors also occur on the nerve terminals of other neurons ("heteroreceptors") where they inhibit the release of the neurotransmitter (monoamines, acetylcholine, glutamate, GABA, peptides) (Haas and Lin 2012). Interestingly, the H3 receptor can signal on its own without an agonist ("constitutive activity") (Arrang et al. 2007). This may explain why inverse agonists are more effective than neutral antagonists in suppressing H3 receptor signalling (Haas and Lin 2012).

4.1.3 Wake-Promoting Histaminergic Activators

H1 Receptor Agonists

Currently available agonists are of little therapeutic significance. Histamine itself injected into the brain of experimental animals has wake-promoting effects (Lin et al. 1988). However, systemically administered histamine cannot pass the blood–brain barrier and can trigger allergic reactions. Betahistine, a derivative of histamine, is a partial H1 receptor agonist that has some central histaminergic effects in humans when administered orally. However, these effects may have been, at least partly, mediated by H3 receptor blockade (Reynolds 2012). At present there are no selective and potent H1 receptor agonists available (Taberan 2013). Although it may be possible to develop more effective centrally acting H1 receptor agonists, recent effort has concentrated on the development of H3 receptor antagonists for the stimulation of the central histaminergic system.

H3 Receptor Antagonists

H3 receptor antagonists and inverse agonists increase histaminergic activity by disinhibiting autoreceptors, thus leading to a wake-promoting effect. This effect is enhanced by the disinhibition of heteroreceptors modulating the release of wake-promoting transmitter substances (monoamines, acetylcholine). There is a large body of evidence demonstrating the wake-promoting propensity of H3 receptor antagonists and inverse agonists (Parmentier et al. 2007). Furthermore, these drugs also have potential cognitive-enhancing effects (Brioni et al. 2011). A large number of H3 receptor antagonists/inverse agonists have been developed over the past 25 years for the treatment of excessive daytime sleepiness (EDS), cognitive disorders and obesity (Gemkow et al. 2009; Berlin et al. 2011). Some early compounds have been superseded by more novel molecules that are at various stages of development. Pitolisant (formerly BF2.649, tiprolisant) has reached the stage of clinical trial for the treatment of EDS in narcolepsy (Schwartz 2011).

4.1.4 Sleep-Promoting Histaminergic Inhibitors

H1 Receptor Antagonists

H1 receptor antagonists are referred to as antihistamines, or more precisely as H1-antihistamines (Church and Church 2013). H1-antihistamines have been used since the 1930s for the treatment of allergic disorders. *First-generation H1-antihistamines* (e.g. diphenhydramine, promethazine, chlorpheniramine, cyclizine) readily cross the blood–brain barrier and have potent sedative/hypnotic effects. These drugs have poor receptor selectivity and often interact with muscarinic cholinoceptors, α-adrenoceptors and serotonin (5HT) receptors, leading to side effects resulting from the blockade of these various receptors. Some of the first-generation H1-antihistamines (e.g. diphenhydramine) are available as over-the-counter hypnotics (Proctor and Bianchi 2012). Some antipsychotic drugs (e.g. chlorpromazine, thioridazine, levomepromazine) and antidepressants (amitriptyline, trazodone, mirtazapine, doxepin) have sedative propensities and prescribed for the treatment of insomnia. Although these drugs interact with a number of receptors, their sleep-promoting effects are likely to be due primarily to the blockade of H1 histamine receptors (Proctor and Bianchi 2012). *Second-generation H1-antihistamines* are chemically different from the first-generation H1-anthistamines, and are more selective for the H1 histamine receptor. These drugs (e.g. desloratadine, levocetirizine, fexofenadine) have an important role in the treatment of allergic disorders: as they do no penetrate into the brain, unwanted sedation is not a problem.

H3 Receptor Agonists

A number of potent and selective H3 histamine receptor agonists (e.g. imetit, R-(α) methylhistamine, BP294) have been developed (Leurs et al. 1998). These drugs, by stimulating release-inhibiting presynaptic H3 histamine receptors, exert an antihistaminergic effect. There is evidence that they promote SWS in experimental animals (Thakkar 2011) and thus have therapeutic potential for the treatment of insomnia. However, they have not reached the stage of clinical development.

4.2 *Orexin*

4.2.1 The Orexinergic Wake-Promoting System

The orexins (orexin A and orexin B, also known as hypocretin-1 and hypocretin-2) are neuropeptides localized in distinct groups of neurons (LH/PF, DMH) in the hypothalamus. Orexinergic neurons of the LH/PF constitute an important wake-promoting system via their projections to the cerebral cortex and other wake-promoting nuclei (TMN, LC, VTA) (Fig. 1). At their projection targets, orexinergic neurons interact with excitatory orexin receptors (OX1 and OX2) (Ohno and Sakurai 2008). The orexinergic neurons of the DMH are involved in circadian regulation and in mediating the effect of light on arousal (Fig. 3).

4.2.2 Narcolepsy: An Orexin-Deficiency State

The loss of orexinergic neurons is responsible for the sleep disorder narcolepsy. The symptoms of narcolepsy are complex and can be related to the intrusion of sleep-related vigilance states into wakefulness. Intrusion of SWS into W leads to EDS and sleep attacks, whereas intrusion of REMS into W results in cataplexy (loss of muscle tone in response to emotions), hypnagogic hallucinations (dream-like visions before falling asleep) and sleep paralysis (inability to move before falling asleep) (Ohno and Sakurai 2008; Sakai 2013).

4.2.3 Wake-Promoting Orexinergic Activators

It would be of great therapeutic importance to replace orexinergic function in narcolepsy. Unfortunately, there are no non-peptide orexin receptor agonists available to date, and the orexin peptides do not readily penetrate into the brain. Administration of orexin peptides into the lateral cerebral ventricle (Mieda et al. 2004) or transplantation of orexin neurons into the LH (Arias-Carrión and Murillo-Rodriguez 2014) have reversed the symptoms of narcolepsy in experimental animals. Furthermore, the administration of orexin A as a nasal spray to narcoleptic patients has been reported to alleviate narcoleptic symptoms (Weinhold et al. 2014).

4.2.4 Sleep-Promoting Orexinergic Inhibitors

A number of orexin receptor antagonists have been developed for the treatment of insomnia (Hoyer and Jacobson 2013). Development has reached the stage of clinical trial with some of these drugs (e.g. almorexant, SB-649868, suvorexant, filorexant) that all antagonize both orexin A and orexin B receptors ("dual orexin receptor antagonists, DORAs"). The first one of these, almorexant, has been withdrawn following Phase III trials, while suvorexant has been submitted for registration. The possible usefulness of selective orexin receptor antagonists is being explored: while OX1 antagonists are probably not good candidates as insomnia drugs, OX2 antagonists have some promise (Hoyer and Jacobson 2013).

An important consideration with the clinical use of orexin receptor antagonists is the risk of provoking the symptoms of narcolepsy, and in particular cataplexy (Tafti 2007). However, clinical trials with DORAs so far have failed to support this possibility (Hoyer and Jacobson 2013).

4.3 Glutamate

4.3.1 The Glutamatergic Wake-Promoting System

The thalamus contains both specific (sensory relay) and non-specific glutamatergic neurons that project diffusely to the cerebral cortex (Brown et al. 2012). The non-specific neurons, located in the intralaminar and reticular nuclei, are wake-promoting and constitute an integral part of the "ascending arousal system" (McCormick and Bal 1997). These neurons receive excitatory inputs from wake-promoting cholinergic and histaminergic neurons (Fig. 1), and their activity is modulated by GABA. The hypnotic effect of gaboxadol has been attributed to the stimulation of extrasynaptic inhibitory $GABA_A$ receptors on thalamocortical neurons (Belelli et al. 2005).

Recently a second glutamatergic wake-promoting system has been identified: glutamatergic neurons in the upper brainstem (parabrachial nucleus and precoeruleus area) evoke wakefulness via an excitatory output to the BF. In fact, in rats, this second system appears to be more important in maintaining alertness than the cortico-thalamic system (Fuller et al. 2011).

Glutamatergic neurons are also involved in the regulation of REMS: they exert a REMS-promoting effect (Sect. 2.2 and Fig. 2).

4.3.2 Glutamate Receptors

There are two classes of glutamate receptor: ionotropic receptors (ligand-gated ion channels) and metabotropic receptors. There are three inotropic receptors defined by their selectivities to some prototype ligands: NMDA, AMPA and kainite receptors (Dingledine et al. 1999). The ionotropic receptors mediate excitatory responses. There are eight metabotropic receptors (mGluR1-8) divided into three groups: I (mGluR1, 5), II (mGluR2, 3), III (mGluR4, 6, 7, 8). The Group I receptors are excitatory, whereas the Group II and Group III receptors are inhibitory. The Group II receptors are mainly release-inhibiting presynaptic receptors (Niswender and Conn 2010).

4.3.3 Wake-Promoting Glutamatergic Activators

An antagonist (negative allosteric modulator) of inhibitory metabotropic mGluR2 glutamate receptors has been reported to increase wakefulness in experimental rats (Ahnaou et al. 2014).

4.3.4 Sleep-Promoting Glutamatergic Inhibitors

Some antiepileptic drugs that act by inhibiting central glutamatergic neurotransmission (Gitto et al. 2012) also possess sedative properties. Thus *topiramate*, an AMPA/kainite receptor antagonist (Poulsen et al. 2004), is a highly sedative drug (Berigan 2002), and *levetiracetam*, a drug that reduces glutamate release by blocking calcium channels (Lee et al. 2009), has marked sleep-promoting effects (Cicolin et al. 2006).

AMN082, a positive allosteric modulator of mGlu7 inhibitory metabotropic glutamate receptors, increases SWS in experimental rats (Cavas et al. 2013b).

4.3.5 REMS-Suppressing Glutamatergic Inhibitors

Stimulation of inhibitory metabotropic mGluR2 (Ahnaou et al. 2009) and mGluR7 (Cavas et al. 2013b) glutamate receptors or antagonism of excitatory mGluR5 metabotropic receptors (Cavas et al. 2013a) have been reported to suppress REMS in experimental animals.

4.4 *Acetylcholine*

4.4.1 The Cholinergic Wake-Promoting System System

There are two major cholinergic arousal systems, one originating in the LD/PPT and another one in the BF. The wake-promoting cholinergic neurons of the LD/PPT project to the cerebral cortex both directly and indirectly via thalamic glutamatergic neurons. The wake-promoting cholinergic neurons of the BF project to the cortex. The wake-promoting cholinergic neurons of both the LD/PPT and the BF receive a facilitatory noradrenergic input from the LC and a facilitatory histaminergic input from the TMN (Sect. 2.1, Fig. 1). Cholinergic neurons are also involved in the regulation of REMS (Sect. 2.2, Fig. 2).

4.4.2 Acetylcholine Receptors (Cholinoceptors)

At their targets, cholinergic neurons interact with two types of receptor: muscarinic and nicotinic receptors. Muscarinic cholinoceptors are G-protein-coupled metabotropic receptors, whereas nicotinic cholinoreceptors are ionotropic receptors (ligand-gated ion channels). There are five types (M_1–M_5) of muscarinic receptor (Brown 2010) and a large number of "subtype assemblies" of the nicotinic receptor (Gotti et al. 2009). Both muscarinic and nicotinic cholinoceptors are present both in the thalamus and the cerebral cortex (McCormick 1992).

4.4.3 Wake-Promoting Cholinergic Activators

Cholinesterase Inhibitors

These drugs, by inhibiting the enzyme responsible for the degradation of acetylcholine, increase cholinergic neurotransmission. *Physostigmine* has marked alerting effect (Votava et al. 1968), and is able to reverse anaesthesia (Meuret et al. 2000). Physostigmine also increases REMS and dreaming (Sitaram et al. 1978), probably due to potentiation of cholinergic mechanisms in the REMS network.

Donepezil, a cholinesterase inhibitor used to improve cognitive function in dementia, increases wakefulness in rats (Abe et al. 2003) and causes insomnia in patients. It also increases REMS (Schredl et al. 2006).

Muscarinic Receptor Agonists

Milameline, a muscarinic receptor agonist, developed for the treatment of dementia, has been reported to increase alertness in experimental animals (Schwarz et al. 1999).

Nicotinic Receptor Agonists

Nicotine is widely used, in the form of inhaled tobacco smoke, for its cognitive-enhancing effect. It increases arousal (Fisher et al. 2012). The enhancement of attention by nicotine is likely to reflect its alerting effect (Myers et al. 2013).

TC-1734, a synthetic nicotinic receptor agonist, has been shown to possess alerting effects (Dunbar et al. 2007).

4.4.4 Sleep-Promoting Cholinergic Inhibitors

Muscarinic cholinoceptor antagonists (e.g. atropine: Santucci et al. 1981; scopolamine: Ebert and Kirch 1998) are sedative. Anticholinergic drugs used to treat overactive bladder cause drowsiness (Scheife and Takeda 2005). The sedative propensities of H1-antihistamines (Church and Church 2013), phenothiazine antipsychotics and tricyclic antidepressants (Proctor and Bianchi 2012) are likely to reflect, to some extent, muscarinic cholinoceptor blockade. In extreme cases, such as administration during anaesthesia (Moos 2007) and poisoning (Parissis et al. 2003), anticholinergic drugs may lead to coma. Physostigmine has proved to be an effective antidote in these cases.

4.4.5 REMS-Suppressing Cholinergic Inhibitors

Anticholinergic drugs, such as scopolamine, suppress REMS (Rao et al. 2004).

4.5 *Noradrenaline*

4.5.1 The Noradrenergic Wake-Promoting System

Central noradrenergic neurons are clustered in seven brainstem nuclei, labelled A1–A7, of which the locus coeruleus (A6) is the largest one (Samuels and Szabadi 2008a; Szabadi 2013). The locus coeruleus (LC) occupies a central position in the sleep-arousal network, collating arousal-related information from all wake-promoting and sleep-promoting nuclei (Fig. 1).

The LC, apart from functioning as a major wake-promoting nucleus, has three additional functions that are intimately linked to its arousal-enhancing effect: autonomic regulation, motor regulation and modulation of the effect of light on arousal ("photomodulation").

Regulation of Arousal

The LC exerts its wake-promoting effect via excitatory projections to the cerebral cortex and some other wake-promoting nuclei (e.g. cholinergic wake-promoting neurons of the PPT/LDT and the BF, serotonergic neurons of the DR) and inhibitory projections to the sleep-promoting GABAergic neurons of the VLPO and BF. The LC also suppresses REM sleep by inhibiting REMS-promoting cholinergic neurons in the PPT/LDT (Figs. 1 and 5; Szabadi 2013). The excitatory projections operate via stimulation of excitatory α_1-adrenoceptors and the inhibitory projections via stimulation of inhibitory α_2-adrenoceptors (Szabadi 2013).

Autonomic Regulation

The LC functions as a premotor autonomic nucleus, sending excitatory projections to sympathetic preganglionic neurons and inhibitory projections to preganglionic parasympathetic neurons (Figs. 1 and 5). LC activation leads to increased arousal, sympathetic activation and parasympathetic deactivation, whereas the opposite pattern occurs as LC activity diminishes. This is beautifully illustrated by the regulation of the diameter of the pupil. It has been shown that fluctuations in pupil diameter, that is under dual sympathetic/parasympathetic control (Szabadi 2013), are closely paralleled by fluctuations in LC activity (Aston-Jones and Cohen 2005; Murphy et al. 2014). Furthermore, both fluctuations in pupil diameter (Wilhelm et al. 2001) and LC activity (Sterpenich et al. 2006; Minzenberg et al. 2008) are related to the level of arousal. Recording pupillary fluctuations forms the basis of the Pupillographic Sleepiness Test (PST) that has been widely used for the assessment of level of arousal in humans (Wilhelm et al. 2001; Samuels et al. 2006).

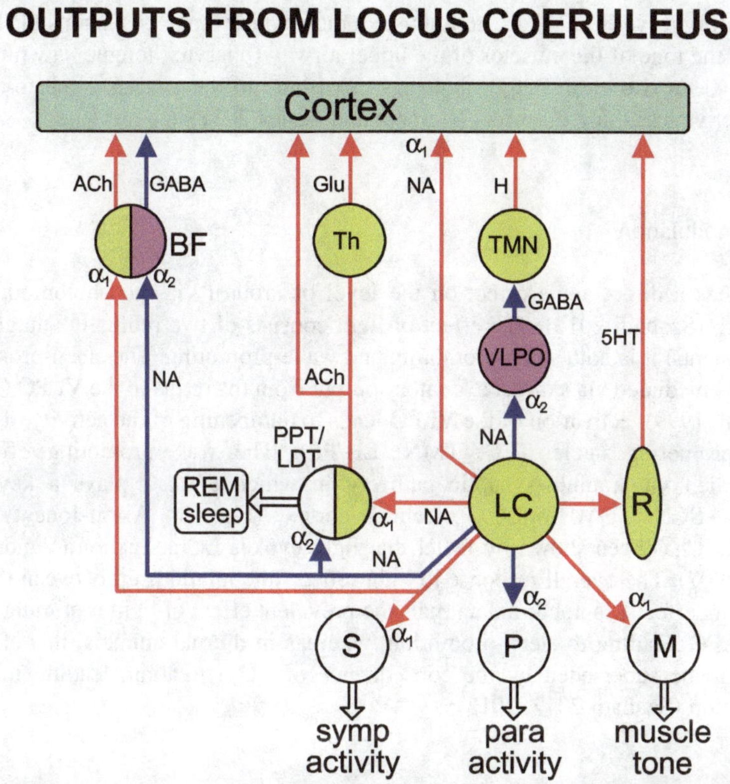

Fig. 5 Schematic diagram of outputs from the noradrenergic locus coeruleus (LC), in relation to the regulation of sleep and arousal. *Wake-promoting nuclei* (*yellow*): BF, basal forebrain; Th, thalamus; TMN, tuberomamillary nucleus; PPT/LDT, pedunculopontine/laterodorsal tegmental nuclei; R, raphe nuclei. *Sleep-promoting nuclei* (*purple*): VLPO, ventrolateral preoptic nucleus; BF, basal forebrain. *Targets outside the sleep/arousal network* (*white*): S, sympathetic preganglionic neurons; P, parasympathetic preganglionic neurons; M, motoneurons. *Connections*: *arrows* (*red*: excitatory; *blue*: inhibitory). *Lettering next to arrow*: neurotransmitter (ACh: acetylcholine; NA: noradrenaline; H: histamine; GABA: γ-aminobutyric acid; 5HT: 5-hydroxytryptamine; Glu: glutamate). *Receptors*: α_1, excitatory α_1-adrenoceptors; α_2, inhibitory α_2-adrenoceptors. See text (Sect. 4.5.1) for details

Motor Regulation

The LC modulates motor activity via excitatory projections to motoneurons (Fig. 5) both in the brainstem and the spinal cord (Szabadi 2013). There is a close link between the modulation of arousal and muscle tone by the LC (McGregor and Siegel 2010). LC activity is reduced during SWS while in REMS there is near complete cessation of LC activity (Gottesmann 2011). The reduction in LC activity during sleep is associated with reduced muscle tone. There is complete cessation of LC activity during attacks of cataplexy in the sleep disorder narcolepsy (Wu et al. 1999), leading to total atonia (Peever 2011). The reduction in the facilitatory drive from the LC to brainstem motor nuclei may be responsible for the loss of the tone of the muscles of the upper airway (pharynx, tongue, jaw muscles) during sleep (Lo et al. 2007), that may contribute to the genesis of sleep-related obstructive breathing disorders (Heinzer and Sériès 2011).

Photomodulation

Light has a direct acute effect on the level of arousal via the photomodulation network (Szabadi 2013). The effect of light consists of two mutually antagonistic components: it is both sleep-promoting and wake-promoting. The sleep-promoting effect is mediated via a direct excitatory output from the retina to the VLPO (Fig. 3; Lu et al. 1999): activation of the VLPO leads to dampening of the activity of major wake-promoting nuclei (LC, TMN, LH/PF). The wake-promoting effect is channelled via a multi-synaptic pathway in which the LC plays a key role: retina → SCN → DMH → LC → cerebral cortex (Fig. 3; Aston-Jones 2005). Indeed, it has been shown by fMRI that light evokes LC activation (Vandewalle et al. 2007). The overall response to light reflects the relationship between the two opposing effects: in nocturnal animals the prevalent effect of light is stimulation of the VLPO, leading to sleep-promotion, whereas in diurnal animals, this effect is likely to be superseded by the consequence of LC activation, leading to wake promotion (Szabadi 2012, 2013).

4.5.2 Cellular Targets of Noradrenergic Drugs

There are three classes of adrenoceptor (α_1, α_2, β): they all have been identified on follower cells ("postsynaptic receptors"). In general, the α_1-adrenoceptors mediate excitatory effects, the α_2-adrenoceptors inhibitory effects and the β-adrenoceptors either excitatory or inhibitory effects. Inhibitory α_2-adrenoceptors have also been identified on the noradrenergic neuron itself ("presynaptic receptors" or "autoreceptors"): in a somato-dendritic location they inhibit the firing of the neuron, whereas located on the nerve terminal they inhibit transmitter release.

Somatodendritic autoreceptors are stimulated by noradrenaline released either from dendrites ("dendritic release") or from recurrent axon collaterals, whereas terminal autoreceptors ("release-modulating receptors") are stimulated by noradrenaline released from the noradrenergic nerve terminals. Stimulation of inhibitory somatodendritic α_2-adrenoceptors dampens the firing of the neuron as activity increases (Huang et al. 2012), providing a mechanism for the autoregulation of central noradrenergic neurons. This mechanism may be responsible for the "switching off" of the central noradrenergic neuron at very high rates of firing (Carter et al. 2010).

Somatodendritic inhibitory α_2-autoreceptors on LC neurons are co-localized with μ opioid receptors: the activation of each receptor leads to cellular inhibition via a shared potassium channel (Christie 1991). Morphine, like the α_2-adrenoceptor agonist clonidine, reduces LC activity (Seutin et al. 1990).

Drugs can also interfere with the release and the reuptake of the transmitter into the nerve terminal or synaptic vesicles. The reuptake of noradrenaline into nerve terminals (Mandela and Ordway 2006) and its accumulation in synaptic vesicles (Zheng et al. 2006) are controlled by specific transporter proteins.

4.5.3 Wake-Promoting Noradrenergic Activators

α_2-Adrenoceptor Antagonists

It is well documented that α_2-adrenoceptor antagonists (yohimbine: Phillips et al. 2000; idazoxan: Glue et al. 1991; atipamezole: Pertovaara et al. 2005) have alerting effects. The alerting effects of these drugs are relatively mild: this is in contrast with the powerful sedative effects of the α_2-adrenoceptor agonists (see Sect. 4.5.4.1).

Noradrenaline Reuptake Inhibitors

These drugs, by impeding the removal of noradrenaline from the synaptic cleft, enhance the effect of noradrenaline. *Selective noradrenaline reuptake inhibitors (NARI)* include the antidepressant reboxetine (Szabadi and Bradshaw 2000) and atomoxetine, a drug indicated for the treatment of attention deficit hyperactivity disorder (ADHD) (Kratochvil et al. 2003). These drugs have relatively mild alerting effects, and insomnia is one of their adverse effects. The NARI have distinct sympathomimetic and parasympatholytic effects, due to the potentiation of the effects of noradrenaline on preganglionic sympathetic and parasympathetic neurons (Szabadi and Bradshaw 2000). *Nonselective catecholamine reuptake inhibitors* (e.g. amphetamine, modafinil), although they also block the reuptake of noradrenaline, act primarily at the dopamine uptake site (see Sect. 4.6.3).

4.5.4 Sleep-Promoting Noradrenergic Inhibitors

α_2-Adrenoceptor Agonists

α_2-Adrenoceptor agonists (e.g. clonidine, medetomidine and dexmedetomidine) are potent sedatives in humans (Hossmann et al. 1980; Scheinin et al. 1987). Dexmedetomidine is used as an anaesthetic (Nelson et al. 2003). The sedative effect of these drugs can be attributed to the stimulation of inhibitory autoreceptors on noradrenergic neurons in the LC leading to the "switching off" of the LC (Abercrombie and Jacobs 1987). The sedation is accompanied by pupil constriction (miosis): this is due to the removal of the excitatory influence of the LC on sympathetic preganglionic neurons and of its inhibitory influence on preganglionic parasympathetic neurons ("disinhibition") (Samuels and Szabadi 2008b). These drugs also cause sedation and miosis in other diurnal species (dog, rabbit); however, they have the opposite effects (i.e. increases in alertness and pupil diameter) in nocturnal animals (cat, rat, mouse). This species difference is likely to reflect the preponderance of the stimulation of α_2-adrenoceptors on LC neurons ("autoreceptors") in diurnal animals versus the stimulation of these receptors on follower neurons ("postsynaptic receptors") in nocturnal animals (Samuels and Szabadi 2008b).

μ Opioid Receptor Agonists

These drugs, including morphine (Paqueron et al. 2002), oxycodone (Verster et al. 2006), tramadol (Lewis and Han 1997) and codeine (Max et al. 1988) are highly sedative in man. The sedation caused by these drugs can be attributed, at least partly, to the inhibition of LC activity via the stimulation of inhibitory μ opioid receptors on the noradrenergic neurons (Seutin et al. 1990). As there is a close association between α_2-adrenoceptors and μ opioid receptors (Sect. 4.5.1), there is a parallelism between the effects of clonidine and morphine on the level of arousal and the diameter of the pupil. Morphine, like clonidine, causes sedation and pupil constriction in diurnal animals (man, dog, rabbit) and increases in alertness and pupil diameter in nocturnal species (cats, rats, mice) (Samuels and Szabadi 2008b).

Apart from its soporific effect, morphine also reduces REM sleep (Cronin et al. 1995). This effect is likely to be due to a decrease in acetylcholine release from central cholinergic neurons involved in REM sleep regulation (Lydic and Baghdoyan 2007).

Monoamine Depletors

Noradrenaline, dopamine and serotonin are accumulated in presynaptic storage vesicles by vesicular monoamine transporter 2, (VMAT2) (Zheng et al. 2006). VMAT2 is inhibited by reserpine and tetrabenazine. Reserpine affects all three monoamines, whereas tetrabenazine shows some selectivity for dopamine. The blockade of VMAT2 leads to emptying ("depletion") of the storage vesicles, resulting in cessation of transmitter release (Shore 1962). The depletion of central noradrenaline stores results in sedation and pupil constriction. Reserpine also depletes peripheral noradrenaline stores leading to a sympatholytic effect (hypotension) (Sulser and Bass 1968). Reserpine is not used clinically any longer. Tetrabenazine is prescribed for the treatment of hyperkinetic movement disorders (e.g. chorea) (Kenney and Jankoovic 2006).

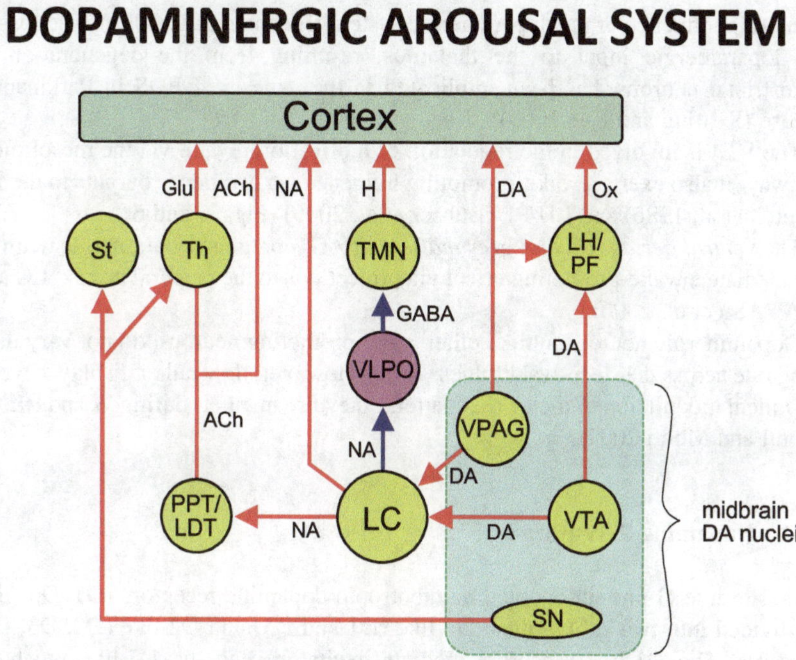

Fig. 6 Schematic diagram of the connections of the midbrain dopaminergic nuclei underlying their wake-promoting propensity. *Wake-promoting nuclei* (*yellow*): midbrain dopaminergic nuclei (SN, substantia nigra; VTA, ventral tegmental area, VPAG: ventral periaqueductal grey matter); TMN, tuberomamillary nucleus; LH/PF, lateral hypothalamic/perifornical area; Th, thalamus; LC, locus coeruleus; PPT/LDT, pedunculopontine/laterodorsal tegmental nuclei; St, striatum. *Sleep-promoting nucleus* (*purple*): VLPO, ventrolateral preoptic nucleus. *Connections*: *arrows* (*red*: excitatory; *blue*: inhibitory). *Lettering next to arrow*: neurotransmitter (ACh: acetylcholine; NA: noradrenaline; H: histamine; Ox: orexin; GABA: γ-aminobutyric acid; DA: dopamine; Glu: glutamate. See text (Sect. 4.6.1). Reproduced, with permission, from Niepel et al. (2013)

4.6 Dopamine

4.6.1 The Dopaminergic Wake-Promoting System

Dopaminergic neurons are localized in distinct nuclei in the midbrain, hypothalamus and olfactory bulb (Niewenhuys 1985). The three nuclei in the midbrain (substantia nigra, VTA, ventral periaqueductal grey matter) are involved in the regulation of motor activity (locomotion), motivation/reward and arousal, whereas the hypothalamic nuclei play a role in neuroendocrine secretion. All three midbrain nuclei make a contribution to the wake-promoting effect of the dopaminergic system, via their direct and indirect projections to the cerebral cortex and other wake-promoting nuclei (Fig. 6; Niepel et al. 2013).

The *substantia nigra (pars compacta)* regulates motor activity via its projection to the striatum (nigrostriatal pathway). This pathway sends excitatory collaterals to the wake-promoting neurons in the thalamus, thus contributing to the wake-promoting influence of the dopaminergic system (Freeman et al. 2001). The loss of the dopaminergic input to the thalamus, resulting from the degeneration of nigrostriatal neurons, has been implicated in the genesis of EDS in Parkinson's disease (Keating and Rye 2003).

The *VTA* is involved in the regulation of motivation/reward via the mesolimbic pathway. It also exerts a wake-promoting influence via excitatory outputs to the LC (Deutch et al. 1986) and LH/PF (Bubser et al. 2005) (Figs. 1 and 6).

The *ventral periaqueductal grey matter (VPAG)* contains dopaminergic neurons that mediate a wake-promoting effect via projections to the cerebral cortex, LC and LH/PF (Lu et al. 2006).

Dopaminergic neurons, unlike other wake-promoting neurons, do not vary their firing rate across the sleep-wakefulness cycle. However, they rather display a cycle-dependent modulation of their firing pattern: they fire in bursts during W and REMS (Monti and Monti 2007).

4.6.2 Dopamine Receptors

There are five G-protein-coupled metabotropic dopamine receptors (D1–D5) that are divided into two subfamilies: D1-like (D1 and D5) and D2-like (D2, D3, D4) receptors. The D1-like receptors mediate excitatory and the D2-like receptors inhibitory cellular responses. The D1-like receptors are post-synaptic, whereas the D2-like receptors are both post-synaptic and presynaptic. The presynaptic receptors inhibit the firing of the neuron and the release of the transmitter (Beaulieu and Gainetdinov 2011).

As D2 receptors occur both presynaptically and postsynaptically, and the presynaptic receptors are more sensitive than the postsynaptic ones, D2 receptor agonists display a dose-dependent dual effect: lower doses decrease and higher doses increase dopamine-mediated functions (Beaulieu and Gainetdinov 2011). It has been proposed that the stimulant effects of D2 receptor agonists on some functions (e.g. locomotion, arousal) may be mediated via inhibition of GABAergic interneurons ("disinhibition") (Monti and Monti 2007).

4.6.3 Wake-Promoting Dopaminergic Activators

Dopamine Reuptake Inhibitors

Blockade of the dopamine transporter (DAT) leads to inhibition of the reuptake of dopamine from the extracellular space, resulting in potentiation of dopaminergic functions. The psychostimulants *amphetamine*, *cocaine* and *methylphenidate* are potent blockers of DAT (Zhu and Reith 2008), and also have marked wake-promoting effects (Nishino et al. 1998). Amphetamine is used to treat EDS/daytime sleep attacks in narcolepsy, and methylphenidate is used for the treatment of ADHD. These drugs also have pro-cognitive effects and a potential for addiction.

Modafinil is a wake-promoting drug licensed for the treatment of EDS in narcolepsy. It also has some procognitive effects (Müller et al. 2013). There is evidence that modafinil, like the psychostimulants amphetamine and cocaine, inhibits dopamine uptake. However, it also differs from the psychostimulants (Wisor 2013). Whereas nonselective psychostimulants (amphetamine, cocaine) enhance all dopaminergic function, including motor activity and motivation/reward, modafinil seems to be relatively selective for the dopaminergic arousal system. This may explain its relatively low addiction potential (Wisor 2013). It has recently been suggested that modafinil may promote wakefulness by selectively potentiating the dopaminergic inhibition of putative sleep-promoting neurons in the nucleus accumbens (Qiu et al. 2012).

Dopamine Release Promoters

The accumulation of monoamines, including dopamine, in presynaptic vesicles is mediated by vesicular monoamine transporter-2 (VMAT-2). Psychostimulants (amphetamine, cocaine, methylphenidate), apart from inhibiting DAT (Sect. 4.6.3.1), also inhibit VMAT-2, thus enhancing the release of dopamine (Riddle et al. 2005).

Nicotine, by stimulating nicotinic receptors on dopaminergic nerve terminals, enhances the release of dopamine (Zhu and Reith 2008). This effect contributes to the wake-promoting propensity of nicotine (see Sect. 4.4.3.3).

Postsynaptic D1 and D2 Dopamine Receptor Agonists

D1 receptor agonists (e.g. SKF 38393) increase wakefulness in experimental animals. D2 receptor agonists (e.g. apomorphine) have a biphasic effect: lower doses decrease alertness, due to stimulation of presynaptic receptors, whereas higher doses increase it, due to the stimulation of postsynaptic receptors (Monti and Monti 2007).

Presynaptic D2 Dopamine Receptor Antagonists

Lower doses of D2 receptor antagonists (e.g. amisulpride) can increase alertness (Patat et al. 1999).

4.6.4 Sleep-Promoting Dopaminergic Inhibitors

Presynaptic D2 Receptor Agonists

D2 receptor agonists (pergolide, pramipexole, ropinirole, bromocriptine) are used to treat Parkinson's disease: they are believed to improve motor deficits through an action at postsynaptic D2 receptors in the striatum. In therapeutic doses these drugs are highly sedative consistent with the stimulation of presynaptic D2 receptors of the wake-promoting dopaminergic neurons in the VTA and VPAG (Samuels et al. 2006). Thus the same dosage of a D2 receptor agonist may stimulate postsysnaptic receptors in the striatum, a structure that is partially denervated in Parkinson's disease, and presynaptic receptors on the anatomically intact wake-promoting neurons of the VTA and VPAG.

Postsynaptic D1 and D2 Receptor Antagonists

D1 receptor antagonists (e.g. SCH 23390) and high doses of D2 receptor antagonists (e.g. haloperidol) have sedative effects (Monti and Monti 2007). The sedation caused by most antipsychotic drugs is only partially attributable to the blockade of postsynaptic D2 dopamine receptors, since many antipsychotics also block other wake-promoting neuroreceptors, such as H1-histamine receptors (Sect. 4.1.4) and cholinoceptors (Sect. 4.4.4)

Monoamine Depletors

Long-term or irreversible blockade of VMAT-2, by reserpine or tetrabenazine, leads to the depletion of monoamine stores, resulting in impairment of central monoaminergic neurotransmission. One of the consequences of central monoamine depletion is sedation (see Sect. 4.5.4).

4.7 Serotonin

4.7.1 The Serotonergic Wake-Promoting System

Serotoninergic neurons are located in nine nuclei (B1–B9) in the midline (raphe) of the brainstem (Niewenhuys 1985). Of these B7 (dorsal raphe nucleus, DR) is involved in the regulation of sleep and arousal (Monti 2010). Serotonergic DR neurons project diffusely to the same areas as LC neurons, including the cerebral cortex, BF and subcortical arousal-modulating nuclei. These neurons vary their activity across the sleep-wakefulness cycle: they are maximally active during W, show reduced activity during SWS and stop firing during REMS. However, mainly due to the distribution of a complex receptor system (Sect. 4.7.2), the serotonergic system does not play a uniform role in the regulation of sleep and arousal: apart from promoting W and SWS, it also initiates SWS (Datta and McLean 2007).

4.7.2 Serotonin Receptors

There are seven classes of serotonin (5-hydroxytryptamine, 5-HT) receptors: 5-HT$_{1-7}$, with a number of subtypes yielding 16 receptors (Hoyer et al. 2002). 5-HT$_3$ receptors are ligand-gated ion channels, whereas all other 5-HT receptors signal via G-proteins. 5-HT$_1$ receptors are negatively, and 5-HT$_{2,4-7}$ receptors are positively coupled to G-proteins. 5-HT$_1$ receptors are inhibitory, and occur both in presynaptic and postsynaptic locations. 5-HT$_2$ and 5-HT$_3$ receptors are excitatory and located mainly postsynaptically. 5-HT$_{2A}$ and 5HT$_{2C}$ receptors are often located on GABAergic interneurons where their activation would mediate an inhibitory influence ("disinhibition") (Monti 2010). In the cerebral cortex, 5-HT$_{2A}$ receptors are co-localized with α_1-adrenoceptors on the same neurons (Santana et al. 2013), indicating synergism between noradrenergic and serotonergic excitatory mechanisms.

4.7.3 Wake-Promoting Serotonergic Activators

Serotonin Receptor Agonists

Systemically administered agonists of 5-HT$_{1A}$ (flesinoxan, 8-OH-DPAT, buspirone), 5-HT$_{1B}$ (GCS 12066B), 5-HT$_{2A/2C}$ (DOI), 5-HT$_3$ (mCPG), 5-HT$_6$ 9SB-399665) and 5-HT$_7$ (LP-44) receptors all increase wakefulness in experimental animals (Monti 2011). In humans, buspirone, used as an anxiolytic, has wake-promoting effects (Manfredi et al. 1991). The mode of action of these different drugs is not known: they are likely to act at different sites in the sleep/arousal network.

Serotonin Reuptake Inhibitors

Released serotonin is removed from the synaptic cleft by the serotonin transporter (SERT): its blockade leads to the potentiation of serotonergic functions (Ramamoorthy et al. 1993). A group of antidepressants (selective serotonin reuptake inhibitors, SSRIs), as their name indicates, act by blocking SERT. These drugs would increase neurotransmission at every serotonergic synapse: in the cerebral cortex increased 5-HT$_{2A}$ receptor stimulation would lead to a wake-promoting effect, whereas in the brainstem 5-HT$_{2C}$ receptor stimulation on GABAergic interneurons, that inhibit wake-promoting noradrenergic neurons in the LC and dopaminergic neurons in the VTA, would lead to a sleep-promoting effect

(Sect. 3.1.1, Figs. 1 and 5). This dual action of the SSRIS may explain their mixed effects on the level of arousal: while they disrupt SWS and cause insomnia, they have little effect on daytime alertness and may even cause sedation (Winokur et al. 2001).

4.7.4 Sleep-Promoting Serotonergic Inhibitors

Serotonin Receptor Antagonists

Since antagonists of 5-HT_{2A} receptors (e.g. ritanserin, ketanserin) increase SWS (Monti 2011), the $5HT_{2A}$ receptor has become a target for the development of anti-insomnia drugs. A number of 5-HT_{2A} receptor antagonists (e.g. volinanserin, eplivanserin) and inverse agonists (e.g. primavanserin) have been developed and are undergoing clinical assessment (Vanover and Davis 2010).

Monoamine Depletors

Inhibitors of VMAT-2, the vesicular membrane pump transporter (Sects. 4.5.4 and 4.6.4), such as reserpine and tetrabenazine, also inhibit the storage of serotonin and lead to the depletion of presynaptic serotonin stores. This leads to a serotonergic contribution to the sleep-promoting effect of these drugs.

4.7.5 REMS-Suppressing Serotonergic Activators

Central serotonergic activation by serotonin receptor agonists and uptake inhibitors leads to the suppression of REMS (Datta and McLean 2007; Monti 2011).

Conclusions

The level of arousal at any one time reflects the intricate interplay between distinct, anatomically and neurochemically defined, wake-promoting and sleep-promoting neuronal systems. These neuronal systems can be grouped into three distinct networks, the sleep/arousal network, the REMS network and the circadian network. In each network there are pharmacologically sensitive sites, such as neuroreceptors and the synaptic machinery: by targeting these sites, drugs can be developed with selective and predictable effects on sleep and arousal. There is complexity at the levels of neuronal connections (e.g. projections to targets via interneurons), cellular localization of neuroreceptors (presynaptic vs postsynaptic) and their signalling mechanisms (excitatory vs inhibitory), and the molecular pharmacology of the drug/

(continued)

receptor interaction (orthosteric vs allosteric). Unravelling the networks at anatomical, cellular and molecular levels has already provided dividends, such as the development of orexin receptor antagonists, $5HT_{2A}$ receptor antagonists and melatonin receptor agonists to treat insomnia and histamine H3 receptor antagonists to treat EDS. The molecular dissection of the $GABA_A$ receptor has already provided subunit selective allosteric modulators with clinical selectivities as hypnotics or anxiolytics. And this is only the beginning of a new area for sleep pharmacology: drugs interacting with the cholinergic, glutamatergic and galaninergic systems provide a hitherto untapped resource.

There is a two-way relationship between the sleep/arousal networks and drugs: while the networks can provide the impetus for the development of novel drugs, drugs can be used as tools for further dissection of the networks. An example for this is modafinil: efforts to discover its mode of action have resulted in important insights into sleep/arousal mechanisms (Qiu et al. 2012).

Finally, new knowledge about the operation of the networks can be obtained from clinical neuropsychiatry and neuropathology. The discovery of the biological bases of the sleep disorders in narcolepsy, RBD and autoimmune encephalitides (Rye 2014), and of the relationship between visual hallucinations and REMS (Manford and Andermann 1998), have yielded valuable insights.

Acknowledgement I am grateful to Rob Langley for drawing the figures.

References

Abe Y, Aoyagi A, Hara T et al (2003) Pharmacological characterization of RS-1259, an orally active dual inhibitor of acetylcholinesterase and serotonin transporter, in rodents: possible treatment of Alzheimer's disease. J Pharmacol Sci 93:95–105

Abercrombie ED, Jacobs BL (1987) Microinjected clonidine inhibits noradrenergic neurons of the locus coeruleus in freely moving cats. Neurosci Lett 76:203–208

Ahnaou A, Drinkenburg WHIM, Bouwknecht JA et al (2008) Blocking melanin-concentrating hormone MCH1 receptor affects rat sleep-wake architecture. Eur J Pharmacol 579:177–188

Ahnaou A, Dautzenberg FM, Geys H et al (2009) Modulation of group II metabotropic glutamate receptor ($mGlu_2$) elecits common changes in rat and mice sleep-wake architecture. Eur J Pharmacol 603:62–72

Ahnaou A, Ver Donck L, Drinkenburg WHIM (2014) Blockade of the metabotropic glutamate (mGluR2) modulates arousal through vigilance states transitions: evidence from sleep-wake EEG in rodents. Behav Brain Res 270:56–67

Argiolas A, Melis MR (2005) Central control of penile erection: role of the paraventricular nucleus of the hypothalamus. Prog Neurobiol 76:1–21

Arias-Carrión O, Murillo-Rodriguez E (2014) Effects of hypocretin/orexin cell transplantation on narcoleptic-like sleep behavior in rats. PLoS One 9:e95342

Arrang JM, Morisset S, Gbahou F (2007) Constitutive activity of the histamine H3 receptor. Trends Pharmacol Sci 28:350–357

Aston-Jones G (2005) Brain structures and receptors involved in alertness. Sleep Med 6(Suppl 1): S3–S7

Aston-Jones G, Cohen JD (2005) An integrative theory of locus coeruleus-norepinephrine function: adaptive gain and optimal performance. Annu Rev Neurosci 28:403–450

Atack JR (2010) Preclinical and clinical pharmacology of the $GABA_A$ receptor α5 subtype inverse agonist α5IA. Pharmacol Ther 125:11–26

Bay T, Eghorn L, Klein AB et al (2014) GHB receptor targets in the CNS: focus on high-affinity binding sites. Biochem Pharmacol 87:220–228

Beaulieu J-M, Gainetdinov RR (2011) The physiology, signaling, and pharmacology of dopamine receptors. Pharmacol Rev 63:182–217

Belelli D, Peden DR, Rosahl TW et al (2005) Extrasynaptic $GABA_A$ receptors of thalamocortical neurons: a molecular target for hypnotics. J Neurosci 25:11513–11520

Benington JH, Kodali SK, Heller HC et al (1995) Stimulation of A_1 adenosine receptors mimics the electroencephalographic effects of sleep deprivation. Brain Res 692:79–85

Berigan TR (2002) Modafinil treatment of excessive daytime sedation and fatigue associated with topiramate. Prim Care Companion J Clin Psychiatry 4:249–250

Berlin M, Boyce CW, Ruiz Mde L (2011) Histamine H3 receptor as a drug discovery target. J Med Chem 54:26–53

Bien CG, Vincent A, Barnett MH et al (2012) Immunopathology of autoantibody-associated encephalitides: clues for pathogenesis. Brain 135:1622–1638

Boscolo-Berto R, Viel G, Montagnese S et al (2012) Narcolepsy and effectiveness of gamma-hydroxybutyrate (GHB): a systematic review and meta-analysis of randomized controlled trials. Sleep Med Rev 16:431–443

Bowery NG (2006) $GABA_B$ receptor: a site of therapeutic benefit. Curr Opin Pharmacol 6:37–43

Brioni JD, Esbenshade TA, Garrison TR et al (2011) Discovery of histamine H_3 antagonists for the treatment of cognitive disorders and Alzheimer's disease. J Pharmacol Exp Ther 336:38–46

Brown DA (2010) Muscarinic acetylcholine receptors (mAChRs) in the nervous system: some functions and mechanisms. J Mol Neurosci 41:340–346

Brown RE, Basheer R, McKenna JT et al (2012) Control of sleep and wakefulness. Physiol Rev 92:1087–1187

Bubser M, Fadel JR, Jackson LL et al (2005) Dopaminergic regulation of orexin neurons. Eur J Neurosci 21:2993–3001

Büttner-Ennever JA, Horn AKE (1997) Anatomical substrates of oculomotor control. Curr Opin Neurobiol 7:872879

Carter ME, Yizhar O, Chikahisha S et al (2010) Tuning arousal with optogenetic modulation of locus coeruleus neurons. Nat Neurosci 13:1526–1533

Cavas M, Scesa G, Navarro JF (2013a) Effects of MPEP, a selective metabotropic glutamate mGlu5 ligand, on sleep and wakefulness in the rat. Prog Neuropsychopharmacol Biol Psychiatry 40:18–25

Cavas M, Scesa G, Navarro JF (2013b) Positive allosteric modulation of mGlu7 receptors by AMN082 affects sleep and wakefulness in the rat. Pharmacol Biochem Behav 103:756–763

Chase MH (2013) Motor control during sleep and wakefulness: clarifying controversies and resolving paradoxes. Sleep Med Rev 17:299–312

Christie MJ (1991) Mechanisms of opioid actions on neurons of the locus coeruleus. Prog Brain Res 88:197–20

Church MK, Church DS (2013) Pharmacology of antihistamines. Indian J Dermatol 58:219–224

Cicolin A, Magliola U, Giordano A et al (2006) Effects of levetiracetam on nocturnal sleep and daytime vigilance in healthy volunteers. Epilepsia 47:82–85

Cronin A, Keier JC, Baghdoyan HA et al (1995) Opioid inhibition of rapid eye movement sleep by a specific mu receptor agonist. Br J Anaesth 74:188–192

Datta S, McLean RR (2007) Neurobiological mechanisms for the regulation of mammalian sleep-wake behavior: reinterpretation of historical evidence and inclusion of contemporary cellular molecular evidence. Neurosci Biobehav Rev 31:775–824

Deutch AY, Goldstein M, Roth RH (1986) Activation of the locus coeruleus induced by selective stimulation of the ventral tegmental area. Brain Res 363:307–314

Dijk DJ, Brunner DP, Aeschbach D et al (1992) The effects of ethanol on human sleep EEG power spectra differ from those of benzodiazepine receptor agonists. Neuropsychopharmacology 7:225–232

Dingledine R, Borges K, Bowie D et al (1999) The glutamate receptor ion channels. Pharmacol Rev 51:7–61

Dodson ER, Zee PC (2010) Therapeutics for circadian rhythm sleep disorders. Sleep Med Clin 5:701–771

Dunbar G, Boeijinga PH, Demazières A et al (2007) Effects of TC-1734 (AZD3480), a selective neuronal nicotinic receptor agonist, on cognitive performance and the EEG of young healthy male volunteers. Psychopharmacology 191:919–929

Ebert U, Kirch W (1998) Scopolamine model of dementia: electroencephalogram findings and cognitive performance. Eur J Clin Invest 28:944–949

España RA, Scammell TE (2011) Sleep neurobiology from a clinical perspective. Sleep 34:845–858

Esplin DW (1970) Drugs acting on the central nervous system – introduction. In: Goodman LS, Gilman A (eds) The pharmacological basis of therapeutics, 4th edn. Macmillan, New York, pp 36–42

Fares A (2011) Night-time exogenous melatonin administration may be a beneficial treatment for sleeping disorders in beta blocker patients. J Cardiovasc Dis Res 2:153–155

Fisher DJ, Daniels R, Jaworska N et al (2012) Effects of acute nicotine administration on resting EEG in nonsmokers. Exp Clin Psychopharmacol 20:71–75

Fort P, Bassetti CL, Luppi PH (2009) Alternating vigilance states: new insights regarding neuronal networks and mechanisms. Eur J Neurosci 29:1741–1753

Freeman A, Ciliax B, Bakay R et al (2001) Nigrostriatal collaterals to thalamus degenerate in parkinsonian animal models. Ann Neurol 50:321–329

Frisullo G, Della Marca G, Mirabella M (2007) A human anti-neuronal autoantibody against $GABA_B$ receptor induces experimental autoimmune agrypnia. Exp Neurol 204:808–818

Froestl W, Gallagher M, Jenkins H et al (2004) SGS742: the first $GABA_B$ receptor antagonist in clinical trials. Biochem Pharmacol 68:1479–1478

Fuller PM, Gooley JJ, Saper CB (2006) Neurobiology of the sleep-wake cycle: sleep architecture, circadian regulation, and regulatory feedback. J Biol Rhythms 21:482–493

Fuller P, Sherman D, Pedersen NP et al (2011) Reassessment of the structural basis of the ascending arousal system. J Comp Neurol 519:933–956

Gallopin T, Luppi PH, Cauli B et al (2005) The endogeneous somnogen adenosine excites a subset of sleep-promoting neurons via A_{2A} receptors in the ventrolateral preoptic nucleus. Neuroscience 134:1377–1390

Gaus SE, Strecker E, Tate BA et al (2002) Ventrolateral preoptic nucleus contains sleep-active, galaninergic neurons in multiple mammalian species. Neuroscience 115:285–294

Gemkow MJ, Davenport AJ, Harich S et al (2009) The histamine H3 receptor as a therapeutic target for CNS disorders. Drug Discov Today 14:9–10

Gitto R, De Luca L, De Grazia S et al (2012) Glutamatergic neurotransmission as molecular target of new anticonvulsants. Curr Top Med Chem 12:971–993

Giuliano F, Rampin O (2004) Neural control of erection. Physiol Behav 83:189–201

Glue P, Wilson S, Lawson C (1991) Acute and chronic idazoxan in normal volunteers: biochemical, physiological and psychological effects. J Psychopharmacol 5:396–403

Gobert A, Rivet J-M, Lejeune F et al (2000) $Serotonin_{2C}$ receptors tonically suppress the activity of mesocortical dopaminergic and adrenergic, but not serotonergic, pathways: a combined dialysis and electrophysiological analysis in the rat. Synapse 36:205–221

Göder R, Seeck-Hirschner M, Stingele K (2011) Sleep and cognition at baseline and the effects of REM sleep diminution after 1 week of antidepressive treatment in patients with depression. J Sleep Res 20:544–551

Gottesmann C (2002) GABA mechanisms and sleep. Neuroscience 111:231–239

Gottesmann C (2011) The involvement of noradrenaline in rapid eye movement sleep mentation. Front Neurol 2:1–10 (Article 81)

Gotti C, Clementi F, Fornari A et al (2009) Structural and functional diversity of native brain neuronal nicotinic receptors. Biochem Pharmacol 78:703–711

Haas HL, Lin J-S (2012) Waking with the hypothalamus. Pflugers Arch Eur J Physiol 463:31–42

Hardeland R (2012) Melatonin in aging and disease – multiple consequences of reduced secretion, options and limits of treatment. Aging Dis 3:194–225

Hassani OK, Lee MG, Jones BE (2009) Melanin-concentrating hormone neurons discharge in a reciprocal manner to orexin neurons across the sleep-wake cycle. Proc Natl Acad Sci USA 106:2418–2422

Heinzer RC, Sériès F (2011) Normal physiology of the upper and lower airways. In: Kryger MH, Roth T, Dement WS (eds) Principles and practice of sleep medicine, 5th edn. Elsevier, St Louis, pp 259–268

Hirshkowitz M, Schmidt MH (2005) Sleep-related erections: clinical perspectives and neural mechanisms. Sleep Med Rev 9:311–329

Hobson JA (2009) REM sleep and dreaming: towards a theory of protoconsciousness. Nat Rev Neurosci 10:803–813

Hoffman EJ, Warren EW (1993) Flumazenil: a benzodiazepine antagonist. Clin Pharm 12:641–656

Hossmann V, Maling TJ, Hamilton CA et al (1980) Sedative and cardiovascular effects of clonidine and nitrazepam. Clin Pharmacol Ther 28:167–176

Hoyer D (2010) Neuropeptide receptor positive allosteric modulation in epilepsy: galanin modulation revealed. Proc Natl Acad Sci USA 107:14943–1494

Hoyer D, Jacobson LH (2013) Orexin in sleep, addiction and more: is the perfect insomnia drug at hand? Neuropeptides 47:477–488

Hoyer D, Hannon JP, Martin GR (2002) Molecular, pharmacological and functional diversity of 5-HT receptors. Pharmacol Biochem Behav 71:533–554

Huang Z-L, Qu W-M, Eguchi N (2005) Adenosine A_{2A}, but not A_1, receptors mediate the arousal effect of caffeine. Nat Neurosci 8:858–859

Huang Z-L, Urade Y, Hayaishi O (2011) The role of adenosine in the regulation of sleep. Curr Top Med Chem 11:1047–1057

Huang H-P, Zhu F-P, Chen X-W et al (2012) Physiology of quantal norepinephrine release from somatodendritic sites of neurons in locus coeruleus. Front Mol Neurosci 5:1–5 (Article 29)

Jego S, Glasgow SD, Gutierrez Herrera C et al (2013) Optogenetic identification of a rapid eye movement sleep modulatory circuit in the hypothalamus. Nat Neurosci 16:1637–1643

Johnson EO, Roehrs T, Roth T et al (1998) Epidemiology of alcohol and medication as aids to sleep in early adulthood. Sleep 21:178–186

Johnston GAR (2013) Advantages of an antagonist: bicuculline and other GABA antagonists. Br J Pharmacol 169:328–336

Johnston GA, Chebib M, Hanrahan JR et al (2003) GABA(C) receptors as drug targets. Curr Drug Targets CNS Neurol Disord 2:260–268

Kalsbeek A, Garidou ML, Palm IF et al (2000) Melatonin sees the light: blocking GABA-ergic transmission in the paraventricular nucleus induces daytime secretion of melatonin. Eur J Neurosci 12:3146–3154

Kalsbeek A, Palm IF, La Fleur SE et al (2006) SCN outputs and the hypothalamic balance of life. J Biol Rhythms 21:458–469

Keating GL, Rye DB (2003) Where you least expect it: dopamine in the pons and modulation of sleep and REM-sleep. Sleep 26:788–789

Kenney C, Jankoovic J (2006) Tetrabenazine in the treatment of hyperkinetic movement disorders. Mov Disord 22:193–197

Ko EM, Eastbrooke IV, McCarthy M et al (2003) Wake-related activity of tubermammillary neurons in rats. Brain Res 992:220–226

Kohlmeier KA, Vardar B, Christensen MH (2013) γ-Hydroxybutyric acid induces actions via the $GABA_B$ receptor in arousal and motor control-related nuclei: implication for therapeutic actions in behavioural state disorders. Neuroscience 248:261–277

Kratochvil CJ, Vaughan BS, Harrington MJ et al (2003) Atomoxetine: a selective noradrenaline reuptake inhibitor for the treatment of attention-deficit/hyperactivity disorder. Expert Opin Pharmacother 4:1165–1174

Krenzer M, Anaclet C, Vetrivelan R et al (2011) Brainstem and spinal cord circuitry regulating REM sleep and muscle atonia. PLoS One 6(10):e24998

Kulisevsky J, Poyurovsky M (2012) Adenosine A_{2A}-receptor antagonism and pathophysiology of Parkinson's disease and drug-induced movement disorders. Eur Neurol 67:4–11

Lal H, Kumar B, Forster MJ (1988) Enhancement of learning and memory in mice by a benzodiazepine antagonist. FASEB J 2:2707–2711

Lambert JL, Cooper MA, Simmons RDJ et al (2009) Neurosteroids: endogenous allosteric modulators of $GABA_A$ receptors. Psychoneuroendocrinology 34S:S48–S58

Lancaster E, Lai M, Peng X et al (2010) Antibodies to the $GABA_B$ receptor in limbic encephalitis with seizures: case series and characterisation of antigen. Lancet Neurol 9:67–76

Landolt H-P (2008) Sleep homeostasis: a role for adenosine in humans? Biochem Pharmacol 75:2070–2079

Lane SD, Green CE, Steinberg JL et al (2012) Cardiovascular and subjective effects of the novel adenosine A_{2A} receptor antagonist SYN115 in cocaine dependent individuals. J Addict Res Ther Suppl 1:1–15

Lazarus M, Shen H-Y, Cherasse Y et al (2011) Arousal effect of caffeine depends on adenosine A_{2A} receptors in the shell of the nucleus accumbens. J Neurosci 31:10067–10075

Le Maître E, Barde SS, Palkovits M et al (2013) Distinct features of neurotransmitter systems in the brain with focus on the galanin system in locus coeruleus and dorsal raphe. Proc Natl Acad Sci USA 110:E536–545

Lee C-Y, Chen C-C, Liou H-H (2009) Levetiracetam inhibits glutamate transmission through presynaptic P/Q-type calcium channels on the granule cells of the dentate gyrus. Br J Pharmacol 158:1753–1762

Lerner JT, Sankar R, Mazarati AM (2008) Galanin and epilepsy. Cell Mol Life Sci 65:1864–1871

Leurs R, Blanidan P, Tedford C (1998) Therapeutic potential of histamine H_3 receptor agonists and antagonists. Trends Pharmacol Sci 19:177–183

Lewis KS, Han NH (1997) Tramadol: a new centrally acting analgesic. Am J Health Syst Pharm 54:643–652

Lin JS, Sakai K, Jouvet M (1988) Evidence for histaminergic arousal mechanisms in the hypothalamus of cat. Neuropharmacology 27:111–122

Lin J-S, Anaclet C, Sergeeva OA et al (2011) The waking brain: an update. Cell Mol Life Sci 68:2499–2512

Lo YL, Jordan AS, Malhorra A et al (2007) The influence of wakefulness on pharyngeal muscle activity. Thorax 62:799–805

Lu J, Shiromani P, Saper CB (1999) Retinal input to the sleep-active ventrolateral preoptic nucleus in the rat. Neuroscience 93:209–214

Lu J, Jhou TC, Saper CB (2006) Identification of wake-active dopaminergic neurons in the ventral periaqueductal gray matter. J Neurosci 26:193–202

Luppi P-H, Clément O, Sapin E et al (2011) The neuronal network responsible for paradoxical sleep and its dysfunctions causing narcolepsy and rapid eye movement (REM) behavior disorder. Sleep Med Rev 15:153–163

Luppi P-H, Clément O, Fort P (2013) Paradoxical (REM) sleep genesis by the brainstem is under hypothalamic control. Curr Opin Neurobiol 23:786–792

Lydic R, Baghdoyan HA (2007) Neurochemical mechanisms mediating opioid-induced REM sleep disruption. In: Lavigne G, Sessle B, Choinière M et al (eds) Sleep and pain. IASP Press, Seattle, pp 99–122
Mandela P, Ordway GA (2006) The norepinephrine transporter and its regulation. J Neurochem 97:310–333
Manford M, Andermann F (1998) Complex visual hallucinations – clinical and neurobiological insights. Brain 121:1819–1840
Manfredi RL, Kales A, Vgontzas AN et al (1991) Buspirone: sedative or stimulant effect? Am J Psychiatry 148:1213–1217
Manns ID, Lee MG, Modirrousta M et al (2003) Alpha 2 adrenergic receptors on GABAergic, putative sleep-promoting basal forebrain neurons. Eur J Neurosci 18:723–727
Martinez-Cué C, Delatour B, Potier M-C (2014) Treating enhanced GABAergic inhibition in Down syndrome: use of GABA α5-selective inverse agonists. Neurosci Biobehav Rev. doi:10.1016/j.neurobiorev.2013.12.008
Massotti M, Lucantoni D, Caporali MG et al (1985) Supraspinal convulsions induced by inverse benzodiazepine agonists in rabbits. J Pharmacol Exp Ther 234:274–279
Matthias S, Wetter TC, Steiger A et al (2001) The GABA uptake inhibitor tiagabine promotes slow wave sleep in normal elderly subjects. Neurobiol Aging 22:247–253
Max MB, Schafer SC, Culnane M et al (1988) Association of pain relief with drug side effects in postherpetic neuralgia: a single-dose study of clonidine, codeine, ibuprofen, and placebo. Clin Pharmacol Ther 43:363–371
McCarley RW (2007) Neurobiology of REM and NREM sleep. Sleep Med 8:302–330
McCormick DA (1992) Neurotransmitter actions in the thalamus and cerebral cortex and their role in neuromodulation of thalamocortical activity. Prog Neurobiol 39:37–388
McCormick DA, Bal T (1997) Sleep and arousal: thalamocortical mechanisms. Annu Rev Neurosci 20:185–215
McGregor R, Siegel JM (2010) Illuminating the locus coeruleus: control of posture and arousal. Nat Neurosci 13:1448–1449
Meuret P, Backman SB, Bonhomme V et al (2000) Physostigmine reverses propofol-induced unconsciousness and attenuation of the auditory steady state response and bispectral index in human volunteers. Anesthesiology 93:708–717
Mieda M, Willie JT, Hara J et al (2004) Orexin peptides prevent cataplexy and improve wakefulness in an orexin neuron-ablated model of narcolepsy in mice. Proc Natl Acad Sci USA 101:4649–4654
Minzenberg MJ, Watrous AJ, Yoon JH et al (2008) Modafinil shifts human locus coeruleus to low-tonic, high-phasic activity during functional MRI. Science 322:1700–1702
Mistlberger RE (2005) Circadian regulation of sleep in mammals: role of the suprachiasmatic nucleus. Brain Res Rev 49:429–454
Monti JM (2010) The structure of the dorsal raphe nucleus and its relevance to the regulation of sleep and wakefulness. Sleep Med Rev 14:307–317
Monti JM (2011) Serotonin control of sleep-wake behavior. Sleep Med Rev 15:269–281
Monti JM, Monti D (2007) The involvement of dopamine in the modulation of sleep and waking. Sleep Med Rev 11:113–133
Monti JM, Torterolo P, Lagos P (2013) Melanin-concentrating hormone control of sleep-wake behaviour. Sleep Med Rev 17:293–298
Moos DD (2007) Central anticholinergic syndrome: a case report. J Perianesth Nurs 22:309–321
Morairty S, Rainnie D, McCarley R et al (2004) Disinhibition of ventrolateral preoptic area sleep-active neurons by adenosine: a new mechanism for sleep promotion. Neuroscience 123:451–457
Müller U, Rowe JB, Rittman T et al (2013) Effects of modafinil on non-verbal cognition, task enjoyment and creative thinking in healthy volunteers. Neuropharmacology 64:490–495
Murck H, Antonijevic IA, Frieboes RM et al (1999) Galanin has REM-sleep deprivation-like effects on the sleep EEG in healthy young men. J Psychiatr Res 33:225–232

Murphy PR, O'Connell RG, O'Sullivan M et al (2014) Pupil diameter covaries with BOLD activity in human locus coeruleus. Hum Brain Mapp. doi:10.1002/hbm.22466
Myers CS, Taylor RC, Salmeron BJ et al (2013) Nicotine enhances alerting, but not executive, attention in smokers and nonsmokers. Nicotine Tob Res 15:277–281
Nelson LE, Guo TZ, Lu J et al (2002) The sedative component of anesthesia is mediated by $GABA_A$ receptors in an endogenous sleep pathway. Nat Neurosci 5:979–984
Nelson LE, Lu J, Guo TZ et al (2003) The α2-adrenoceptor agonist dexmedetomidine converges on an endogenous sleep-promoting pathway to exert its sedative effects. Anesthesiology 98:428–436
Niepel G, Bibani RH, Vilisaar J et al (2013) Association of a deficit of arousal with fatigue in multiple sclerosis: effect of modafinil. Neuropharmacology 64:380–388
Niewenhuys R (1985) Chemoarchitecture of the brain. Springer, Berlin
Nishino S, Mao J, Sampathkumaran R et al (1998) Increased dopaminergic transmission mediates the wake-promoting effects of CNS stimulants. Sleep Res Online 1:49–61
Niswender CM, Conn PJ (2010) Metabotropic glutamate receptors: physiology, pharmacology, and disease. Annu Rev Pharmacol Toxicol 50:295–322
Nutt DJ, Stahl SM (2010) Searching for perfect sleep: evolution of $GABA_A$ receptor modulators as hypnotics. J Psychopharmacol 24:1601–1612
Ögren SO, Kuteeva E, Hökfelt T et al (2006) Galanin receptor antagonists: a potential novel pharmacological treatment for mood disorders. CNS Drugs 20:633–654
Ohno K, Sakurai T (2008) Orexin neuronal circuitry: role in the regulation of sleep and wakefulness. Front Neuroendocrin 29:70–87
Pandi-Perumal SR, Srinivasan V, Mastroni GJM et al (2006) Melatonin: nature's most versatile signal? FEBS J 273:2813–2838
Paqueron X, Lumbroso A, Mergoni P et al (2002) Is morphine-induced sedation synonymous with analgesia during intravenous morphine titration? Br J Anaesth 89:697–701
Parissis D, Mellidis C, Boutis A et al (2003) Neurological findings in a case of coma secondary to *Datura stramonium* poisoning. Eur J Neurol 10:745–746
Parmentier AC, Guhennec C et al (2007) The brain H_3 receptor as a novel therapeutic target for vigilance and sleep-wake disorders. Biochem Pharmacol 73:1157–1171
Patat A, Rosenzweig P, Miget N et al (1999) Effects of 50 mg amisulpride on EEG, psychomotor and cognitive functions in healthy sleep-deprived subjects. Fundam Clin Pharmacol 13:582–594
Peever J (2011) Control of motoneuron function and masseter muscle tone during REM sleep, REM sleep behaviour disorder and cataplexy/narcolepsy. Arch Ital Biol 149:454–466
Peigneux P, Laureys S, Fuchs S et al (2001) Generation of rapid eye movements during paradoxical sleep in humans. NeuroImage 14:701–708
Pertovaara A, Haapaalinna A, Sirviö J et al (2005) Pharmacological properties, central nervous system effects, and potential therapeutic application of atipamezole, a selective alpha2-adrenoceptor antagonist. CNS Drug Rev 11:273–288
Petit-Pedrol M, Armangue T, Peng X et al (2014) Encephalitis with refractory seizures, status epilepticus, and antibodies to the $GABA_A$ receptor: a case series, characterisation of the antigen, and analysis of the effects of antibodies. Lancet Neurol 13:276–286
Phillips MA, Szabadi E, Bradshaw CM (2000) Comparison of the effects of clonidine and yohimbine on spontaneous pupillary fluctuations in healthy human volunteers. Psychopharmacology 150:85–89
Pinard A, Seddick R, Bettler B (2010) $GABA_B$ receptors: physiological functions and mechanisms of diversity. Adv Pharmacol 58:231–255
Porkka-Heiskanen T (2013) Sleep homeostasis. Curr Opin Neurobiol 23:799–805
Poulsen CF, Simeone TA, Mar TE et al (2004) Modulation by topiramate of AMPA and kainite mediated calcium influx in cultured cerebral cortical, hippocampal and cerebellar neurons. Neurochem Res 29:275–282

Proctor A, Bianchi MT (2012) Clinical pharmacology in sleep medicine. ISRN Pharmacol (Article ID 914168). doi:10.5402/2012/914168
Qiu M-H, Liu W, Qu W-M (2012) The role of nucleus accumbens core/shell in sleep-wake regulation and their involvement in modafinil-induced arousal. PLoS One 7:e45471
Rajarao SJ, Platt B, Sukoff SJ et al (2007) Anxiolytic-like activity of the non-selective galanin receptor agonist, galnon. Neuropeptides 41:307–320
Ramamoorthy S, Bauman AL, Moore KR et al (1993) Antidepressant- and cocaine-sensitive human serotonin transporter: molecular cloning, expression, and chromosomal localization. Proc Natl Acad Sci USA 90:2542–2546
Rao U, Lin KM, Schramm P et al (2004) REM sleep and cortisol responses to scopolamine during depression and remission in women. Int J Neuropsychopharmacol 7:265–274
Reynolds GP (2012) Histamine and antipsychotic drug-induced weight gain. J Psychopharmacol 26:1608–1609
Riddle EL, Fleckenstein AE, Hanson GR (2005) Role of monoamine transporters in mediating psychostimulant effects. AAPS J 7:E847–851
Roth T, Lines C, Vandormael K et al (2010) Effect of gaboxadol on patient-reported measures of sleep and waking function in patients with primary insomnia: results from two randomized, controlled, 3-month studies. J Clin Sleep Med 6:30–39
Rudolph U (2004) GABAergic system. In: Offermanns S, Rosehthal W (eds) Encyclopedic reference of molecular pharmacology. Springer, Berlin, pp 373–378
Rudolph U, Knoflach F (2011) Beyond classical benzodiazepines: novel therapeutic potential of $GABA_A$ receptor subtypes. Nat Rev Drug Discov 10:685–697
Rye D (2014) Paralysis lost: a new cause for a common parasomnia? Lancet Neurol 13:531–532
Sakai T (2013) Orexin deficiency and narcolepsy. Curr Opin Neurobiol 23:760–766
Samuels ER, Szabadi E (2008a) Functional neuroanatomy of the noradrenergic locus coeruleus: its roles in the regulation of arousal and autonomic function. Part I: principles of functional organisation. Curr Neuropharmacol 6:235–253
Samuels ER, Szabadi E (2008b) Functional neuroanatomy of the noradrenergic locus coeruleus: its roles in the regulation of arousal and autonomic function. Part II: physiological and pharmacological manipulations and pathological alterations of locus coeruleus activity in humans. Curr Neuropharmacol 6:254–285
Samuels ER, Hou RH, Langley RW et al (2006) Comparison of pramipexole and modafinil on arousal, autonomic and endocrine functions in healthy volunteers. J Psychopharmacol 20:756–770
Sánchez-López A, Escudero M (2011) Tonic and phasic components of eye movements during REM sleep in the rat. Eur J Neurosci 33:2129–2138
Santana N, Mengod G, Artigas F (2013) Expression of α_1-adrenergic receptors in rat prefrontal cortex: cellular co-localization with 5-HT_{2A} receptors. Int J Neuropsychopharmacol 16:1139–1151
Santucci V, Glatt A, Demieville H et al (1981) Quantification of slow wave EEG induced by atropine: effects of physostigmine, amphetamine and haloperidol. Eur J Pharmacol 73:113–122
Saper CB, Fuller PM, Pedersen NP et al (2010) Sleep state switching. Neuron 68:1023–1042
Scammell TE, Gershchenko DY, Mochizuki T et al (2001) An adenosine A2a agonist increases sleep and induces Fos in ventrolateral preoptic neurons. Neuroscience 107:653–663
Scheife R, Takeda M (2005) Central nervous system safety of anticholinergic drugs for the treatment of overactive bladder in the elderly. Clin Ther 27:144–153
Scheinin M, Kallio A, Koulu M (1987) Sedative and cardiovascular effects of medetomidine, a novel selective alpha 2-adrenoceptor agonist, in healthy volunteers. Br J Clin Pharmacol 24:443–451
Schredl M, Hornung O, Regen F (2006) The effect of donepezil on sleep in elderly, healthy persons: a double-blind placebo-controlled study. Pharmacopsychiatry 39:205–208

Schwartz J-C (2011) The histamine H3 receptor: from discovery to clinical trials with pitolisant. Br J Pharmacol 163:713–721

Schwarz RD, Callahan MJ, Coughenour LL (1999) Milameline (CI-979/RU35926): a muscarinic receptor agonist with cognition-activating properties: biochemical and in vivo characterization. J Pharmacol Exp Ther 291:812–822

Seutin V, Franchimont N, Massotte L et al (1990) Comparison of the effect of morphine on locus coeruleus noradrenergic and ventral tegmental area dopaminergic neurons *in vitro*. Life Sci 46:1879–1885

Shore PA (1962) Release of serotonin and catecholamines by drugs. Pharmacol Rev 14:531–550

Sigel E, Steinmann ME (2012) Structure, function, and modulation of $GABA_A$ receptors. J Biol Chem 287:40224–40231

Sitaram N, Moore AM, Gillin JC (1978) The effect of physostigmine on normal human sleep and dreaming. Arch Gen Psychiatry 35:1239–1243

Stäubli U, Scafidi J, Chun D (1999) $GABA_B$ receptor antagonism: facilitatory effects on memory parallel those on LTP induced by TBS but not HFS. J Neurosci 19:4609–4615

Steiger A (2007) Neurochemical regulation of sleep. J Psychiatr Res 41:537–552

Sterpenich V, D'Argembeau A, Desseilles M et al (2006) The locus ceruleus is involved in the successful retrieval of emotional memories in humans. J Neurosci 26:7416–7423

Sulser F, Bass AD (1968) Pharmacodynamic and biochemical considerations on the mode of action of reserpine-like drugs. In: Efron DH, Cole JO, Levine J et al (eds) Psychopharmacology: a review of progress 1957–1967. Public Health Service Publication No 1836. Washington, DC, pp 1093–1100

Szabadi E (2012) Modulation of physiological reflexes by pain: role of the locus coeruleus. Front Integr Neurosci 6:1–15 (Article 94)

Szabadi E (2013) Functional neuroanatomy of the central noradrenergic system. J Psychopharmacol 27:659–693

Szabadi E (2014) Selective targets for arousal-modifying drugs: implications for the treatment of sleep disorders. Drug Discov Today 19:701–708

Szabadi E, Bradshaw CM (2000) Mechanisms of action of reboxetine. Rev Contemp Pharmacother 11:267–282

Taberan IV (2013) Functional pharmacology of histamine H_1 receptors expressed in mouse preoptic/anterior hypothalamic neurons. Br J Pharmacol 170:415–425

Tafti M (2007) Reply to 'Promotion of sleep by targeting the orexin system in rats, dogs and humans'. Nat Med 13:525–526

Thakkar MM (2011) Histamine in the regulation of wakefulness. Sleep Med Rev 15:65–74

Torterolo P, Morales FR, Chase MH (2002) GABAergic mechanisms in the pedunculopontine tegmental nucleus of the cat promote active (REM) sleep. Brain Res 944:1–9

Triarhou LC (2006) The percipient observations of Constantin von Economo on encephalitis lethargica and sleep disruption and their lasting impact on contemporary sleep research. Brain Res Bull 69:244–258

Tsunematsu T, Ueno T, Tabuchi S et al (2014) Optogenetic manipulation of activity and temporally controlled cell-specific ablation reveal a role for MCH neurons in sleep/wake regulation. J Neurosci 34:6896–6909

Van den Heuvel CJ, Reid KJ, Dawson D (1997) Effect of atenolol on nocturnal sleep and temperature in young men: reversal by pharmacological doses of melatonin. Physiol Behav 61:795–802

Vandewalle G, Schmidt C, Albouy G et al (2007) Brain responses to violet, blue and green monochromatic light exposures in humans: prominent role of blue light and the brainstem. PLoS One 2(11):e1247

Vanover KE, Davis RE (2010) Role of 5-HT_{2A} receptor antagonists in the treatment of insomnia. Nat Sci Sleep 2:139–150

Venault P, Chapouthier G (2007) From behavioural pharmacology of beta-carbolines to seizures, anxiety, and memory. ScientificWorldJournal 7:204–223

Verster JC, Veldhuijzen DS, Volkerts ER (2006) Effects of an opioid (oxycodone/paracetamol) and an NSAID (bromfenac) on driving ability, memory functioning, psychomotor performance, pupil size, and mood. Clin J Pain 22:499–504

Votava Z, Benesová O, Bohdanecký Z et al (1968) Influence of atropine, scopolamine and benactyzine on the physostigmine arousal reaction in rabbits. Prog Brain Res 28:40–47

Webling KE, Runesson J, Bartfai T et al (2012) Galanin receptors and ligands. Front Endocrinol (Lausanne) 3:146

Weinhold SL, Seeck-Hirschner NA et al (2014) The effect of intranasal orexin-A (hypocretin-1) on sleep, wakefulness and attention in narcolepsy with cataplexy. Behav Brain Res 262:8–13

Wilhelm B, Giedke H, Lüdtke H et al (2001) Daytime variations in central nervous system activation measured by a pupillographic sleepiness test. J Sleep Res 10:1–7

Wilson SJ, Nutt DJ, Alford C et al (2010) British Association for Psychopharmacology consensus statement on evidence-based treatment of insomnia, parasomnias and circadian rhythm disorders. J Pschopharmacol 24:1577–1600

Winokur A, Gary KA, Rodner S et al (2001) Depression, sleep physiology, and antidepressant drugs. Depress Anxiety 14:19–28

Winsky-Sommerer R (2009) Role of $GABA_A$ receptors in the physiology and pharmacology of sleep. Eur J Neurosci 29:1779–1794

Wisor J (2013) Modafinil as a catecholaminergic agent: empirical evidence and unanswered questions. Front Neurol 4:139 (eCollection 2013)

Wu MF, Gulyani SA, Yau E et al (1999) Locus coeruleus neurons: cessation of activity during cataplexy. Neuroscience 91:1389–1399

Zant JC, Rozov S, Wigren HK et al (2012) Histamine release in the basal forebrain mediates cortical activation through cholinergic neurons. J Neurosci 32:13244–13254

Zecharia AY, Yu X, Götz T et al (2012) GABAergic inhibition of histaminergic neurons regulates active waking but not the sleep-wake switch or propofol-induced loss of consciousness. J Neurosci 32:13062–13075

Zhang J-P, Qi X, Yuan X-S et al (2013) Projections of nucleus accumbens adenosine A_{2A} receptor neurons in the mouse brain and their implications in medicating sleep-wake regulations. Front Neuroanat 7:1–15 (Article 43)

Zheng G, Dwoskin LP, Crooks PA (2006) Vesicular monoamine transporter 2: role as a novel target for drug development. AAPS J 8:E682–E692

Zhu J, Reith MEA (2008) Role of dopamine transporter in the action of psychostimulants, nicotine, and other drugs of abuse. CNS Neurol Disord Drug Targets 7:393–409

Classification of Sleep Disorders

Michael Thorpy

Abstract The *International Classification of Sleep Disorders* (ICSD-3) produced by the American Academy of Sleep Medicine is a major revision of the prior classification and was published in 2014. It is a major advance over previous versions but it is unfortunate that some of the diagnostic criteria differ from that of the American Psychiatric Association's revised version of the *Diagnostic and Statistical Manual of Mental Disorders fifth edition (DSM-V)* published in 2013 which includes a smaller section entitled "Sleep Wake Disorders." The DSM-V serves as an entry level classification, mainly for psychiatrists, and it is to be hoped that in the future the two classifications will be merged into one that will cause less confusion not only for clinicians but also for agencies that reimburse for health care and provide for treatment options. This review discusses the main features of the diagnostic entries in the ICSD-3 and presents the pharmacological treatment focus for the disorders.

The *International Classification of Sleep Disorders* (ICSD-3) produced by the American Academy of Sleep Medicine was a major revision of the prior classification and was published in 2014 [1 ICSD3]. In 2013 the American Psychiatric Association published the revised version of the *Diagnostic and Statistical manual of mental Disorders fifth edition (DSM-V)* which includes a section entitled "Sleep Wake Disorders," an update of the DSM-IV section [2 DSMV]. This more simplified classification system results in a classification for mental health and general medical clinicians who are not experts in sleep medicine. *The International Classification of Diseases* modified version, the ICD-10-CM, that will be adopted in the USA in 2014 contains a classification that more closely conforms to the ICSD-3 [3 ICD10]. The ICSD3 will be used for this review.

The *International Classification of Sleep Disorders* (ICSD-3) is a major revision of the ICSD-2 and was published in March of 2014 (Table 1). The main change was

M. Thorpy, M.D. (✉)
Sleep-wake Disorders Center, Montefiore Medical center, Albert Einstein College of Medicine, Bronx, NY, USA
e-mail: thorpy@aecom.yu.edu

A. Guglietta (ed.), *Drug Treatment of Sleep Disorders*, Milestones in Drug Therapy,
DOI 10.1007/978-3-319-11514-6_3

Table 1 ICSD-3 (Adapted from the International Classification of Sleep Disorders, Third Revision 2014)

	ICD-9-CM code:	ICD-10-CM code:
Insomnia disorders:		
Chronic insomnia disorder	342	F51.01
Short-term insomnia disorder	307.41	F51.02
Other insomnia disorder	307.49	F51.09
Isolated symptoms and normal variants:		
Excessive time in bed		
Short sleeper		
Sleep-related breathing disorders:		
Obstructive sleep apnea disorders:		
Obstructive sleep apnea, adult	327.23	G47.33
Obstructive sleep apnea, pediatric	327.23	G47.33
Central sleep apnea syndromes:		
Central sleep apnea with Cheyne–Stokes breathing	786.04	R06.3
Central apnea due to a medical disorder without Cheyne-Stokes breathing	327.27	G47.37
Central sleep apnea due to high altitude periodic breathing	327.22	G47.32
Central sleep apnea due to a medication or substance	327.29	G47.39
Primary central sleep apnea	327.21	G47.31
Primary central sleep apnea of infancy	770.81	P28.3
Primary central sleep apnea of prematurity	770.82	P28.4
Treatment-emergent central sleep apnea	327.29	G47.39
Sleep-related hypoventilation disorders:		
Obesity hypoventilation syndrome	278.03	E66.2
Congenital central alveolar hypoventilation syndrome	327.25	G47.35
Late-onset central hypoventilation with hypothalamic dysfunction	327.26	G47.36
Idiopathic central alveolar hypoventilation	327.24	G47.34
Sleep-related hypoventilation due to a medication or substance	327.26	G47.36
Sleep-related hypoventilation due to a medical disorder	327.26	G47.36
Sleep-related hypoxemia disorder:		
Sleep-related hypoxemia	327.26	G47.36
Isolated symptoms and normal variants:		
Snoring		
Catathrenia		
Central disorders of hypersomnolence:		
Narcolepsy Type 1	347.01	G47.411

(continued)

Table 1 (continued)

	ICD-9-CM code:	ICD-10-CM code:
Narcolepsy Type 2	347.00	G47.419
Idiopathic hypersomnia	327.11	G47.11
Kleine–Levin syndrome	327.13	G47.13
Hypersomnia due to a medical disorder	327.14	G47.14
Hypersomnia due to a medication or substance	292.85 (drug-induced);	F11-F19
	291.82 (alcohol-induced)	
Hypersomnia associated with a psychiatric disorder	327.15	F51.13
Insufficient sleep syndrome	307.44	F51.12
Isolated symptoms and normal variants:		
Long sleeper		
Circadian rhythm sleep–wake disorders:		
Delayed sleep–wake phase disorder	327.31	G47.21
Advanced sleep–wake phase disorder	327.32	G47.22
Irregular sleep–wake rhythm disorder	327.33	G47.23
Non-24-h sleep–wake rhythm disorder	327.34	G47.24
Shift work disorder	327.36	G47.26
Jet lag disorder	327.35	G47.25
Circadian sleep–wake disorder not otherwise specified (NOS)	327.30	G47.20
Parasomnias:		
NREM-related parasomnias:		
Disorders of arousal (From NREM sleep)		
Confusional arousals	327.41	G47.51
Sleepwalking	307.46	F51.3
Sleep terrors	307.46	F51.4
Sleep-related eating disorder	327.40	G47.59
REM-related parasomnias		
REM sleep behavior disorder:	327.42	G47.52
Recurrent isolated sleep paralysis	327.43	G47.51
Nightmare disorder	307.47	F51.5
Other parasomnias:		
Exploding head syndrome	327.49	G47.59
Sleep-related hallucinations	368.16	H53.16
Sleep enuresis	788.36	N39.44
Parasomnia due to a medical disorder	327.44	G47.54
Parasomnia due to a medication or substance	292.85 (drug-induced)	F11-F19
	291.82 (alcohol-induced)	
Parasomnia, unspecified	327.40	G47.50

(continued)

Table 1 (continued)

	ICD-9-CM code:	ICD-10-CM code:
Isolated symptoms and normal:		
Sleep talking		
Sleep-related movement disorders:		
Restless legs syndrome	333.94	G25.81
Periodic limb movement disorder	327.51	G47.61
Sleep-related leg cramps	327.52	G47.62
Sleep-related bruxism	327.53	G47.63
Sleep-related rhythmic movement disorder	327.59	G47.69
Benign sleep myoclonus of infancy	327.59	G47.69
Propriospinal myoclonus at sleep onset	327.59	G47.69
Sleep-related movement disorder due to a medical disorder	327.59	G47.69
Sleep-related movement disorder due to a medication or substance	292.85 (drug-induced)	F11-F19
	291.82 (alcohol-induced)	
Sleep-related movement disorder, unspecified	327.59	G47.69
Isolated symptoms and normal variants:		
Excessive fragmentary myoclonus		
Hypnagogic foot tremor and alternating leg muscle activation		
Sleep starts (Hypnic Jerks)		
Other sleep disorder	327.8	G47.8
Appendix A:		
Fatal familial insomnia	046.8	A81.83
Sleep-related epilepsy	345	G40.5
Sleep-related headaches	784.0	R51
Sleep-related laryngospasm	787.2	J38.5
Sleep-related gastroesophageal reflux	530.1	K21.9
Sleep-related myocardial ischemia	411.8	I25.6
Appendix B:		
ICD-10-CM coding for substance-induced sleep disorders		F10-F19

the simplification of the Insomnia disorders and an expansion of the sleep-related breathing disorders.

The organization of the ICSD-3 produced a greater degree of standardization between disorder texts. It includes information in all the following categories where available: Alternate Names, Diagnostic Criteria, Essential Features, Associated Features, Clinical and Pathophysiological Subtypes, Demographics: Prevalence, Gender Bias, Racial/Ethnic Bias, Cultural Issues, Predisposing and Precipitating Factors: Risk factors, Familial Pattern (Genetics, Familial clusters), Onset, Course, and Complications: Medical, Neurological, Psychiatric/social, Developmental

Issues (Pediatric, Geriatric), Pathology and Pathophysiology, Objective Findings; Sleep logs, Actigraphy, Questionnaires, Polysomnography, Multiple sleep latency test, Neurological (Electroencephalogram, Cerebrospinal Fluid, Neuroimaging, Electromyogram, Autonomic), Endocrine, Genetic Testing, Physical Findings (Respiratory, Arterial Blood Gas, Pulmonary Function, Ventilatory Response), Cardiac (Electrocardiogram, Echocardiogram, Cardiac Catheterization), and Serum Chemistry. Several disorders are now classified as Isolated Symptoms and Normal Variants, which includes: Excessive Time in Bed, Short Sleeper, Snoring, Catathrenia, Long Sleeper, Sleep Talking, Excessive Fragmentary Myoclonus, Hypnagogic Foot Tremor and Alternating Leg Muscle Activation, and Sleep Starts (Hypnic Jerks).

1 Insomnia Disorders

The Insomnia disorders are characterized by one major disorder termed Chronic Insomnia Disorder. This recognizes the fact that the clinical features of insomnia can be the result of a primary or secondary process but the consequences are similar no matter what the etiology (Edinger et al. 2011). The diagnosis rests upon a sleep symptom such as difficulty initiating sleep that occurs three times per week for at least 3 months and has daytime consequences. Psychophysiological insomnia and insomnia disorders of the ICSD-2 are mentioned as subtypes of chronic insomnia disorder. The inclusion of Short-Term Insomnia Disorder with similar diagnostic criteria applies to insomnia that is less than 3 months in duration. Excessive time in bed and short sleeper are included as isolated symptoms and normal variants, not as specific disorders.

Treatment of the insomnia disorders is by both behavioral as well as pharmacological means. Cognitive behavior therapy for insomnia (CBT-I) is a well recognized effective means of treating insomnia that is often combined with pharmacological therapy. Most often the hypnotics are used in conjunction with CBT-I. Those FDA approved include the benzodiazepines such as triazolam, flurazepam, and temazepam which have largely been replaced with the newer benzodiazepine receptor agonists (BZRAs) such as zolpidem, eszopiclone, and zaleplon (Schwartz and Goradia 2013). The choice of agent depends upon the half-life of the medication, side effect profile, and reimbursement issues. Alternative approved hypnotics include a melatonin agonist, ramelteon, and a sedating antidepressant medication doxepin. Sedating antidepressants are not FDA approved for insomnia but are often used due to their lack of drug scheduling and potential for habituation, such as trazodone or amitriptyline (Bertisch et al. 2014). New hypnotic agents based upon hypocretin receptor antagonism are likely to be available in 2015 (Zisapel 2012).

2 Sleep-Related Breathing Disorders

The Sleep-Related Breathing Disorders are organized into four main categories: obstructive sleep apnea (OSA) disorders, central sleep apnea (CSA) syndromes, sleep-related hypoventilation disorders, and sleep-related hypoxemia disorder. The central sleep apnea syndromes are divided into eight types: two related to Cheyne–Stokes breathing (CSB), high altitude, substance, three primary CSA disorders of which one is infancy and the other prematurity, and a new entity entitled treatment-emergent central sleep apnea. The latter category applies to central apnea that follows CPAP administration.

Obstructive sleep apnea syndrome maintains the criterion of five or more respiratory events per hour of sleep when studied in a sleep center or by out of center sleep studies (OCST), so long as typical symptoms are present, otherwise 15 or more predominantly obstructive respiratory events are sufficient to make the diagnosis. The OSA disorders are divided into adult and pediatric types. In the pediatric criteria, for those less than 18 years of age, only one obstructive event is required per hour of sleep so long as respiratory symptoms or sleepiness are present; alternatively, obstructive hypoventilation along with symptoms is required.

Central sleep apnea with Cheyne–Stokes breathing (CSA-CSB) is five or more central apneas or hypopneas per hour of sleep with a pattern that meets criteria for CSB. Central sleep apnea without CSB is diagnosed as central sleep apnea due to a medical disorder without Cheyne–Stokes breathing that occurs as a consequence of a medical or neurological disorder. Central sleep apnea due to high altitude periodic breathing is central apnea attributable to high altitude of at least 1,500 m but usually above 2,500 m. Central sleep apnea due to a medicine or substance most typically due to an opioid or respiratory depressant is not associated with CSB. Primary central sleep apnea (CSA) is five or more central apneas or central hypopneas per hour of sleep in the absence of CSB and of unknown etiology. Primary central sleep apnea of infancy occurs in an infant with greater than 37 weeks conceptional age with recurrent, prolonged (>20 s duration) central apneas and periodic breathing for more than 5 % of total sleep time during sleep. Primary central sleep apnea of prematurity occurs in an infant of less than 37 weeks conceptional age with similar respiratory events.

Treatment-emergent central sleep apnea is diagnosed when five or more obstructive events occur during a PSG with continuous positive airway pressure (CPAP) that shows resolution of obstructive events and presence of central apneas or hypopneas (Westhoff et al. 2012).

The treatment of the sleep-related breathing disorders is mainly based upon improved ventilation during sleep with devices such as a CPAP, bilevel positive airway (BiPAP) device, or an adaptive servoventilation (ASV) device. Medication is rarely used but may be considered in central sleep apnea syndrome where medroxyprogesterone has been shown to be helpful especially in patients with an obesity-hypoventilation component to their disease. Rarely a respiratory stimulant antidepressant such as protriptyline may also be useful, but its usefulness is impaired by significant adverse effects (Hudgel 1995).

3 Central Disorders of Hypersomnolence

The central disorders of hypersomnolence comprise eight disorders. Narcolepsy has undergone a major revision with elimination of the disorder name terms, with or without cataplexy. Type 1 narcolepsy is that presumed to be due to hypocretin loss with either measured reduction in csf hypocretin or cataplexy with associated electrophysiological findings. Narcolepsy Type 2 is that which is confirmed by electrophysiological studies in the absence of cataplexy or with a normal csf hypocretin level. A major change in the narcolepsy criteria is the addition of including a SOREMP on the nocturnal PSG as one of the two required to meet the MSLT criteria of two SOREMPs for diagnosis. This is based upon a study that indicates that the positive predictive value of a SOREMP on the nocturnal PSG for narcolepsy is 92 % (Andlauer et al. 2013). Approximately 50 % of patients with narcolepsy will have a SOREMP less than 15 min on the nocturnal PSG.

Narcolepsy is treated by medications that either treat both major symptoms of EDS and cataplexy, such as sodium oxybate, or by medications that target each symptom separately. Wake promoting agents such as modafinil or armodafinil are the preferred agents for EDS and can be combined with a norepinephrine antagonist such as venlafaxine for the treatment of cataplexy. Alternative medications for sleepiness include the amphetamines such as dextroamphetamine, combination amphetamine salts (adderall) or rarely methamphetamine, and methylphenidate (Ahmed and Thorpy 2010). Mazindol is available in France and has been used successfully for treating the sleepiness and cataplexy in narcolepsy (Nittur et al. 2013).

Idiopathic hypersomnia (IH) is a single entity with elimination of the two ICSD-2 hypersomnia disorders that had specific sleep duration criteria. The new idiopathic hypersomnia disorder requires either an MSLT mean sleep latency of 8 m or less, or a nocturnal sleep duration of at least 660 min. The ICSD-2 category of recurrent hypersomnia has been reduced to a single entry Kleine–Levin syndrome (KLS) with a subtype of menstrual-related Kleine–Levin syndrome (Arnulf et al. 2008). The sleepiness must persist for 2 days to 5 weeks and at least once every 18 months. There can be only one symptom with the sleepiness of cognitive dysfunction, altered perception, eating disorder, or disinhibited behavior.

Treatment of IH is similar to that for narcolepsy without cataplexy. Wake promoting agents such as modafinil or armodafinil, amphetamines, and methylphenidate are the preferred agents (Lavault et al. 2011). Sodium oxybate can be considered but little data exists on its effectiveness for IH. Mazindol is available in France and has been used successfully (Nittur et al. 2013). KLS treatment is generally disappointing but medications that have been tried include lithium carbonate and amantadine (Arnulf et al. 2008).

Insufficient sleep syndrome is the new term for the previous more cumbersome term of behaviorally induced insufficient sleep syndrome. The reduced sleep must be present most days for at least 3 months. Extension of sleep time must result in resolution of symptoms. The other three items in the hypersomnia disorders section are hypersomnia related to a medical disorder, medication or substances, or psychiatric disorder.

Long sleeper is no longer regarded as a disorder but as a normal variant. There are no diagnostic criteria but a total sleep time of 10 or more hours is suggested as being usually accepted.

4 Circadian Rhythm Sleep–Wake Disorders

The circadian rhythm sleep–wake disorders comprise six specific disorders including delayed sleep–wake phase disorder (DSWPD), advanced sleep–wake phase disorder (ASWPD), irregular sleep–wake rhythm disorder, non-24-h sleep–wake rhythm disorder, shift work disorder, and jet lag disorder. These disorders arise when there is a substantial misalignment between the internal circadian rhythm and the desired sleep–wake schedule. Specific general diagnostic criteria are given for circadian rhythm sleep–wake disorder (CRSWD). A 3-month duration of symptoms is a requirement for diagnosing all these disorders except for jet lag disorder which has a requirement of jet travel across at least two time zones. A circadian rhythm disorder not otherwise specified (NOS) is listed for patients who have a circadian rhythm sleep–wake disorder who meet all the criteria for CRSWD but not the specific types.

The CRSWDs are treated by considering chronobiological principles. Most require some alteration and management of the sleep–wake schedule by behavioral means. Medications can be helpful, particularly melatonin to stabilize or to help advance the sleep pattern. Chronobiotics include circadian regulators capable of entraining desynchronized or misaligned circadian rhythms as one might observe in patients with delayed sleep phase disorder or non-24-h sleep–wake disorder (Thorpy and Roth 2013). These medications can reset the circadian clock in the suprachiasmatic nucleus of the hypothalamus, resulting in alignment of circadian rhythms with the day/night cycle (Touitou and Bogdan 2007). The only FDA approved chronobiotic for a CRSD is the melatonin agonist, tasimelteon, which is approved for the treatment of non-24-h sleep–wake rhythm disorder (Dhillon and Clarke 2014). Patients with other CRSWDs may be helped by the use of hypnotics at night to help with the quality of nocturnal sleep and/or the use of wake promoting agents during the day, such as modafinil, to prevent sleep episodes at inappropriate times of the day (Morgenthaler et al. 2007).

5 Parasomnias

The parasomnias are divided into three groups: the NREM-related parasomnias, REM-related parasomnias, and an other parasomnia category. They are defined as undesirable physical events or experiences that occur during entry into sleep, within sleep, during arousal from sleep.

The NREM-related parasomnias comprise general diagnostic criteria for the group heading of disorders of arousal (from NREM sleep). Specific general diagnostic criteria are given for disorders of arousal (DA) and the detailed text applies to all of the DA's as no text is presented in each of the specific DAs except for diagnostic criteria. Sleep-related abnormal sexual behaviors are listed as a subtype to be classified under confusional arousals. Diagnostic criteria are given for three disorders: confusional arousals, sleepwalking, and sleep terrors. The final NREM-related parasomnia is sleep-related eating disorder (SRED) that requires an arousal from the main sleep period to distinguish it from night eating syndrome (NES) disorder which is excessive eating between dinner and bedtime, and SRED requires an adverse health consequence from the disorder.

Treatment of the non-REM-related parasomnias is primarily by ensuring the patient is protected in the environment by removing any sharp objects such as furniture from near the bed. Sometimes hypnosis or cognitive behavioral therapy (CBT) has been helpful. Medications have rarely been effective for sleepwalking or sleep terrors but some patients have been helped with benzodiazepines, such as clonazepam, which reduce slow-wave sleep or the use of antidepressant medications such as paroxetine (Attarian and Zhu 2013).

The REM-related parasomnias include REM sleep behavior disorder (RBD), recurrent isolated sleep paralysis (RISP), and nightmare disorder. RBD which is repeated episodes of vocalizations and/or complex motor behaviors requires the polysomnographic evidence of REM sleep without atonia (RWA) (Schenck and Howell 2013). RISP is the recurrent inability to move the trunk and all of the limbs at sleep onset or upon awakening from sleep that causes distress or fear of sleep. Nightmare disorder is repeated occurrence of extended, extremely dysphoric, and well-remembered dreams that usually involve threats to survival, security, or physical integrity.

The REM-related parasomnias, such as nightmares or sleep paralysis, are usually treated by the use of REM suppressing medications which typically are antidepressants that are sedative such as imipramine. Behavioral management is also important as well as psychiatric management in the case of nightmares. However in RBD, clonazepam is the most effective medication, and less effective but useful when clonazepam cannot be taken is melatonin in high doses up to 12 mg (McCarter et al. 2013). Rivastigmine has been shown to be helpful in refractory RBD patients with mild cognitive impairment (Brunetti et al. 2014).

The other parasomnia section includes three specific disorders: exploding head syndrome (EHS), sleep-related hallucinations, and sleep enuresis. EHS is a complaint of a sudden noise or sense of explosion in the head either at the wake–sleep transition or upon awakening during the night associated with abrupt arousal. Sleep-related hallucinations are predominantly visual hallucinations that are experienced just prior to sleep onset or upon awakening during the night or in the morning. Sleep enuresis is involuntary voiding during sleep at least twice a week in people older than 5 years of age. Parasomnias associated with medical disorders, and medication or substance and unspecific parasomnia, comprise the other entries in this category. Sleep talking is a normal variant that can occur in both NREM or REM sleep and can be associated with parasomnias such as RBD or DAs.

EHS usually does not require treatment although topiramate has been reported to be helpful (Palikh and Vaughn 2010). Sleep-related hallucinations may be helped by REM suppressant medications. Sleep enuresis is largely a maturational problem that requires no treatment other than behavioral management; however, sometimes desmopressin may be helpful. There is a report that sertraline can be effective in desmopressin nonresponders (Mahdavi-Zafarghandi and Seyedi 2014).

6 The Sleep-Related Movement Disorders

The sleep-related movement disorders (SRMD) comprises seven specific disorders; restless legs syndrome (RLS), periodic limb movement disorder (PLMD), sleep-related leg cramps, sleep bruxism, sleep-related rhythmic disorder (RMD), benign sleep myoclonus of infancy (BSMI), and propriospinal myoclonus at sleep onset (PSM). SRMD are relatively simple, usually stereotyped movements that disturb sleep or its onset.

Restless legs syndrome (also known as Willis–Ekbom disease) is an urge to move the legs, usually accompanied by or thought to be caused by uncomfortable and unpleasant sensations in the legs. The ICSD-3 criteria do not include any frequency or duration criteria as is contained in the DSM-V criteria.

PLMD is defined as the polysomnography demonstration of periodic limb movements (PLMS) of >5/h in children and >15/h in adults that cause significant sleep disturbance or impairment of functioning. Sleep-related leg cramps are painful sensations that occur in the leg or foot with sudden, involuntary muscle hardness or tightness. Sleep-related bruxism is tooth grinding during sleep that is associated with tooth wear or morning jaw muscle pain or fatigue. RMD is repetitive, stereotyped, and rhythmic motor behaviors involving large muscle groups that are sleep related. BSMI is repetitive myoclonic jerks that involve the limbs, trunk, or whole body that occurs from birth to 6 months of age during sleep. As PSM mainly occurs during relaxed wakefulness and drowsiness as the patient attempts to sleep, the term "at sleep onset" has been added to the propriospinal myoclonus name. The three final categories are related to a medical disorder, medication of substance, and an unspecified parasomnia.

Treatment of RLS is mainly by dopamine agonists such as pramipexole or rotigitine; however, because of augmentation there is a move to use alpha-2-delta ligands, such as gabapentin enacarbil or pregabalin, as first-line medications

(Silber et al. 2013). Opioids, such as oxycodone or tramadol, can be used when dopamine agonists or alpha-2-delta agonists are not effective. PLMD, when treatment is required, is usually treated with the same medications as RLS. RMD is fairly unresponsive to medications but clonazepam has been used but behavioral treatments or hypnosis may be more effective (Chisholm and Morehouse 1996). BSMI and PSM usually require no medication treatment.

7 Isolated Symptoms and Normal Variants

Isolated symptoms and normal variants include excessive fragmentary myoclonus (EFM), hypnagogic foot tremor and alternating muscle activation, and sleep starts (hypnic jerks). EFM is now regarded as a normal variant found on polysomnographic EMG recordings that are characterized by small movements of the corners of the mouth, fingers, or toes or without visible movement. Hypnagogic foot tremor (HFT) is rhythmic movement of the feet or toes that occurs in the transition between wake and sleep or in light NREM sleep, and alternating muscle activation (ALMA) is brief activation of the anterior tibialis in one leg with alternation in the other leg. Sleep starts (hypnic jerks) are brief, simultaneous contractions of the body, or one or more body segments occurring at sleep onset.

The isolated symptoms and normal variants usually do not require any specific treatment.

The final category in the ICSD-3 is a general other sleep disorder category for disorders that cannot be classified elsewhere.

Conclusion

The new ICSD-3 is a major advance over previous versions, but it is unfortunate that some of the diagnostic criteria differ from that of DSM-V, for example, the criteria for narcolepsy. However, the DSM-V serves as an entry level classification, mainly for psychiatrists, and it is to be hoped that in the future the two classifications will be merged into one that will cause less confusion not only for clinicians but also for agencies that reimburse for health care and provide for treatment options.

Appendix A lists several disorders that are coded in other sections of ICD 10 other than the sleep sections and include: Fatal Familial Insomnia, Sleep-Related Epilepsy, Sleep-Related Headaches, Sleep-Related Laryngospasm, Sleep-Related Gastroesophageal Reflux, and Sleep-Related Myocardial Ischemia. Appendix B lists the ICD sleep-related substance-induced sleep disorders.

References

Ahmed I, Thorpy M (2010) Clinical features, diagnosis and treatment of narcolepsy. Clin Chest Med 31(2):371–381

American Academy of Sleep Medicine (2014) International classification of sleep disorders, 3rd edn. American Academy of Sleep Medicine, Darien, IL

American Psychiatric Association (2013) Diagnostic and statistical manual of mental disorders, 5 DSM-Vth edn. American Psychiatric Association, Washington, DC

Andlauer O, Moore H, Jouhier L et al (2013) Nocturnal REM sleep latency for identifying patients with narcolepsy/hypocretin deficiency. JAMA Neurol 6:1–12

Arnulf I, Lin L, Gadoth N et al (2008) Kleine-Levin syndrome: a systematic study of 108 patients. Ann Neurol 63:482–493

Attarian H, Zhu L (2013) Treatment options for disorders of arousal: a case series. Int J Neurosci 123(9):623–625

Bertisch SM, Herzig SJ, Winkelman JW, Buettner C (2014) National use of prescription medications for insomnia: NHANES 1999–2010. Sleep 37(2):343–349

Brunetti V, Losurdo A, Testani E et al (2014) Rivastigmine for refractory REM behavior disorder in mild cognitive impairment. Curr Alzheimer Res 11(3):267–273

Chisholm T, Morehouse RL (1996) Adult headbanging: sleep studies and treatment. Sleep 19 (4):343–346

Dhillon S, Clarke M (2014) Tasimelteon: first global approval. Drugs 74(4):505–511

Edinger JD, Wyatt JK, Stepanski EJ et al (2011) Testing the reliability and validity of DSM-IV-TR and ICSD-2 insomnia diagnoses: results of a multi-method/multi-trait analysis. Arch Gen Psychiatry 68:992–1002

Hudgel DW (1995) Pharmacologic treatment of obstructive sleep apnea. J Lab Clin Med 126 (1):13–18

International Classification of Diseases (2010) Tenth Revision, Clinical Modification (ICD-10-CM), National Center for Health Statistics. Centers for Disease Control and Prevention (CDC), December 20, 2010

Lavault S, Dauvilliers Y, Drouot X et al (2011) Benefit and risk of modafinil in idiopathic hypersomnia vs. narcolepsy with cataplexy. Sleep Med 12(6):550–556

Mahdavi-Zafarghandi R, Seyedi A (2014) Treatment of monosymptomatic nocturnal enuresis: sertraline for non-responders to desmopressin. Iran J Med Sci 39(2):136–139

McCarter SJ, Boswell CL, St Louis EK et al (2013) Treatment outcomes in REM sleep behavior disorder. Sleep Med 14(3):237–242

Morgenthaler TI, Lee-Chiong T, Alessi C et al (2007) Standards of practice committee of the American academy of sleep medicine. Practice parameters for the clinical evaluation and treatment of circadian rhythm sleep disorders. An American academy of sleep medicine report. Sleep 30(11):1445–1459

Nittur N, Konofal E, Dauvilliers Y et al (2013) Mazindol in narcolepsy and idiopathic and symptomatic hypersomnia refractory to stimulants: a long-term chart review. Sleep Med 14 (1):30–36

Palikh GM, Vaughn BV (2010) Topiramate responsive exploding head syndrome. J Clin Sleep Med 6(4):382–383

Schenck CH, Howell MJ (2013) Spectrum of RBD (overlap between RBD and other parasomnias). Sleep Biol Rhythms 11(Suppl 1):27–34

Schwartz TL, Goradia V (2013) Managing insomnia: an overview of insomnia and pharmacologic treatment strategies in use and on the horizon. Drugs Context 2013:212257. doi:10.7573/dic.212257

Silber MH, Becker PM, Earley C et al (2013) Medical advisory board of the Willis-Ekbom disease foundation. Willis-Ekbom disease foundation revised consensus statement on the management of restless legs syndrome. Mayo Clin Proc 88(9):977–986

Thorpy MJ, Roth T (2013) Toward a classification of medications for sleep and circadian rhythm disorders. Nat Sci Sleep 5:143–145

Touitou Y, Bogdan A (2007) Promoting adjustment of the sleep-wake cycle by chronobiotics. Physiol Behav 90(2–3):294–300

Westhoff M, Arzt M, Litterst P (2012) Prevalence and treatment of central sleep apnoea emerging after initiation of continuous positive airway pressure in patients with obstructive sleep apnoea without evidence of heart failure. Sleep Breath 16:71–78

Zisapel N (2012) Drugs for insomnia. Expert Opin Emerg Drugs 17(3):299–317

Development of New Therapeutical Agents for Treatment of Insomnia and Other Sleeps Disorders

Gloria Martin and Antonio Guglietta

Abstract The development of a new therapeutic agent for human use is a complex, highly regulated, risky, and expensive process which, in the case of insomnia and other sleep disorders, presents some special challenges typical of this group of diseases. The development of a new therapeutic agent consists of a series of Nonclinical and clinical phases in which the new drug candidate has to demonstrate, among other things, a favorable benefit/risk ratio in the treatment of a given disease before receiving market authorization and be used to treat patients. In the Nonclinical development phase chemical development, formulation development, analytical methods development, pharmacology development, pharmacokinetic development, and Nonclinical safety evaluation of the new drug candidate are carried out. The main goal of these studies, conducted both in vitro and in vivo in laboratory animals, is to collect sufficient information to authorize the evaluation of the new therapeutic agent in humans. The clinical development phase of a new therapeutic agent consists of several steps in which the new drug candidate is administered to humans in a stepwise progression starting with administration to a reduced number of healthy volunteers and then progressing to patients in order to assess its safety and efficacy. In addition, in the case of hypnotics, specific points of concerns such as development of tolerance, potential for drug abuse, interaction with other drugs, and next-day performance should also be addressed and evaluated promptly and properly. Several guidelines issued by regulatory agencies are available to guide investigators throughout the entire development process and should be consulted to correctly design and execute the individual studies of the development plan. Nevertheless, during the development of new therapeutic agents for insomnia and other sleep disorders, specific issues can arise which are not covered by the available guidelines. Therefore, it is always highly advisable to consult with regulatory agencies on how to address these specific points. A better understanding on the drug regulatory framework and rules can maximize the chances of a

G. Martin (✉)
Grupo Ferrer Internacional, Joan Buscallá 1, 08173 Sant Cugat del Valles, Barcelona, Spain
e-mail: gloriamartin@ferrer.com

A. Guglietta
Independent Pharmaceutical Consultant, Barcelona, Spain
e-mail: consultant@guglietta.es

A. Guglietta (ed.), *Drug Treatment of Sleep Disorders*, Milestones in Drug Therapy,
DOI 10.1007/978-3-319-11514-6_4

successful development of a new therapeutic agent for treatment of insomnia and other sleep disorders.

1 Introduction

Sleep disorders consist of a group of disorders that have in common a disturbance of the normal sleep process such as difficulty of falling asleep, inadequate duration of sleep, etc. Several distinct sleep disorders are known today which cause a major impact on the quality of life of a person and have important social and economic consequences (see chapter "Classification of Sleep Disorders"). Relatively few drugs are available on the market to treat some of these disorders (Equihua et al. 2013; Kryger et al. 2011) and it is likely that new therapeutic agents with an improved efficacy and safety profile will be developed in the future to replace the existing treatments and to provide treatments for sleep disorders conditions for which, presently, there is no specific treatment.

The development and approval of a new drug for human use is a highly regulated process. In order to be commercialized, a new drug has to receive a market authorization by the competent government body of a country. This authorization is granted after evaluation of all the scientific data available for the new therapeutic agent including data on efficacy, safety, and manufacturing process of the active molecule and its formulations. In order to be approved, a new drug candidate has to be manufactured reproducibly under strict quality conditions and has to demonstrate, among other things, in well-designed and well-conducted studies a statistically beneficial effect in a given disease as well as to be safe to humans which, in broader terms, means to demonstrate a favorable benefits/risks ratio. Each country has its own government body that grants a market authorization for new drugs. In the United States the agency that reviews the information and scientific data on a new drug and grants its market authorization is the Food and Drug Administration (FDA). In the European Community the situation is somehow more complex since there is a central European agency (European Medicines Agency or EMA) which coexists and shares some responsibilities with country-specific agencies (for instance the Medicines and Healthcare Products Regulatory Agency in United Kingdom or the "Agence National de Sécurité du Medicament et des Produits de Santé" in France) and several review and approval regulatory procedures coexist.

The rate of failure to get market authorization for a new drug may be quite high although this rate varies depending on several factors such as the therapeutic area and indication and development phase of the drug (DiMasi et al. 2010). Reasons for failure of drug development are multiple and include scientific, technical, regulatory, and commercial issues (Table 1). Although the risks associated with drug development failure cannot be completely eliminated, they can be reduced by a careful selection of the project to develop, by providing adequate development

Table 1 Main reasons for drug development failure

Group	Reason	Example
Scientific	Efficacy	Lack of efficacy in Nonclinical and clinical studies
	Safety	Safety concerns in Nonclinical and clinical studies
Technical	API issues	Technical problem in the synthesis of the API
	Formulation issues	Stability of the formulation
Regulatory	Regulatory requirements	Development plan did not meet regulatory requirements
	Regulatory decisions	Unfavorable benefit/risk ratio or quality issues
Commercial	Development cost	High development cost
	Patent issues	Lack of IP protection
	Market value	Change in the market value of the product

funding and by designing a proper development plan. In this regard, it is very important to understand what kind of data are required by the regulatory agencies to approve a new drug and to comply with their requirements in terms of number and type of studies to be conducted, quality of the data, statistics, etc. While it may be acceptable, although painful, to stop the development of a drug candidate due, for instance, to lack of efficacy or safety concerns after a well-designed development plan, it may be quite frustrating to abandon the development of a new therapeutic agent due to bad planning and/or to poorly designed development studies. In order to facilitate development of a new drug, in many cases, the regulatory agencies issue specific guidelines to guide investigators throughout this process. Regulatory guidelines however are limited in scope. They tell you what to do but not how to do it. There is always room for interpretation and following these guidelines does not guarantee the approval of a new drug. Therefore, it is always highly advisable during the entire development process to have frequent meetings with the regulatory agencies to address specific issues that might have arisen during the process and to discuss with them results and ways to proceed. To this end, some agencies have established specific consultations programs such as scientific advices, pre-IND meetings, etc. to meet with the investigators to review the available data and address their concerns.

In the area of insomnia and sleep disorders, despite some new drugs have been recently successfully developed, several other molecules have, for a variety of reasons, failed to reach the market place (Food and Drug Administration 2013, Table 1). However, there are several still unmet medical needs in this area that require specific and better treatments. This represents a unique opportunity from a scientific, medical, and commercial standpoint to develop more specific therapeutic agents to treat these conditions. In this regard, a better understanding of the drug development and approval process and the peculiarities of the sleep disorders area may help to maximize the chance of success.

2 Drug Discovery in Insomnia and Other Sleep Disorders

The process that leads to the approval and commercialization of a new drug, including those to treat sleep disorders, starts with the identification of a new therapeutic molecule such as a small molecule, a peptide, a protein (Fig. 1), etc. Similarly to other therapeutic areas, new therapeutic molecules to treat sleep disorders can be discovered using different approaches such as:

Target-Based Approach The starting point of this approach is the identification of a molecular target that plays a key role in a given medical condition such as receptors or enzymes whose activity is then modified by molecules that bind to it. The target chosen may be a "new target" against which no drugs have ever been developed, at least for a specific disease, or an "old target" which is one that has been known for some time to play a role in a given disease and against which drugs have already been developed to treat the medical condition that is object of our investigation. Most of the time, the goal of this approach is to identify a molecule that can block the activity of a receptor (receptor antagonists) or inhibit the activity of an enzyme (enzyme inhibitors). These molecules can be identified by a variety of methods such as a process of medicinal chemistry or through the screening of library of molecules.

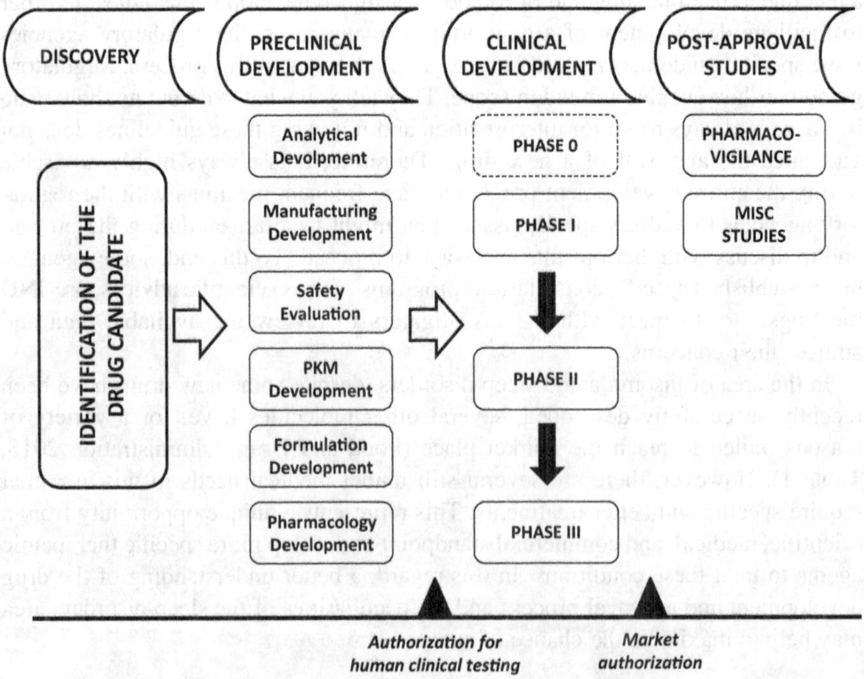

Fig. 1 The drug development process. The *dotted box* indicates an optional step of the clinical development phase

In the area of sleep disorders, an example of a new target is represented by the orexin receptors which were recently discovered (de Lecea et al. 1998; Sakurai et al. 1998) and their involvement in sleep demonstrated (de Lecea and Huerta 2014; Krystal et al. 2013; Sellayah and Sikder 2013). Several orexin antagonists have been synthesized and are currently in development for treatment of insomnia (Equihua et al. 2013). If successful, the orexin antagonists will be the first molecules of a new class of hypnotics (orexin antagonists) that will reach the market.[1] An example of an "old target" in sleep disorders is represented by the $GABA_A$ receptor against which at least three generation of hypnotics have been developed (barbiturates, benzodiazepines, z-drugs) (López-Muñoz et al. 2005; Wick 2013; Nutt and Sathl 2010). Although potentially very rewarding, the new target approach is also very risky since no previous drug acting on this target has ever been developed to treat the disease object of our research or any other medical condition. The synthesis of molecules that bind to the new target may not be technically or commercially feasible or may be associated with target-specific toxicology and adverse effects. The old target approach on the other hand may be less rewarding but is also less risky since a lot of information may be available on target-specific unwanted effects and chemical structures that bind to it. This previous available information therefore may be very useful to the investigators to design the discovery phase of a new molecule.

Molecule-Based Approach Another approach to discover new drugs, including those to treat sleep disorders, is the molecule-based approach in which the starting point of the discovery program is a molecule known to have the desired biological effects and which is either usually already on the market or in advanced phase of development. In this kind of approach, the molecular structure of an existing active molecule is modified in an attempt to identify a new chemical structure that improves the benefit/risk ratio of the reference molecule. This approach may be less risky than other approaches; however sometimes it may not be possible to synthesize a new molecule structure that improves the benefit/risk ratio of the reference molecule also because several chemical modifications of the reference molecule may have been already explored and may not be patentable.

Endogenous Molecules Replacement Approach A new therapeutic agent can be identified based on a better understanding on the physiology of endogenous molecules. For instance, the discovery that some narcoleptic patients have low level of orexin in their brain (Mignot et al. 2002; Kanbayashi et al. 2002; Nakamura et al. 2010; Nishino et al. 2000) suggests that, at least in theory, the exogenous administration of this peptide may restore its physiological levels and be beneficial in the treatment of this condition.

[1] After this book went to press, on Aug 13th 2014, the US Food and Druf Administration approved Suvorexant (Belsomra) tablets to treat difficulty in falling and stying alseep (insomnia). Suvorexant is the first approved drug of the orexin antagonists class.

In the discovery phase of a new drug such as a new therapeutic agent to treat sleep disorders, some preliminary chemical and biological characterization of the new molecule is carried out. This characterization may include biochemical in vitro assay such as receptor binding, enzyme inhibition assay, as well as in vivo pharmacology and pharmacokinetic characterization and toxicology tests both in vitro and in vivo using laboratory animals. Typically, to obtain the final molecule to be developed, it is necessary to go through a cycling process in which initially identified molecular structures are tested in a variety of chemical and biological assays and then modified on the basis of the results obtained. This process may be repeated several times until a molecule with the final desired characteristics is identified. For instance, to identify new orexin antagonists that may become potential new drugs to treat insomnia, the new small molecules synthesized may be screened for their ability to bind to the orexin receptors and act as receptor antagonists while the in vivo activity, pharmacokinetics, metabolism, etc. of these new molecules may also be tested in animals.

When, based on the data collected during the discovery phase and other considerations (market size, development cost, time to market, cost of goods, etc.), a decision is made to develop the new therapeutic agent, the Nonclinical development phase of a new drug begins.

3 Nonclinical Development in Insomnia and Other Sleep Disorders

The Nonclinical development phase of a new therapeutic agent consists of different groups of studies that may be carried out sequentially or in parallel whose goal is to collect sufficient data on the new molecule to make a judgment on whether its activity and safety can also be evaluated in humans (Fig. 1).

3.1 Definition of the Manufacturing Process of the Drug Substance

Throughout the Nonclinical development, the process of synthesis and manufacturing of the new therapeutic agent is defined (chemical synthesis, biotechnology, etc.) and characterized in terms of presence and levels of impurities, stability of the drug substance, physical properties, yield of the manufacturing process, etc. The process of manufacturing is scaled up both to assure that sufficient quantity of the drug substance of the required quality is available to carry out the Nonclinical development studies and to define a large scale manufacturing process. The final goal of the manufacturing development is to have a well-characterized, reliable, and reproducible method of synthesis of the new therapeutic agent of the required quality suitable for human use and commercialization.

The results and conclusions of this development step are summarized in a regulatory document which will be part of the application submitted prior to human testing to the competent government agencies. Specific guidelines issued/adopted by several regulatory agencies are available to guide the manufacturing company throughout this process (ICH Harmonized Tripartite Guideline 2012). However, it may always be a good idea to discuss specific issues that may arise, during the process with the competent regulatory agencies.

The definition and characterization of the manufacturing process of the drug substance usually runs in parallel with other Nonclinical development phases. This implies that the drug substance used in the different studies of the Nonclinical development phase may come from different batches whose characteristics may be slightly different. This may be acceptable providing that no major differences such as in the presence and level of impurities or critical physical properties (polymorphism, etc.) exist between different batches of drug substance. Generally speaking, it is always a good idea to use the closest to final product drug substance available for all Nonclinical development studies and in particular for key studies such as the toxicology and other safety studies. Changes in the manufacturing process that lead to major changes of the characteristics of a new therapeutic agent (i.e., different impurity profile) may have a major impact on the development program and may require running additional studies that, ultimately, result in longer development time and increased development cost. Although during the development of a new drug, a change of the manufacturing process of the drug substance is often necessary, sometimes even in the late phases of development, for instance to optimize the process of synthesis and reduce costs, the risk to significantly modify the characteristics of the drug substance, and its impact on the development process should be carefully considered.

3.2 *Analytical Methods Development*

Several studies that need to be conducted during the Nonclinical and clinical development of a new therapeutic agent require the identification and quantification of several chemical products such as the drug substance, its metabolites, degradation products, etc. both in vitro (cell cultures, etc.) and in biological samples (blood levels, organs, and tissue levels, etc.). The manufacturing of the drug substance as well as toxicology, safety pharmacology studies, pharmacokinetic, metabolism, formulation, and clinical studies cannot be carried out according to regulations without the availability of specific methods of analysis for the drug substance, drug product, and related compounds. Therefore, it is necessary that specific and validated analytical methods are developed and are available at the time of running all clinical as well as specific nonclinical studies.

3.3 Nonclinical Safety Evaluation

In order to be administered to humans, the safety of a new therapeutic agent has to be extensively characterized in the nonclinical development phase both in vitro and in vivo using a variety of laboratory animals. To characterize the toxicology profile of a new drug candidate, a series of properly designed studies are conducted. As in the case of manufacturing development of drug substance, several regulatory guidelines (ICH, FDA, or EMA) are available to help investigators to decide the type and design of toxicology studies needed to support the human testing of the new molecule. The main purpose of this evaluation is to collect enough data to support the proposed clinical studies and to conclude that the new therapeutic agent can be administered to humans, under the experimental conditions of the study(ies) proposed, with an acceptable risk of having some undesirable effects (ICH Harmonised Tripartite Guideline 2009). More specifically, a well-designed toxicology plan should allow the identification and characterization (incidence, severity, reversibility, etc.) of possible adverse effects induced by a new therapeutic agent and to establish the doses at which such events occur. Toxicology studies may include in vitro studies such as those using isolated cells in cultures and in vivo animal studies which are generally conducted in mice, rats, rabbits, guinea pigs, minipigs, dogs, and primates. The choice of a given animal to conduct toxicology studies should be justified mainly in terms of their relevance to humans (i.e., same metabolites profile, etc.). Examples of toxicology in vitro studies include, but are not limited to, the AMES test (Martelmans and Zeiger 2000; ICH Harmonised Tripartite Guideline 2011) and the chromosomal aberration test (Clare 2012; OECD 1997) to assess the carcinogenicity potential of a new molecule (ICH Harmonised Tripartite Guideline 1995) or the hERG test which detects potential cardiotoxicity of the new drug (Katchman et al. 2005; Johannesen et al. 2014; ICH Harmonised Tripartite Guideline 2000). Examples of in vivo toxicology studies include general toxicology studies (acute and repeated doses studies) (ICH Harmonised Tripartite Guideline 1998; European Medicines Agency 2010a) and reproductive and developmental studies (ICH Harmonised Tripartite Guideline 2005a; Food and Drug Administration 2011; European Medicines Agency 2008). In addition, depending on the type of product to be developed, some special safety studies may be required such as immunotoxicity studies in the case of proteins, peptides, or other molecules that have the potential to induce antibody formation (ICH Harmonised Tripartite Guideline 2005b; Food and Drug Administration 2002) or drug abuse studies in the case of compounds known to cause drug addiction (European Medicines Agency 2006; Food and Drug Administration 2010). The toxicology studies sometimes are not limited to the new drug substance but are extended also to their formulations, impurities, and metabolites (Food and Drug Administration 2008).

In toxicology studies, it is generally expected to find toxicological effects. Therefore, the dose of the drug under evaluation should be increased until such effects appear. The results obtained from these studies will allow to determine the maximum initial dose that should be administered in humans (ICH Harmonised

Tripartite Guideline 2009; European Medicines Agency 2007; FDA 2005b) and to establish the Nonclinical therapeutic window of the new drug which is the ratio between the doses at which an undesirable effect occurs and those that induce a desirable effect. The larger this ratio the safer the new drug is. A toxicology effects-free study may not be conclusive and may generate doubts on whether the animals have been exposed to the new therapeutic agent. Therefore, it is advisable, at least in key toxicology studies, to also measure the blood level and, sometimes tissue levels, of the molecule administered (toxicokinetic) (ICH Harmonised Tripartite Guideline 1994a). The presence and level of the molecule under evaluation in the blood would prove that the animals have been exposed to the new therapeutic agent and will allow to relate a toxicological effect to a given blood level of the molecule.

In evaluating the results of the toxicology studies, one should always keep in mind the pharmacological activity of the molecule under testing and not to classify as toxic an effect that is actually the desired pharmacological activity expected. For instance, when evaluating the toxicology effects of hypnotics, it is expected to detect some sedation of the animals which, in the case of other class of drugs, could be considered a toxic effect. However, in the case of hypnotics, sedation is part of their desired pharmacologic activity and thus it should not be considered a toxic effect but actually a demonstration of its therapeutic activity.

Safety evaluation of a new therapeutic agent also includes safety pharmacology studies whose goal is to investigate the potential undesirable effect of a new therapeutic agent on specific organs or systems after exposure to therapeutic or higher dose of the molecule. Safety pharmacology studies include evaluation of the effects of the new therapeutic agent on vital organs or systems such as cardiovascular, respiratory, and central nervous system. In some cases, evaluation of the effects on other systems such as renal and gastrointestinal systems may also be required (European Medicines Agency 2010b). The safety pharmacology plan and studies should be designed following the recommendations of specific guidelines and on the basis of the specific properties of the new molecule and the pharmacologic class to which it belongs (ICH Harmonised Tripartite Guideline 2000; European Medicines Agency 2001).

In the safety evaluation of a new molecule intended to be used for treatment of insomnia and other sleep disorders, the following special aspects should be considered:

3.3.1 Tolerance, Physical Dependence, and Addiction

Tolerance causes a decrease of effectiveness of a therapeutic agent such as a hypnotic over time which can be overcome by administering higher concentrations of the drug (Heit 2003). Although tolerance may develop to all effects of a drug, the rate and the degree at which it occurs may be different and specific for each effect (Bateson 2002; Griffiths and Johnson 2005; Vinkers and Oliver 2012; Rabbani et al. 1995). While the development of tolerance to an undesirable effect may not be relevant and actually may be beneficial since it widens the therapeutic window of the drug, development of tolerance to a desirable effect may have negative therapeutic impacts since higher doses of the drug are needed to be administered to have the same degree of a therapeutic effect with the consequences of narrowing the therapeutic window and increasing the risk of producing toxic effects. Physical dependence, on the other hand, is a physical reaction characterized by symptoms and signs such as anxiety, vomiting convulsions, etc. that occur after abrupt discontinuation of a chronically administered drug (Heit 2003). Finally addiction can be defined as a compulsive need to use the drug despite harmful consequences (Heit 2003).

It is known that development of tolerance, physical dependence, and addiction occur, although to different degrees, with drugs that act by modulating the activity of the $GABA_A$ receptor such as benzodiazepines and non-benzodiazepine hypnotics (Greenblatt and Roth 2012; Hajak et al. 2003; O'brien 2005; Uzun et al. 2010; Zisapel 2012). Although the development of these undesirable effects of a drug ultimately needs to be evaluated in humans, Nonclinical studies in animals may provide useful insights of the risk of developing them after administration of a new drug candidate to humans and allow to design the clinical studies accordingly. A variety of animals models have been developed to evaluate these effects in species such as rats, dogs, and monkeys ranging from chronic administration of a drug to precipitating a withdrawal syndrome to self-administration of drugs (Toth and Bhargava 2013).

3.3.2 Risk of Falling

Falling may be a dangerous side effect of several drugs including hypotensive, antipsychotics, muscle relaxants, etc. The mechanism of action of this effect is often not clear and may be related to mechanisms such as drowsiness, dizziness, and ataxia (Cardario et al. 2011). An increased risk of number of falls has been reported in people taking hypnotics, particularly those acting on the $GABA_A$ receptor (Kolla et al. 2013; Obayashi et al. 2013; Mets et al. 2010) and motor impairment and muscle relaxation have been associated with modulation of the activity of selected α subunits that make up this receptor (Milić et al. 2012; Rowlett et al. 2005). Therefore, one possible mechanism for the increased number of falls in patients taking $GABA_A$-acting hypnotics may be related to motor impairment muscle relaxation and coordination. Addressing this point at a Nonclinical stage in animals using appropriate tests (Deacon 2013; Meyer et al. 1979; Nordmann 1985; Milić et al. 2012) may provide useful information on the new molecule and its potential to cause an increased risk of number of falls.

3.3.3 Next-Day Effects

A good hypnotic should be able to promptly induce physiological sleep of good quality and of adequate duration. In addition, the use of a hypnotic should not be associated with any residual undesirable effects such as memory impairment and performance deficits the day after drug administration and the individual should be able to carry out daily tasks such as driving normally. The next-day effects constitute an important part of the clinical evaluation of a hypnotic which however should also be studied during the Nonclinical development phase.

3.3.4 Interaction with Other Drugs

Other drugs that a patient takes concomitantly with sleep disorders medications may produce drug–drug interactions that may result in some unexpected and undesirable effects. In the case of hypnotics, interaction with alcohol deserves a special mention. It has been reported that consumption of alcohol in patients that take hypnotics is quite frequent particularly in certain age-groups (Ilomäki et al. 2013). Alcohol may interact with other drugs by modifying their metabolism and excretion of and/or by producing pharmacodynamics interactions (Saunders 1986). The consequences of interactions of hypnotics with alcohol may result in excessive CNS depression, impaired psychomotor performance, driving impairment, etc. which frequently lead to intensive care unit admission (Gunja 2013; Hesse et al. 2003; Zosel et al. 2011). Therefore, already in the Nonclinical development of a hypnotic it is important to evaluate its interaction with other drugs and in particular alcohol.

3.4 Pharmacokinetic and Metabolism Characterization

The pharmacokinetic profile of a sleep disorder medication allows to predict its effects in humans. Generally speaking, an ideal sleep disorder drug has to be absorbed quickly, cross the blood–brain barrier and reach the CNS, bind to its molecular target in sufficient amount and for enough time to exert its action for the desired duration of time, and then be eliminated completely from the body. Nonclinical evaluation of the pharmacokinetic profile of a new molecule in several animal species allows to predict how the new therapeutic agent will behave in humans and will give insights on how to design clinical studies. These evaluations need to be conducted mainly by the intended route of administration in humans both on the drug substance and on its formulated product(s) developed for human use (ICH Harmonised Tripartite Guideline 1994b; European Medicines Agency 1994)

3.5 Formulation Development

For any new molecule being developed for human use, at least one formulation suitable for human administration has to be developed before the new therapeutic agent can be administered to people. The formula of the new formulation has to be carefully selected making sure that all its components are suitable for use in humans and has to be characterized in terms of degradation products, stability, storage conditions, release of the active component(s), and in vivo pharmacokinetic profile (ICH Harmonised Tripartite Guideline 1996). In the case of sleep disorders medications, in most cases the formulation developed is an immediate release oral tablet. However, in selected cases, some special formulations such as extended release tablet forms, sublingual formulations, oral spray formulations, and intranasal formulations have been developed or proposed (Costantino et al. 2007; Owen 2009)

3.6 Pharmacology Development

The purpose of this phase is to gather additional data on the primary pharmacologic effect of the new drug to complement the information already collected during the discovery phase of the new therapeutic agent. The primary pharmacology evaluation plan varies greatly depending on the characteristics of the new therapeutic agent and the specific points that need to be addressed. Both in vitro and in vivo studies may be carried out during this phase. In vitro studies may include binding studies to specific receptors and receptor subtypes to determine important variables of the drug–receptor interaction such as affinity, selectivity, potency, and efficacy of the new therapeutic agent for a variety of receptors as well as its agonist or antagonist activity (Nutt and Sathl 2010; Mould et al. 2014; Winrow and Renger 2014). In vivo pharmacologic evaluation may include studies such as activity of the new therapeutic agent in animal models of sleep disorders such as models of insomnia, sleep apnea, restless legs syndrome, and narcolepsy (De la Herrán-Arita and Druken-Colin 2012; Toth and Bhargava 2013). The choice of the right animal model depends on many factors such as indication, type of molecule, and specific points to address.

It is important to stress that the extent of the Nonclinical development phase of a new therapeutic agent may vary greatly depending on a variety of factors. While for a new therapeutic agent never previously tested in humans, the development phase may be very comprehensive, in specific cases such as for instance a new indication for an already marketed compound, it may be less extensive. The purpose of the Nonclinical development phase is to collect enough data on a new therapeutic agent to receive authorization by the competent government body to start its evaluation in humans. Although the extent of this phase may initially be limited to those studies required to support the first human study, Nonclinical development of a new therapeutic agent does not stop when the new molecule enters clinical phase but continues throughout the entire development process to support additional clinical studies and market authorization.

The data collected in the development Nonclinical phase along with the proposed clinical plan and study(ies) are submitted to the competent regulatory body and ethical committees of the country/medical centers where the clinical studies are planned. In case of a favorable evaluation, an authorization to conduct the clinical study(ies) proposed is granted. Because of the different data required by different regulatory agencies and different criteria of evaluation, sometimes the permission to initiate clinical studies may be granted in a country and denied in another.

4 Clinical Development in Insomnia and Other Sleep Disorders

The clinical development phase of new therapeutic agents, including those intended for treatment of sleep disorders, consists of different studies of increasing complexity whose goal is to assess the efficacy and safety of the new molecule in humans (Fig. 1).

4.1 Early Phase Clinical Studies

These studies are initial clinical evaluation of a new therapeutic agent intended to collect data on safety, pharmacokinetics, and pharmacodynamics of the new molecule. The information gathered in this phase will then be used to design the large confirmatory studies required prior to market authorization of the new drug. Typically, early phase studies are grouped in different phases known as phase 0, phase I, and phase II studies.

Phase 0 Studies (Exploratory IND Studies) In selected cases, initial clinical evaluation of a new therapeutic agent may be authorized very early in the development process and with only limited Nonclinical data availability. In the phase 0, the new therapeutic agent is administered to a small number of subjects at subtherapeutic doses (microdoses) to collect preliminary data on the pharmacokinetic and pharmacodynamics of the drug (Food and Drug Administration 2006). Although phase 0 is not mandatory and often may not even be recommended or feasible, in selected cases, it may be useful to gather preliminary human data which can then be used to make decision on further development steps of drug candidates.

Phase I Studies These are small studies in healthy human volunteers usually conducted in a single medical center which typically involve around 20–100 subjects. The main purpose of these studies is to collect safety data on the new therapeutic agent. These studies evaluate pharmacokinetics, pharmacodynamics, and tolerability after single and repeated doses of the new drug and are usually conducted in a stepwise manner starting with the administration of a low dose of the new agent, determined on the basis of the Nonclinical data, and then progressively increasing the dose to be administered. Patients are closely monitored for safety and studies are stopped as soon as serious safety concerns arise.

Phase II Studies Contrary to phase I, phase II studies are conducted in patients and may involve up to a few hundred subjects. In addition to further explore the safety of the drug, the goal of these studies is to determine the therapeutic activity of the new molecule on a given medical condition and to identify the doses at which it occurs. To achieve these goals, typically in phase II studies several doses of the drug under study are tested in order to determine a dose–response curve for a given pharmacologic effect. To insure that the right patients are entering these studies, an accurate diagnosis of the sleep disorder being evaluated is necessary. The selection of a well-defined and well-characterized patient population, particularly in early clinical evaluations, minimizes the risk of introducing confounding factors in the studies and helps to reduce the variability of the results allowing a better interpretation of the data. In the case of insomnia, the diagnosis of primary or secondary insomnia of acute or chronic duration should be established and the patient population should be well characterized through, for example, a well-conducted clinical interview and neurophysiological data such as polysomnography. In insomnia it is recommended to first assess the clinical activity of new therapeutic agent in primary insomnia of short duration and then, as the investigation of the new therapeutic agent progresses, explore the activity of the new drug on secondary and chronic insomnia and in insomnia in specific age-groups.

Although the design of these studies follows a similar pattern independently of the type of drug or indication, special features may apply in specific situations. In evaluating new therapeutic agents for treatment of sleep disorders and particularly in the case of insomnia, already in early clinical studies, special emphasis should be given to the evaluation of pharmacokinetic parameters in subsets of the general population such as children and elderly people and to the accumulation and circadian variation of the drug.

Proof-of-Concept Studies These are no mandatory small exploratory studies whose goal is to have a proof of a therapeutic activity in humans. These studies, usually conducted after phase I studies and demonstration of safety of the drug under investigation in humans, include a small number of patients suffering from a given disease and are very useful to demonstrate a clinical beneficial activity and make a decision on whether continuing the development of the molecule and confirm its activity in larger clinical studies.

4.2 Late Phase Clinical Studies

Phase III Studies These are large clinical studies which may involve up to several thousand patients. Typically, they are multicenter and often multinational studies whose goal is to confirm and further characterize the beneficial activity of the drug on a given condition as well as to evaluate its safety in a large patient population.

It is recommended, whenever possible, to include, at least in some of the studies, an active marketed comparator which allows a useful benchmark to evaluate the new therapeutic agent.

4.2.1 Late Phase Clinical Studies in Insomnia

In the case of insomnia, specific regulatory guidelines exist for these kinds of studies (Food and Drug Administration 1977; European Medicines Agency 2011) to help investigators to design and conduct the right clinical studies. In its guideline, the European Medicines Agency stresses the importance of documenting the effect of a new therapeutic agent for the treatment of insomnia both on subjective endpoints in the "natural setting" and objective endpoints (polysomnography). The suggested design for these studies is a randomized, double-blind, placebo-controlled, parallel group studies. At minimum, the activity of the new drug should be assessed on sleep onset latency, sleep continuity, sleep duration, feeling of restorative sleep, quality of sleep, and subsequent daytime functioning in the natural setting. Ideally, the new therapeutic agent should improve all these aspects of sleep. In the case that only some of these aspects are improved after treatment with a new therapeutic agent, then the approved indication may be restricted to a subset of the insomnia patient population (i.e., insomnia characterized by difficulty to fall asleep) providing that the improvement is robust and clinically relevant. In any case, the measurement of sleep architecture by polysomnography is considered helpful for phase III studies and the importance of an improvement in the quality of next-day time functioning is stressed (European Medicines Agency 2011). In order to receive a market authorization, the results of the phase III studies should be "robust and clinically meaningful" and therefore a responder/remitter analysis at least for the primary endpoint is considered essential by the EMA to assess the clinical relevance of the data in insomnia (European Medicines Agency 2011).

The studies performed in a sleep laboratory with polysomnography and psychometric methods are very useful to evaluate the effect of a new drug on sleep since they allow an objective quantification of several sleep parameters such as sleep onset latency, number and duration of awakenings, and total sleep time. However, these measurements are obtained in an artificial setting which may have an impact on the outcome of the studies and therefore may not be completely representative of the results obtained in a normal setting. Therefore, the importance to demonstrate a beneficial effect of a new therapeutic agent in insomnia in a natural setting should be again emphasized (European Medicines Agency 2011). Ambulatory polysomnography, in which a device that records several parameters of insomnia and other sleep disorders (Doering et al. 2008) is worn by the patient, may reduce the impact of the artificial setting of a sleep lab allowing the test to be conducted at home. Actimetry, which measures the gross motor activity usually by means of a wrist device worn by the patient, also reduces the impact of a sleep lab. However, the data obtained with actimetry in sleep (Wang et al. 2008) are considered useful by the EMA only to complement the polysomnographic data but are unacceptable when used alone (European Medicines Agency 2011).

As mentioned, since insomnia can negatively impact the quality of life, the effect of a new therapeutic agent on this parameter should be assessed using, for instance, questionnaires and semi- or structured interviews.

In the first late phase clinical studies of a new hypnotic agent, a duration of treatment of 2–4 weeks is recommended. However, unless there are safety concerns or unless only a short-term indication is sought, the activity of the new hyphotic should later on be also evaluated in long-term studies of at least 6-month duration. The preferred design for long-term studies is a double-blind placebo-controlled randomized withdrawal design in which, after a trial of sufficient duration, responders to the new drug under study are randomized to continue treatment or switch to placebo and the number of patients worsening and the time to this effect chosen as primary endpoints (European Medicines Agency 2011). As an alternative to the randomized withdrawal design, a double-blind placebo-controlled extension study could be chosen (European Medicines Agency 2011).

It is recognized that insomnia may present special features in pediatric and elderly population (Owens 2006; Kamel and Gammack 2006; Bain 2006) and that children and elderly people may respond to drugs differently than young adults (Stephenson 2005; Noble 2003). Therefore, data on drug efficacy and safety obtained in young adults cannot be extrapolated to these groups and, if an indication in these special populations is sought, specific clinical efficacy and safety data should be collected (European Medicines Agency 2011).

In pediatric population, a specific three-arm study of the same duration as in adults including a placebo and an active comparator should be conducted once the efficacy and safety of the new drug has been proven for adults (European Medicines Agency 2011). In children, clinical outcome endpoints are preferred in phase III studies and next-day performance included as co-primary endpoint should be assessed by evaluating school performance (European Medicines Agency 2011).

In the elderly, frequency of both primary and secondary insomnia is higher than in young adults and sleep architecture tends to have a different pattern in this age-group. Furthermore, the pharmacokinetic and sensitivity of the elderly population to a drug may be different than in young adults and often elderly people also take other drugs for a variety of medical conditions which can interfere with the new investigational drug. For all these reasons, assessment of efficacy and safety of a new hypnotic drug in elderly requires conducting specific studies or alternatively, elderly patients should be included in general phase III studies in sufficient number to allow a subset analysis for this age-group (European Medicines Agency 2011).

Safety data of a new drug candidate are collected throughout the entire development process including large phase III studies. In the case of hypnotics, specific adverse events particularly relevant in this class of drugs should be monitored throughout the entire development phase in both short-term and long-term studies. These include hangover, rebound, and withdrawal effects as well as development of tolerance and dependence to the new therapeutic agent (European Medicines Agency 2011). In addition, particular attention should be given to the development of CNS adverse effects such as cognition, reaction time, and driving (European Medicines Agency 2011).

At the end of Phase III studies, all the data are submitted to the competent regulatory body of a country for evaluation (New Drug Application or NDA in United States). The application of the new drug can be rejected, approved, or sometimes the agency may request additional data before a final decision is made. If the regulatory agency thinks that the new drug has a positive benefit/risk profile and that it fulfills all their requirements, a market authorization in a given country is granted. The indication that the new drug receives may be restricted to certain conditions or age-groups (for instance short-term insomnia, certain age-groups, etc.) based on the data and/or the results submitted to the agency. It is not uncommon that a new drug is authorized to be commercialized in a country and the same application is rejected in another country. This may happen for instance because the requirement for approval may vary between countries.

Once a drug receives a market authorization by a regulatory agency it can be commercialized. However, in order to be sold, somebody (the patient himself, an insurance company, or the government of a country) has to be willing to pay for it. The approval of a new drug by the regulatory agency does not guarantee that the drug will be paid by, for instance, an insurance company or by the government of a country. Particularly in countries with a strong socialized medicine, the opinion of independent agencies such as the National Institute for Health and Care Excellence (NICE) (formerly National Institute for Health and Clinical Excellence) in the United Kingdom which evaluates parameters not required for approval but equally important to assure efficient and high quality health care to citizens is taken into account to make this decision. Among others, one parameter that is evaluated to make a decision on whether or not to pay for the new drug is the relationship between benefits and costs of therapeutic agents (Chalkidou 2009). Therefore, given the high importance of this relationship, pharmacoeconomic studies are often included by pharmaceutical companies in the development plan of a new drug candidate (Bodrogi and Kaló 2010; International Society for Pharmacoeconomics and Outcome Research)

5 Post-approval Studies in Insomnia and Other Sleep Disorders

After a new therapeutic agent has received the first approval to be commercialized, the research and development on the activity on the new therapeutic agent does not stop but continues with the so-called post-approval studies. These studies are aimed to gather more information of the new drug to, for instance, better clarify some aspects of its development and/or expand its indications and monitoring its safety. Several types of post-approval studies exist (Fig. 1).

Studies Requested by the Regulatory Agency These might be additional Nonclinical or clinical studies that are requested by a regulatory agency to evaluate or better understand certain aspects of the activity and/or safety of the new drug. A new therapeutic agent may be approved by a regulatory agency with the condition that additional and specific studies will be conducted by the pharmaceutical company. Based on the results of these post-approval studies, the regulatory agency may then make some changes to the approval conditions for instance by putting some restriction on the use of the new therapeutic agent.

Studies to Expand the Indication that the New Drug Has Received After receiving the first approval, a pharmaceutical company may decide to run additional studies to expand the indication of the new drug, to seek a completely new indication, or to get an approval in another country. For instance, a new hypnotic may be studied in a special population such as elderly people to receive authorization to its use in this age-group, or long-term studies in insomnia can be conducted to receive an indication for chronic insomnia, or specific studies in Asian countries and population may be conducted to receive an approval in these geographic areas.

Studies to Develop New Formulation of the Same Active Molecule After a drug is approved, the pharmaceutical company may decide to develop new formulations of the same active molecule. This may be done just to offer more choices to the physician and patients or in response to some concern that has arisen after the approval of the drug. For instance, a slow release formulation or a fast-absorbed formulation may be developed in order to overcome a short and inadequate duration of the activity of a molecule or to speed up its absorption. The development of controlled release of and sublingual formulations of zolpidem are examples of these strategies (see chapter “Zolpidem Sublingual Formulations”).

Safety Monitoring (Pharmacovigilance) One of the criteria for approval of a new drug to be commercialized is its safety to humans. The safety of a new investigational drug is evaluated throughout the entire development process and approval is granted if a favorable benefit/risk ratio is demonstrated. Obviously, the correctness and the value of this assessment is a function of the total number of subjects exposed to the new drug in the development phases. In the best case scenario, throughout the entire development process, only a few thousand people have received the new drug candidate. Simply because of statistic considerations, this number maybe insufficient to detect certain adverse events such as rare effects that occur only in a small subset of the general population. When a drug is commercialized, suddenly a much larger population becomes exposed to it therefore increasing the likelihood of detecting adverse events not identified during the drug development process. Therefore, it is mandatory for the pharmaceutical company to continuously monitor for the entire life of the new drug product the appearance and the degree of adverse events after a drug is commercialized and to report them to the competent government body. This process is regulated by specific guidelines that assure a prompt detection and reporting of any adverse event associated with the use of a new drug (European Medicines Agency 2014;

Food and Drug Administration 2005a). Based on the results of this continuous surveillance activity, several actions may be taken by the regulatory agency such as restriction of the use of new drug, request of additional studies, or even withdrawal of the drug from the market.

Miscellaneous Studies Additional studies that a pharmaceutical company may decide to run after approval of the drug may also include a variety of other studies such as pharmacoeconomic studies and promotional marketing-driven studies.

Conclusions

The discovery and development of new therapeutic agents for treatment of sleep disorders is a complex, lengthy, and expensive process which, as in the case of any drug, is highly regulated by the government of a country. In order to maximize the chances for approval, it is necessary to understand the rules that govern this process and design a proper development plan which addresses all the points and concerns of the regulatory agency. Although there are several regulatory guidelines that help the investigators throughout this process, each development can be quite unique and may present issues not covered by any guideline. Therefore, it is always recommended to consult with the regulatory agency to discuss specific points that might have arisen during the development. In addition, the development of new therapeutic agents for sleep disorders presents some specific features typical of this group of disorders. In the case of development of hypnotics for example, issues such as tolerance, potential for abuse, interaction with other drugs, and next-day performance need to be addressed properly during the development. Despite that several different sleep disorders are known, very few specific medications are available to treat them. Therefore, with a better understanding of the physiology and pathogenesis of the different sleep disorders it can be expected that new, more specific, and better therapeutic agents to treat them may be discovered and developed in the future.

References

Bain KT (2006) Management of chronic insomnia in elderly persons. Am J Geriatr Pharmacother 4:168–192

Bateson AN (2002) Basic pharmacologic mechanism involved in benzodiazepine tolerance and withdrawal. Curr Pharm Des 1:5–21

Bodrogi J, Kaló Z (2010) Principles of pharmacoeconomics and their impact on strategic imperatives of pharmaceutical research and development. Br J Pharmacol 159:1367–1373

Cardario B et al (2011) Drugs and the risk of falling. Guidance document. http://www.health.gov.bc.ca/prevention/pdf/medications-and-the-risk-of-falling.pdf

Chalkidou K (2009) Comparative effectiveness review within the U.K.'s National Institute for Health and Clinical Excellence. The Commonwealth fund 59: pub 1296

Clare G (2012) The in-vitro mammalian chromosomal aberration test. Methods Mol Biol 817:69–91

Costantino HR, Illum L, Brandt G et al (2007) Intranasal delivery: physicochemical and therapeutic aspects. Int J Pharm 337:1–24

De la Herrán-Arita AK, Druken-Colin R (2012) Models for narcolepsy with catalepsy drug discovery. Expert Opin Drug Discov 7:152–164

de Lecea L, Huerta R (2014) Hypocretin (orexin) regulation of sleep-to-wake transition. Front Pharmacol 5:16

de Lecea L, Kilduff TS, Peyron C et al (1998) The hypocretins: hypothalamus-specific peptides with neuroexcitatory activity. Proc Natl Acad Sci USA 95:322–327

Deacon RM (2013) Measuring the strength of mice. J Vis Exp 76:2610

DiMasi JA, Feldman J, Seckler A et al (2010) Trends in risks associated with new drug development: success rates for investigational drugs. Clin Pharmacol Ther 87:272–277

Doering S, Boeckmann JA, Hugger S et al (2008) Ambulatory polysomnography for the assessment of sleep bruxism. J Oral Rahabil 35:572–576

Equihua AC, De la Herrán-Arita AK, Drucker-Colin R (2013) Orexin receptor antagonists as therapeutic agents for insomnia. Front Pharmacol 4:1–10

European Medicines Agency (EMA) (1994) Pharmacokinetics and metabolic studies in the safety evaluation of new medicinal products in animals. 3BS 11A

European Medicines Agency (EMA) (2001) Safety pharmacology studies for human pharmaceuticals. CPMP/ICH/539/00

European Medicines Agency (EMA) (2006) Guideline on the non-clinical investigation of the dependence potential of medicinal products. EMEA/CHMP/SWP/94227/2004

European Medicines Agency (EMA) (2007) Guideline on strategy to identify and mitigate risks for first in human clinical trials with investigational medicinal products. EMEA/CHMP/SWP/28367/07

European Medicinal Agency (EMA) (2008) Guidelines on risk assessment of medicinal products on human reproduction and lactation: from data to labelling. EMEA/CHMP/203927/2005

European Medicines Agency (EMA) (2010a) Guideline on repeated dose toxicity. CPMP/SWP/1042/99 Rev 1 Corr

European Medicines Agency (EMA) (2010b) Reflection paper on non-clinical evaluation of drug-induced liver injury (DILI). EMEA/CHMP/SWP/150115/2006

European Medicines Agency (EMA) (2011) Medicinal products for the treatment of insomnia. CHMP/16274/2009 Rev 1

European Medicines Agency (EMA) (2014) Guideline on medicinal products for the treatment of insomnia. EMA/213031/2014

Food and Drug Administration (1977) Guideline for the clinical evaluation of hypnotic drugs

Food and Drug Administration (2002) Guidance for industry immunotoxicology evaluation of investigational new drugs

Food and Drug Administration (2005a) Good pharmacovigilance practices and pharmacoepidemiologic assessment

Food and Drug Administration (2005b) Estimating the maximum safe starting dose in initial clinical trials for therapeutics in adult healthy volunteers

Food and Drug Administration (2006) Exploratory IND studies

Food and Drug Administration (2008) Safety testing of drug metabolites

Food and Drug Administration (2010) Assessment of abuse potential of drugs

Food and Drug Administration (2011) Reproductive and developmental toxicities–integrating study results to assess concern

Food and Drug Administration (2013) FDA-approved drugs to treat sleep disorders. J Psychosoc Nurs Ment Health Serv 51:9–10

Greenblatt DJ, Roth T (2012) Zolpidem for insomnia. Expert Opin Pharmacother 13:879–893

Griffiths RR, Johnson NW (2005) Relative abuse liability of hypnotic drugs: a conceptual framework and algorithm for differentiating among compounds. J Clin Psychiatry 66 (S9):31–41

Gunja N (2013) In the Zzz zone: the effects of Z-drugs on human performance and driving. J Med Toxicol 9:163–171

Hajak G, Müller WE, Wittchen HU et al (2003) Abuse and dependence potential for the non-benzodiazepine hypnotics zolpidem and zopiclone: a review of case reports and epidemiological data. Addiction 98:1371–1378

Heit HA (2003) Addiction, physical dependence, and tolerance: precise definition to help clinicians evaluate and treat chronic patients. J Pain Palliat Care Pharmacother 17:15–29

Hesse LM, von Moltke LL, Greenblatt DJ (2003) Clinically important drug interaction with zopiclone, zolpidem and zaleplon. CNS Drugs 17:513–532

ICH Harmonised Tripartite Guideline (1994a) Note for guidance on toxicokinetics: the assessment of systemic exposure in toxicity studies. Guideline S3A

ICH Harmonised Tripartite Guideline (1994b) Guidance for repeated dose tissue distribution studies. Guideline S3B

ICH Harmonised Tripartite Guideline (1995) Guideline on the need for carcinogenicity studies of pharmaceuticals. Guideline S1A

ICH Harmonised Tripartite Guideline (1996) Stability for new dosage forms. Guideline Q1C

ICH Harmonised Tripartite Guideline (1998) Duration of chronic testing in animals (rodent and non rodent toxicity testing). Guideline S4

ICH Harmonised Tripartite Guideline (2005a) Detection of toxicity to reproduction for medicinal products and toxicity to male fertility. Guideline S5(R2)

ICH Harmonised Tripartite Guideline (2005b) Immunotoxicity studies for human pharmaceuticals. Guideline S8

ICH Harmonised Tripartite Guideline (2000) Safety pharmacology studies for human pharmaceuticals. Guideline S7A

ICH Harmonised Tripartite Guideline (2009) Guideline on non-clinical safety studies for the conduct of human clinical trials and marketing authorization for pharmaceuticals. Guideline M3(R2)

ICH Harmonised Tripartite Guideline (2011) Guideline on genotoxicity testing and data interpretation for pharmaceuticals intended for human use. Guideline S2(R1)

ICH Harmonised Tripartite Guideline (2012) Development and manufacture of drug substance (chemical entities and biotechnological/biological entities). Guideline Q11

Ilomäki J, Paljärvi T, Korhonen MJ et al (2013) Prevalence of concomitant use of alcohol and sedative-hypnotic drugs in middle and older aged persons: a systematic review. Ann Pharmacother 47:257–268

International Society for Pharmacoeconomics and Outcome Research. Pharmacoeconomic guidelines around the world. www.ispor.org/peguidelines/index.asp

Johannesen L, Vicente J, Gray RA et al (2014) Improving the assessment of heart toxicity for all new drugs through translational regulatory science. Clin Pharmacol Ther 95:501–508

Kamel NS, Gammack JK (2006) Insomnia in the elderly: cause, approach, and management. Am J Med 119:463–469

Kanbayashi T, Inoue Y, Chiba S et al (2002) CSF hypocretin-1 (orexin A) concentrations in narcolepsy with and without catalepsy and idiopathic hipersomnia. J Sleep Res 11:91–93

Katchman AN, Koerner J, Tosaka T et al (2005) Comparative evaluation of HERG currents and QT intervals following challenge with suspected torsadogenic and non torsadogenic drugs. J Pharmacol Exp Ther 316:1098–1106

Kolla BP, Lovely JK, Mansukhani MP et al (2013) Zolpidem is independently associated with increased risk of inpatient falls. J Hosp Med 8:1–6

Kryger MH, Roth T, Dement WC (eds) (2011) Principles and practice of sleep medicine, 5th edn. Elsevier Saunders, St. Louis, MO

Krystal AD, Benca RM, Kilduff TS (2013) Understanding the sleep-wake cycle: sleep, insomnia and the orexin system. J Clin Psychiatry S1:3–20

López-Muñoz F, Ucha-Udabe R, Alamo C (2005) The history of barbiturates a century after their clinical introduction. Neuropsychiatr Dis Treat 1:329–343

Martelmans K, Zeiger E (2000) The Ames salmonella/microsome mutagenicity assay. Mutat Res 455:29–60
Mets MA, Volkerts ER, Oliver B (2010) Effect of hypnotic drugs on body balance and standing steadiness. Sleep Med Rev 14:259–267
Meyer OA, Tilson HA, Byrd WC (1979) A method for the routine assessment of fore- and hindlimb grip strength of rats and mice. Neurobehav Toxicol 1:233–236
Mignot E, Lammers GJ, Ripley B et al (2002) The role of cerebrospinal fluid hypocretin measurement in the diagnosis of narcolepsy and other hypersomnias. Arch Neurol 59:1553–1562
Milić M, Divljaković J, Rapalli S et al (2012) The role of α_1 and α_5 subunits containing $GABA_A$ receptors in motor impairment induced by benzodiazepine in rats. Behav Pharmacol 23:191–197
Mould R, Brown J, Marshall FH et al (2014) Binding kinetics differentiates functional antagonism of orexin-2 receptor ligands. Br J Pharmacol 171:351–363
Nakamura M, Kanbayashi T, Sugiura T et al (2010) Relationship between clinical characteristics of narcolepsy and CSF orexin-A levels. J Sleep res 20:45–49
Nishino S, Ripley B, Overeem S et al (2000) Hypocretin (orexin) deficiency in human narcolepsy. Lancet 355:39–40
Noble RE (2003) Drug therapy in the elderly. Metabolism 52(S2):27–30
Nordmann H (1985) Grip strength test and infrared thermometry as non-invasive methods to complement acute toxicity data in mice. Arch Toxicol Suppl 8:435–441
Nutt DJ, Sathl SM (2010) Searching for perfect sleep: the continuing evolution of $GABA_A$ receptor modulators as hypnotics. J Psychopharmacol 24:1601–1612
O'brien CP (2005) Benzodiazepines use, abuse and dependence. J Clin Psychiatry 2:28–33
Obayashi K, Araki T, Nakamura K et al (2013) Risk of falling and hypnotic drugs: retrospective drugs of inpatients. Drugs R D 13:159–164
OECD (1997) In-vitro mammalian chromosomal Aberration test. Test No: 473
Owen RT (2009) Novel zolpidem formulations. Drugs Today 45:395–400
Owens JA (2006) Pediatric insomnia. Sleep Med Clin 1:423–435
Rabbani M, Wright EJ, Little HJ (1995) Tolerance of competitive NMDA antagonists, but not cross tolerance with barbiturates. Pharmacol Biochem Behav 50:9–15
Rowlett JK, Platt DM, Lelas S et al (2005) Different $GABA_A$ receptor subtypes mediate the anxiolytic, abuse-related and motor effects of benzodiazepine-like drugs in primates. Proc Natl Acad Sci USA 102:915–920
Sakurai T, Amemiya A, Ishii M et al (1998) Orexin and orexin receptors: a family of hypothalamic neuropeptides and G-protein coupled receptors that regulate feeding and feeding behavior. Cell 4:573–585
Saunders JB (1986) Drugs and alcohol—their interactions and how to avoid them. Curr Ther 27:31–53
Sellayah D, Sikder D (2013) Food for thought: understanding the multifaceted nature of orexins. Endocrinology 154:2990–3999
Stephenson T (2005) How children's responses to drugs differ from adults. Br J Clin Pharmacol 59:670–673
Toth LA, Bhargava P (2013) Animal models of sleep disorders. Comp Med 63:91–104
Uzun S, Kozumplik O, Jakovljevic M et al (2010) Side effects of treatment with benzodiazepines. Psychiatr Danub 22:90–93
Vinkers CH, Oliver B (2012) Mechanism underlying tolerance after long-term benzodiazepine use: a future for subtype-selective $GABA_A$ receptor modulators? Adv Pharmacol Sci 2012:416864
Wang D, Wong KK, Dungan GC et al (2008) The validity of wrist actimetry assessment of sleep with and without sleep apnea. J Clin Sleep Med 4:450–455
Wick JY (2013) The history of benzodiazepines. Consult Pharm 28:538–548

Winrow CJ, Renger JJ (2014) Discovery and development of orexin receptor antagonists as therapeutic for insomnia. Br J Pharmacol 17:283–293

Zisapel N (2012) Drugs for insomnia. Expert Opin Emerg Drugs 17:299–317

Zosel A, Osterberg EC, Mycyk MB (2011) Zolpidem miuse with other medications or alcohol frequently results in intensive cure unit admission. Am J Ther 18:305–208

Part II
Drugs to Treat Insomnia

Low-Dose Doxepin for Insomnia

Dimitri Markov and Karl Doghramji

Abstract A tricyclic antidepressant first approved in 1969, doxepin has long been available in larger doses (10-, 25-, 50-, 75-, 100-, and 150-mg capsules) to treat depression and anxiety and as a topical preparation (5 % cream) for pruritus, but not in dosages <10 mg. Low-dose doxepin—3 and 6 mg—has demonstrated efficacy for insomnia characterized by frequent or early morning awakenings and an inability to return to sleep. FDA-approved in March 2010, doxepin (3 and 6 mg) is only the second insomnia medication not designated as a controlled substance and thus may be of special value in patients with sleep maintenance insomnia with a history of substance use disorders.

1 Introduction

Sleep consists of two strikingly different states: Rapid eye movement (REM) sleep and nonrapid eye movement (NREM) sleep, which alternate in a cyclical fashion. NREM sleep can be subdivided further into three stages (N1, N2, and N3). Stage N1 of NREM is viewed as a "shallow" sleep, during which an individual can be easily aroused. With the onset of stage N2, the arousal threshold increases, and a more intense stimulus is needed to arouse a sleeper. Stage N3 of NREM sleep (also known as deep sleep, delta sleep, or slow wave sleep) predominate during the first third of the night. Slow-wave sleep is associated with a higher arousal threshold than "lighter" stages of NREM sleep. Sleep begins with Stage N1 of NREM and "deepens" to NREM Stages N2 and N3, which are followed by the first brief

D. Markov, M.D.
Department of Psychiatry and Human Behavior, Sleep Disorders Center, Jefferson Medical College, Thomas Jefferson University, Philadelphia, PA, USA
e-mail: Dimitri.Markov@jefferson.edu

K. Doghramji, M.D. (✉)
Department of Psychiatry and Human Behavior, Sleep Disorders Center, Jefferson Medical College, Thomas Jefferson University, 211 South Ninth Street, Suite 500, Philadelphia, PA, USA
e-mail: Karl.Doghramji@jefferson.edu

A. Guglietta (ed.), *Drug Treatment of Sleep Disorders*, Milestones in Drug Therapy,
DOI 10.1007/978-3-319-11514-6_5

episode of REM in approximately 90 min. After the first sleep cycle, NREM and REM sleep continue alternating in a cyclical fashion. The duration of each cycle is approximately 90 min. REM sleep episodes become longer as the night progresses, and the longest REM periods are found in the last third of the night (Sinton and McCarley 2004; Iber et al. 2007; Doghramji 1998; Siegel 2005; Carskadon and Dement 2005; Walczak and Chokroverty 1999).

2 Major Neurotransmitters Involved in Sleep and Wakefulness

2.1 *Adenosine*

Adenosine is believed to be an endogenous sleep-producing substance. A breakdown product of adenosine triphosphate (ATP), adenosine is believed to mediate the transition from prolonged wakefulness to NREM sleep. Other substances hypothesized to be involved in promoting sleep include proinflammatory cytokines (interleukin-1), prostaglandin D2, and growth hormone-releasing hormone (Pace-Schott and Hobson 2002; Reiter 2003; Borbely and Achermann 2005; Mignot et al. 2002; Porkka-Heiskanen et al. 2002; McGinty and Szymusiak 2005).

2.2 *Acetylcholine*

Cholinergic neurons have a dual role: Some promote sleep, and others promote wakefulness (McCarley 1999).

2.3 *Serotonin and Norepinephrine*

Serotonin, norepinephrine, and histamine are believed to be neurotransmitters which promote wakefulness. The serotoninergic, noradrenergic, and histaminergic wakefulness-promoting neurons have a discharge pattern nearly opposite to that of the cholinergic sleep-promoting neurons. The discharge rate of serotoninergic, noradrenergic, and histaminergic neurons is fastest during wakefulness, decreases during NREM sleep, and virtually stops firing during REM sleep.

Serotonin and norepinephrine neurons promote cortical activation during wakefulness by rapid firing. During the NREM sleep period, at the beginning of the first sleep cycle, the serotoninergic and noradrenergic neurons significantly reduce their firing rate. This removes the inhibition from the REM-on cholinergic neurons, leading to the first REM sleep period approximately 90 min later (Pace-Schott and Hobson 2002; Jones 2005; Siegel 2004).

2.4 *Histamine*

Antihistaminergic drugs that cross the blood–brain barrier are known to produce sedation. The neurotransmitter histamine plays a key role in maintenance of wakefulness. Histaminergic neurons originate from the posterior hypothalamus and project diffusely throughout the brain. In the cortex, histamine facilitates cortical arousal. Histaminergic neurons fire most rapidly during cortical activation in the wake state and turn off during REM sleep (Pace-Schott and Hobson 2002; Jones 2005; Siegel 2004).

2.5 *Hypocretin (Orexin)*

The newly discovered peptides called hypocretins (also known as orexins) are thought to regulate wakefulness by interacting with histaminergic, aminergic, and cholinergic systems. Hypocretins consist of two neuropeptides, hypocretin 1 and hypocretin 2, and have key roles in regulation of arousal and metabolism. They bind to their corresponding receptors (Hcrtr1 and Hcrtr2) throughout the brain and spinal cord. These peptides are produced by hypothalamic neurons that surround the fornix bilaterally and in the dorsolateral hypothalamus. These hypothalamic regions are implicated in control of nutritional balance, blood pressure, and temperature regulation and endocrine secretion and arousal. Hypocretins likely play a role in all of these functions (Sutcliffe and de Lecea 2002; Peyron et al. 1998).

3 Insomnia

Insomnia is the complaint of an inability to fall or stay asleep or unrefreshing sleep. It represents the second most commonly expressed complaint (after pain) in clinical settings (Mahowald et al. 1997).

3.1 *Prevalence and Impact of Insomnia*

Of the general adult population, 35 % experience insomnia during the course of 1 year (Mellinger et al. 1985). Half experience the problem as severe, and 20 % of adults are dissatisfied with their sleep or take medication for sleeping difficulties

(Ohayon 1996). Insomnia is more common in women by a factor of 1.5:1 (Sutton et al. 2001). In women, its prevalence peaks during pregnancy and in the peri- and postmenopausal years. It is also more common in adolescents (ages 11–14) than in younger girls (30.4 % versus 16.8 %) (Camhi et al. 2000). The prevalence of insomnia increases with advancing age. Insomnia affects more than one-third of the population aged 65 and older (Ohayon et al. 2001). Insomnia causes a significant burden for the healthcare system and also for employers of insomniacs in both direct and indirect expenses including medical expenses, ramifications of accidents, and reduced productivity due to absenteeism and decreased work efficiency (Leger et al. 2001; Weissman et al. 1997; Mallon et al. 2002).

4 Doxepin

Low-dose doxepin—3 and 6 mg—has demonstrated efficacy for insomnia characterized by frequent or early morning awakenings and an inability to return to sleep. FDA approved in March 2010, doxepin (3 and 6 mg) is only the second insomnia medication not designated as a controlled substance and thus may be of special value in patients with a history of substance use disorders (Silenor 2010). Ramelteon, the other hypnotic that is not a controlled substance, is indicated for sleep initiation insomnia (i.e., inability to fall asleep). In contrast, low-dose doxepin is for patients with sleep maintenance insomnia, which consists of the complaint of waking up frequently or early in the morning and not falling back asleep (Silenor 2010; Goforth 2009). A tricyclic antidepressant first approved in 1969, doxepin has long been available in larger doses (10-, 25-, 50-, 75-, 100-, and 150-mg capsules) to treat depression and anxiety and as a topical preparation (5 % cream) for pruritus, but not in dosages <10 mg. An inexpensive generic doxepin oral solution (10 mg/ml) is available and can be titrated to smaller dosages by a dropper. Liquid doxepin costs 10–20 cents per dose. A pharmacist can provide a dropper, and patients should mix the medication in 4 ounces of water, milk, or juice; 0.3 ml of liquid doxepin contains 3 mg of active ingredient, and 0.6 ml of solution contains 6 mg of doxepin. These other preparations of doxepin, however, are not FDA-approved for insomnia.

4.1 *Mechanism of Action*

Doxepin's mechanism of action for treating depression and insomnia remains unknown. The antidepressant effect of doxepin is thought to result from inhibition of serotonin and norepinephrine reuptake at the synaptic cleft. Animal studies have shown anticholinergic and antihistaminergic activity with doxepin (Goforth 2009). Doxepin is a potent histamine antagonist—predominantly at the H1 receptor—and its binding potency to the H1 receptor is approximately 100-times higher than its binding potency for monoamine transporters (serotonin and norepinephrine) (Goforth 2009; Stahl 2008). Brain histamine is believed to be 1 of the key elements in maintaining wakefulness, and the activation of the H1 receptor is thought to play an important role in mediating arousal. Blockade of the H1 receptor by doxepin likely plays a role in reducing wakefulness. Typically, therapeutic doses of antidepressants with antihistaminergic properties, such as doxepin at antidepressant doses, amitriptyline, or desipramine, do not selectively block H1 receptors, but act at cholinergic, serotonergic, adrenergic, histaminergic, and muscarinic receptors, which can cause adverse effects (Stahl 2008). However, low doses of doxepin (1, 3, and 6 mg) can achieve selective H1 blockade (Roth et al. 2007; Scharf et al. 2008). Patients taking >25 mg/day of doxepin may report clinically significant anticholinergic effects.

4.2 *Pharmacokinetics*

When doxepin, 6 mg, was administered to healthy, fasting patients, time to maximum concentration (Tmax) was 3.5 h. Peak plasma concentration (Cmax) increased in a dose-related fashion when doxepin was increased from 3 to 6 mg. Doxepin, 6 mg, taken with a high-fat meal resulted in area under the curve increase of 41 %, Cmax increase of 15 %, and almost 3-h delay in Tmax. Therefore, to prevent a delay in onset of action and to minimize the likelihood of daytime sedation, doxepin should not be taken within 3 h of a meal (Silenor 2010; Goforth 2009; Stahl 2008). Doxepin is metabolized primarily by the liver's cytochrome P450 (CYP) 2C19 and CYP2D6 enzymes; CYP1A2 and CYP2D6 are involved to a lesser extent. If doxepin is coadministered with drugs that inhibit these isoenzymes, such as fluoxetine and paroxetine, doxepin blood levels may increase. Doxepin does not seem to induce CYP isoenzymes. This medication is metabolized by demethylation and oxidation; the primary metabolite is nordoxepin (N-desmethyldoxepin), which later undergoes glucuronide conjugation. The half-life is 15 h for doxepin and 31 h for nordoxepin. Doxepin is excreted in urine primarily as glucuronide conjugate (Silenor 2010; Goforth 2009; Stahl 2008). Coadministration with cimetidine, an inhibitor of CYP isoenzymes, could double the doxepin plasma concentration; therefore, patients taking cimetidine should not exceed 3 mg/day of doxepin.

4.3 *Efficacy*

Doxepin reduced insomnia symptoms in 3 pilot studies at doses of 10, 25, and 50 mg, and in 2 phase III randomized, double blind, placebo-controlled clinical trials using 1, 3, and 6 mg (Roth et al. 2007; Scharf et al. 2008). Clinical studies lasted up to 3 months (Silenor 2010; Goforth 2009; Stahl 2008; Hajak et al. 1996,

2001; Rodenbeck et al. 2003). In the first phase III trial, 67 patients, age 18 to 64 with chronic primary insomnia, were randomly assigned to placebo or 1, 3, or 6 mg of doxepin for 2 nights. All patients received all treatments, and each treatment was followed by 8 h of polysomnography (PSG) evaluation in a sleep laboratory (Roth et al. 2007). In this study, patients taking doxepin at all doses achieved improvement in objective (PSG-defined) and subjective (patient-reported) measures of sleep duration and sleep maintenance. Wake after sleep onset (WASO), total sleep time (TST), and sleep efficiency (SE) improved with all doxepin doses, and wake time during sleep (WTDS)—which was the primary study endpoint—decreased with 3 and 6 mg doses, but not with 1 mg or placebo. In addition, PSG indicators of early-morning awakenings (terminal insomnia) were reduced, as shown by an increase in SE during the final third of the night and the 7th and 8th h of sleep (1, 3, and 6 mg doses) and a reduction in wake time after sleep (WTAS) during the final third of the night (6 mg only). The effects on sleep duration and maintenance were more robust with 3 and 6 mg doses. Improved sleep onset was seen only with the 6 mg dose. Next-day alertness was assessed using the Visual Analogue Scale (VAS) for sleepiness and the Digit-Symbol Substitution Test (DSST) and the Symbol-Copying Task (SCT) for psychomotor function. No statistically significant differences were found among placebo and any of the doxepin doses on the VAS, DSST, or SCT. Doxepin was well tolerated. Reported adverse events were mild or moderate. Headaches and somnolence were reported by >2 % of patients. The incidence of adverse events, including next-day sedation, was similar to that of placebo. Additionally, there were no spontaneous reports of anticholinergic side effects, which are associated with higher doxepin doses. The second phase III trial examined safety and efficacy of 1, 3, and 6 mg doxepin in patients age ≥65 (Scharf et al. 2008). Seventy-six adults with primary insomnia were randomly assigned to receive placebo or doxepin for 2 nights; all patients received all treatments, and each treatment was followed by an 8-h PSG. Patients taking any doxepin dose achieved objective and subjective improvement in sleep duration and sleep maintenance, which lasted into the final hours of the night. WTDS (primary study endpoint), WASO, TST, and overall SE improved at all doxepin doses compared with placebo, and WTAS and SE at hours 7 and 8 improved at doxepin doses of 3 and 6 mg compared with placebo. These findings suggest that doxepin, 3 and 6 mg, can help older insomnia patients with early morning awakenings. In this study, no statistically significant differences were found among placebo and any doxepin doses on VAS, DSST, or SCT or next-day residual sedation. The incidence of side effects was low and similar to that of placebo. Adverse events were mild or moderate; 1 incident of chest pain was reported, but it was determined not to be of cardiac origin and not related to study drug. There were no spontaneous reports of anticholinergic side effects associated with higher doses of doxepin. There were no reports of memory impairment (Scharf et al. 2008).

4.4 Tolerability

Clinical studies that evaluated the safety of doxepin lasted up to 3 months. Somnolence/sedation, nausea, and upper respiratory tract infection were reported by >2 % of patients taking doxepin and were more common than in patients treated with placebo (Silenor 2010). All reported adverse events were mild to moderate. Doxepin appears to be better tolerated at hypnotic doses (3 and 6 mg) than at antidepressant doses (50–300 mg/day), although direct comparative studies are not available. Additionally, psychomotor function assessed using DSST and SCT and next-day sedation assessed using VAS in patients receiving hypnotic doses of doxepin (1 and 3 mg) were the same as placebo. Two studies noted small-to-modest decreases in DSST, SCT, and VAS when doxepin, 6 mg, was administered.

Patients taking doxepin at antidepressant doses report significant anticholinergic side effects, including sedation, confusion, urinary retention, constipation, blurred vision, and dry mouth. Hypotension also has been reported at antidepressant doses, and there seems to be a dose-dependent cardiotoxicity, with higher incidence of adverse effects occurring at higher doses of the drug. Severe toxicity or death from overdose is presumably less likely with hypnotic doses of doxepin than with higher doses, although this has not been systematically explored. If an insomniac overdosed on a 30-day supply of an hypnotic dose (3 or 6 mg), he or she would take only 90–180 mg of doxepin, which would be unlikely to cause severe toxicity or death (Goforth 2009; Stahl 2008; Roth et al. 2007). Symptoms of withdrawal and rebound insomnia—an increase in WASO compared with baseline after discontinuing the medication—were assessed in a 35-day double-blind study of adults with chronic insomnia (Silenor 2010). There was no evidence of withdrawal syndrome as measured by Tyler's Symptom Checklist after doxepin 3 and 6 mg was discontinued. Discontinuation period-emergent nausea and vomiting was noted in 5 % of patients taking 6 mg of doxepin, but not in those taking placebo or 3 mg of doxepin. There was no evidence of rebound insomnia after doxepin 3 and 6 mg was discontinued (Silenor 2010).

4.5 Contraindications

Doxepin is contraindicated in patients with hypersensitivity to doxepin hydrochloride, with severe urinary retention, with narrow angle glaucoma, and who have used monoamine oxidase inhibitors (MAOIs) within the previous 2 weeks. Serious adverse effects, including hypertensive crisis and death, have been reported with coadministration of MAOIs and certain drugs, such as serotonergic antidepressants and some opioids derivatives. There are no reports of concomitant use of doxepin with MAOIs (Silenor 2010).

4.6 Dosing

In adults, the recommended hypnotic dose for doxepin is 6 mg taken 30 min before bedtime. For patient's age ≥65, the recommended starting hypnotic dose is 3 mg 30 min before bedtime, which can be increased to 6 mg if indicated (Silenor 2010).

References

Borbely AA, Achermann P (2005) Sleep homeostasis and models of sleep regulation. In: Kryger MH, Roth T, Dement WC (eds) Principles and practice of sleep medicine. Elsevier Saunders, Philadelphia, pp 405–417

Camhi SL, Morgan WJ, Pernisco N, Quan SF (2000) Factors affecting sleep disturbances in children and adolescents. Sleep Med 1(2):117–123

Carskadon MA, Dement WC (2005) Normal human sleep: an overview. In: Kryger MH, Roth T, Dement WC (eds) Principles and practice of sleep medicine. Elsevier Saunders, Philadelphia, pp 13–23

Doghramji K (1998) The evaluation and management of sleep disorders. In: Stoudemire A (ed) Clinical psychiatry for medical students. J.B. Lippincott Company, Philadelphia, pp 783–818

Goforth HW (2009) Low-dose doxepin for the treatment of insomnia: emerging data. Expert Opin Pharmacother 10(10):1649–1655

Hajak G, Rodenbeck A, Adler L et al (1996) Nocturnal melatonin secretion and sleep after doxepin administration in chronic primary insomnia. Pharmacopsychiatry 29:187–192

Hajak G, Rodenbeck A, Voderholzer U et al (2001) Doxepin in the treatment of primary insomnia: a placebo-controlled, double-blind, polysomnographic study. J Clin Psychiatr 62:453–463

Iber C, Anvoli-Israel S, Chesson AL (2007) The AASM manual for the scoring of sleep and associated events. The American Academy of Sleep Medicine, Weschester

Jones BE (2005) Basic mechanisms of sleep-wake states. In: Kryger MH, Roth T, Dement WC (eds) Principles and practice of sleep medicine. Elsevier Saunders, Philadelphia, pp 136–153

Leger D, Scheuermaier K, Philip P, Paillard M, Guilleminault C (2001) SF-36: evaluation of quality of life in severe and mild insomniacs compared with good sleepers. Psychosom Med 63:49–55

Mahowald MW, Kader G, Schenck CH (1997) Clinical categories of sleep disorders I. Continuum 3(4):35–65

Mallon L, Broman JE, Hetta J (2002) Sleep complaints predict coronary artery disease mortality in males: a 12-year follow-up study of a middle-aged Swedish population. J Intern Med 251 (3):207–216

McCarley RW (1999) Sleep neurophysiology: basic mechanisms underlying control of wakefulness and sleep. In: Chokroverty S (ed) Sleep disorders medicine: basic science, technical considerations, and clinical aspects. Butterworth Heinemann, Boston, pp 21–50

McGinty D, Szymusiak R (2005) Sleep-promoting mechanisms in mammals. In: Kryger MH, Roth T, Dement WC (eds) Principles and practice of sleep medicine. Elsevier Saunders, Philadelphia, pp 169–184

Mellinger GD, Balter MB, Uhlenhuth EH (1985) Insomnia and its treatment. Prevalence and correlates. Arch Gen Psychiatr 42(3):225–232

Mignot E, Taheri S, Nishino S (2002) Sleeping with the hypothalamus: emerging therapeutic target for sleep disorders. Nat Neurosci 5(Suppl):1071–1075

Ohayon M (1996) Epidemiological study on insomnia in the general population. Sleep 19(Suppl 3):S7–S15

Ohayon MM, Zulley J, Guilleminault C, Smirne S, Priest RG (2001) How age and daytime activities are related to insomnia in the general population: consequences for older people. J Am Geriatr Soc 49:360–366
Pace-Schott EF, Hobson JA (2002) The neurobiology of sleep: genetics, cellular physiology and subcortical networks. Nat Rev Neurosci 3:591–605
Peyron C, Tighe DK, van den Pol AN et al (1998) Neurons containing hypocretin (orexin) project to multiple neuronal systems. J Neurosci 18:9996–10015
Porkka-Heiskanen T, Alanko L, Kalinchuk A et al (2002) Adenosine and sleep. Sleep Med Rev 6 (4):321–332
Reiter RJ (2003) Melatonin: clinical relevance. Best Pract Res Clin Endocrinol Metab 17 (2):273–285
Rodenbeck A, Cohrs S, Jordan W et al (2003) The sleep-improving effects of doxepin are paralleled by a normalized plasma cortisol secretion in primary insomnia. A placebo-controlled, double-blind, randomized, cross-over study followed by an open treatment for 3 weeks. Psychopharmacology 170:423–428
Roth T, Rogowski R, Hull S et al (2007) Efficacy and safety of doxepin 1 mg, 3 mg, and 6 mg in adults with primary insomnia. Sleep 30(11):1555–1561
Scharf M, Rogowski R, Hull S et al (2008) Efficacy and safety of doxepin 1 mg, 3 mg, and 6 mg in elderly patients with primary insomnia: a randomized, double-blind, placebo controlled cross-over study. J Clin Psychiatr 69:1557–1564
Siegel JM (2004) The neurotransmitters of sleep. J Clin Psychiatr 65(Suppl 16):4–7
Siegel JM (2005) REM sleep. In: Kryger MH, Roth T, Dement WC (eds) Principles and practice of sleep medicine. Elsevier Saunders, Philadelphia, pp 120–135
Silenor (2010) Somaxon, San Diego [package insert]
Sinton CM, McCarley RW (2004) Neurophysiological mechanisms of sleep and wakefulness: a question of balance. Semin Neurol 24(3):211–223
Stahl SM (2008) Selective histamine H1 antagonism: novel hypnotic and pharmacologic actions challenge classical notions of antihistamines. CNS Spectr 13(12):1027–1038
Sutcliffe JG, de Lecea L (2002) The hypocretins: setting the arousal threshold. Nat Rev Neurosci 3:339–349
Sutton DA, Moldofsky H, Badley EM (2001) Insomnia and health problems in Canadians. Sleep 24(6):665–670
Walczak T, Chokroverty S (1999) Electroencephalography, electromyography, and electro-oculography: general principles and basic technology. In: Chokroverty S (ed) Sleep disorders medicine: basic science, technical considerations, and clinical aspects. Butterworth Heinemann, Boston, pp 175–203
Weissman MM, Greenwald S, Nino-Murcia G, Dement WC (1997) The morbidity of insomnia uncomplicated by psychiatric disorders. Gen Hosp Psychiatry 19:245–250

Lorediplon: A New $GABA_A$ Modulator Drug for Treatment of Insomnia

Fabiana d' Aniello, Benjamin Santos, and Antonio Guglietta

Abstract Drugs acting on the $GABA_A$ receptor such as the so-called Z-drugs Zolpidem, Zaleplon, and Eszopiclone are widely used in the treatment of insomnia. Despite their efficacy and good safety profile however, none of these drugs represent the perfect solution to the treatment of this disorder. Therefore, there is still room for improvement of this class of drugs and for addressing still unmet clinical needs. Lorediplon is a new promising drug acting on the $GABA_A$ receptor currently under clinical evaluation which selectively binds to the α_1 subunit of the $GABA_A$ receptor. In animals, Lorediplon showed potent sedative effects and much less potent activity on inhibition of muscular tone. No development of tolerance was shown to the sedative effect of Lorediplon after 5 days continuous daily treatment. Preclinical in vivo studies also demonstrated that Lorediplon at hypnotic dose has a low risk of development of amnesia, next-day hangover effects, or development of physical dependence. The pharmacological activity and pharmacokinetic profile of Lorediplon were evaluated in three clinical studies. After oral administration, the molecule is absorbed quickly with a dose-related linear profile and T_{max} around 2 h. In a phase advance model of insomnia in healthy human volunteers, Lorediplon, compared to Zolpidem, showed a longer lasting maintenance of sleep. The sleep induced by Lorediplon was rated by the subjects participating in the study as of improved quality while EEG recording in mice demonstrated that Lorediplon-induced sleep is less fragmented and more physiological than that induced by Zolpidem. Lorediplon also shows an excellent safety profile with a comparable to Zolpidem number of adverse events reported in clinical studies. Therefore, Lorediplon has the potential to become the best-in-class $GABA_A$ acting hypnotic and a major player in the treatment of insomnia and related disorders.

F. d' Aniello (✉)
Grupo Ferrer International, Diagonal 549, 08029 Barcelona, Spain
e-mail: fdaniello@ferrer.com

B. Santos
Spherium Biomed, Barcelona, Spain
e-mail: bsantos@spheriumbiomed.com

A. Guglietta
Independent Pharmaceutical Consultant, Barcelona, Spain
e-mail: consultant@guglietta.es

A. Guglietta (ed.), *Drug Treatment of Sleep Disorders*, Milestones in Drug Therapy,
DOI 10.1007/978-3-319-11514-6_6

1 Introduction

Sleep disorders are a group of medical conditions characterized by abnormality of normal sleep which includes approx. a hundred different disorders ranging from difficult falling and staying asleep to excessive daytime sleepiness to alterations of sleep patterns to unusual behavior during sleep (see chapter "Classification of Sleep Disorders").

Insomnia is a very common medical condition characterized by difficult to fall and/or maintain sleep. It affects people worldwide and, although its prevalence increases with age, it can occur in all age groups. Insomnia can be transient or chronic or primary or secondary to other medical conditions and can have serious daytime consequences such as fatigue, difficulty in concentrating, memory impairment, irritability, anxiety, and depression (Wafford and Eber 2008; Taylor et al. 2003). All this can lead to problems such as low productivity, increased risk of automobile and industrial accidents, and suicide (Schenck et al. 2003). Patients with insomnia also have an increased incidence of medical and psychiatric problems and reduced quality of life (Roth 2005).

The social and economic impact of insomnia can be enormous and it has been estimated that its total cost including treatment, lost productivity, and insomnia-related accidents may exceed 100 billion USD per year in the United States alone (Fullerton 2006).

2 The Role of GABA Receptors in Sleep

Two major classes of GABA receptors are known: the $GABA_A$ receptor which belongs to the class of ionotropic receptors and whose activation results in regulation of ion flow through the cell membrane and the $GABA_B$ receptor which belongs to the family of metabotropic receptors and is coupled to G-protein (Perfilova and Tiurenkov 2010; Benarroch 2012; Sharman et al. 2013; Bormann 2000; Olsen and Sieghart 2008). A key role in the regulation of sleep is played by the $GABA_A$ receptors and chemical agents that modulate the activity of this receptor have been developed for the treatment of sleep disorders and insomnia (Gottesmann 2002).

The ionotropic $GABA_A$ receptor is part of the family of ligand-gated ion channels known as the Cys loop receptors. Members of the same family include the nicotinic acetylcholine receptors, glycine receptors, and 5-HT_{3A} receptors (Nys et al. 2013). Structurally the $GABA_A$ receptor is a transmembrane protein made of five protein subunits arranged around a central channel permeable to Cl^- through which the ions can enter the cell. In humans, the five proteins of the receptor are made of several different protein subunits which can be divided into several classes some of which can be further divided into subclasses giving origin to a total of at least 19 possible combinations (Table 1). Despite all possible combinations, by far, the five subunits of $GABA_A$ receptors are most commonly made of two copies of

Table 1 Protein subunits of the $GABA_A$ receptor present in humans

Subunits	Subunits Subdivision	Most frequent arrangement
α	α_1, α_2, α_3, α_4, α_5, α_6,	$\alpha_1\beta_2\gamma_2$ (60 %)
β	β_1, β_2, β_3	
γ	γ_1, γ_2, γ_3	$\alpha_2\beta_3\gamma_2$ (15–20 %)
δ		
ε		$\alpha_3\beta_n\gamma_2$ (10–15 %)
γ		
ρ	ρ_1, ρ_2, ρ_3	

Whole brain. Numbers in parentheses represent the estimated percentages of the most frequent subunits. The homopentameric ρ receptor has also been referred as the $GABA_C$ receptor (Bormann 2000; Perfilova and Tiurenkov 2011)

the α_1, two copies of the β_2, and one copy of γ_2 subunit ($\alpha_1\beta_2\gamma_2$) with the $\alpha_2\beta_3\gamma_2$ and the $\alpha_3\beta_n\gamma_2$ receptors also highly prevalent (Table 1) (Whiting 2003; Nutt 2006; Möhler et al. 2002; Möjler 2007).

The main endogenous ligand for the $GABA_A$ receptor is the γ-aminobutyric acid although other endogenous ligands such as neurosteroids and endocannabinoids have also been shown to bind to this receptor (Sigel and Steinmann 2012; Belelli and Lambert 2005; Siegel et al. 2011; Nutt and Sathl 2010). The γ-aminobutyric acid binds at the receptor at two specific sites located at the α/β subunit interface (Rudolph and Knoflach 2011; Sigel and Steinmann 2012). Activation of the $GABA_A$ receptor by GABA leads to a change in the conformation of the proteins of the receptor opening the Cl^- channel and allowing chloride ions to enter the cell (Sigel and Steinmann 2012).

The $GABA_A$ receptor modulation is associated with several biological effects such as sedation, myorelaxation, sleep, anxiety, epilepsy, and memory (Rudolph et al. 1999). The exact contribution of the different subunits to specific biologic effects is far from being completely understood. The majority of studies have focused on the role of the α subunit of the $GABA_A$ receptor. Existing data, based mainly on absence of a specific benzodiazepine-induced effect in mice lacking a specific subunit, suggest that the α_1 subunit is associated with sedation, convulsions, and memory, the α_2 with anxiety, electroencephalogram (EEG) changes, and sleep/wake rhythms, the α_3 with anxiety, sleep, and depression, and the α_5 subunit with sleep, learning, and memory (Nutt 2006; Nutt and Sathl 2010; Sigel and Steinmann 2012) (Table 2). The α_1, α_2, α_3, and α_5 are the subunits most involved in regulation of sleep, mood, and cognition (Nutt and Sathl 2010).

Table 2 Function of the α subunits most involved in regulation of sleep of the $GABA_A$ receptor

Subunit	Role
α_1	Sedation, convulsions, memory
α_2	Sleep/wake, EEG changes, anxiety
α_3	Sleep, anxiety, depression
α_5	Sleep, learning, memory

3 Hypnotic Drugs Acting on the $GABA_A$ Receptor

Several drugs also target the $GABA_A$ receptor and bind and/or modulate its action to elicit a biological response. The binding of benzodiazepines to the $GABA_A$ receptor is well known. The classical binding site for benzodiazepines is different from the GABA-binding site and it is located at the α/γ subunit interface of the receptor (Sigel and Steinmann 2012). In addition to this site, a new binding site for benzodiazepines has recently been described which is located at the α/β interface (Baur et al. 2012; Ramerstorfer et al. 2011). Benzodiazepines do not open the Cl^- channel of the $GABA_A$ receptor but instead increase the apparent affinity for channel gating by GABA (Sigel and Steinmann 2012). In addition to benzodiazepines, several other drugs have been shown to bind to the $GABA_A$ receptor and/or modulate its activity. These include volatile anesthetics such as halothane, intravenous anesthetics such as propofol and etomidate, barbiturates, gabapentin, and pregabalin, among others (Whiting 2003; Sieghart et al. 2012; Vedula et al. 2009).

The $GABA_A$ receptor is involved in the regulation of sleep (Nutt and Sathl 2010) which is mainly mediated through its α subunits. More specifically the α_1 subunit, in addition to be involved in memory, has been demonstrated to have a sedative and anticonvulsant effect and thus to contribute to regulation of sleep (Nutt and Sathl 2010). However, this subunit does not mediate the benzodiazepine-induced changes of EEG (Tobler et al. 2001) and therefore it seems to be involved more in sedation than in regulation of sleep (Nutt and Sathl 2010). The α_2 subunit, on the other hand, seems to mediate the EEG activity in nonrapid eye movement, sleep, and theta activity in rapid eye movements (Kopp et al. 2004). The α_3 subtype, which has a unique distribution in the thalamus, is also involved in the regulation of sleep mainly through the thalamo-cortical projections (Browne et al. 2001). Finally, the α_5 subunit may have a sedative effect and thus may also contribute to regulation of sleep (Van Rijnsoever et al. 2004; Savič et al. 2008).

3.1 *Barbiturates and Benzodiazepines*

The key role of the $GABA_A$ receptor in the regulation of sleep has been widely exploited to develop hypnotic drugs. The barbiturates, introduced clinically about a century ago, were perhaps the first pharmaceutical agents to have demonstrated a real efficacy as hypnotics and, in particular, the short-acting and intermediate-

Table 3 Commercialized Z-drugs

Z-drugs	Common commercial names	FDA approval
Zopiclone	Imovane, Zimovane	1986[a]
Zolpidem	Ambien, Stilnox	1992
Zaleplon	Sonata	1999
Eszopiclone	Lunesta	2004

[a]Not approved by the FDA but available in other countries other than USA

acting barbiturates were largely used to treat this condition in the last century until the introduction of the benzodiazepines in the 1960s (López-Muñoz et al. 2005). The hypnotic activity of these drugs is mainly due to modulation of the activity of GABA$_A$ receptor to which they bind (Whiting 2003). Benzodiazepines were first introduced as hypnotic in 1960 and largely replaced barbiturates in the treatment of insomnia (Wick 2013). As mentioned above, benzodiazepines also act by binding to two different sites of the GABA$_A$ receptor and by modulating its activity (Sigel and Steinmann 2012).

3.2 *Non-benzodiazepines Hypnotics*

The last class of hypnotic drugs acting through the GABA$_A$ receptor are the non-benzodiazepine hypnotics (the so-called Z-drugs) which are today the most widely drugs used to treat insomnia and which also act by binding to specific GABA$_A$ subunits. Currently there are four non-benzodiazepine hypnotic drugs approved for treatment of insomnia: Zopiclone, Zolpidem, Zaleplon, and Eszopiclone. They all have been approved in the period 1986–2004 and several new formulations of these molecules were developed and commercialized in the following years (Table 3). As the benzodiazepines, the Z-drugs have low activity at α_4 and α_6, but on the contrary they show significant differences in their affinity and activity for the other α units (Nutt and Sathl 2010).

Other GABA$_A$ modulators such as Indiplon underwent clinical development as hypnotic but, for a variety of reasons, so far they all have failed to reach the market.

Despite that all the drugs that have been developed for insomnia, including those that act on the GABA$_A$ receptor, have significantly improved the treatment of this disorder, we are still far from having found the ideal hypnotic and all the hypnotic drugs on the market have drawbacks that have to be taken into account when using them.

4 Lorediplon

Despite being good effective drugs, the hypnotics acting on GABA$_A$ receptor, including the Z-drug, can still be improved in terms of both efficacy and safety. Therefore, the idea behind the discovery and development of Lorediplon was the

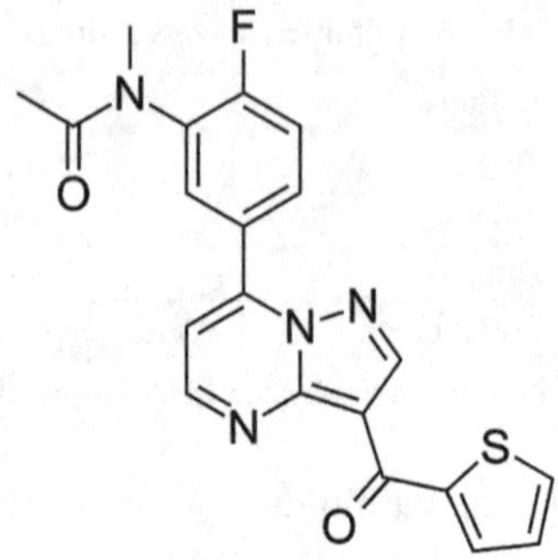

Fig. 1 Chemical structure of Lorediplon

search for a better $GABA_A$ hypnotic which would induce physiologic sleep quickly of adequate duration while displaying reduced undesirable effects.

4.1 Chemical Structure

Chemically, Lorediplon (Fig. 1) is a pyrazolopyrimidine and therefore belongs to the same chemical class of the hypnotics Zaleplon and Indiplon. More specifically, Lorediplon is chemically related to Indiplon from which it differs in the presence of a Fluorine atom in position 2. The introduction of fluorine atom in a molecule may change its biological activity. For instance, atorvastatin a drug used to lower blood levels of cholesterol contains fluorine in its molecule which is essential for its activity. The substitution of fluorine with an atom of hydrogen in the atorvastatin structure leads to a compound that is more than 800 % less active (Roth 2002; Ruth et al. 1990; Istvan and Deisenhofer 2001). It is believed that due to the high stability of the carbon–fluorine bond, fluorinated compounds display a delayed metabolism and elimination and thus different pharmacokinetics properties (Hagmann 2008). Based on its chemical structure and biological data, Lorediplon was found to fulfill all the criteria for patentability and a composition of matter patent has already been granted in some countries while being under review in others.

4.2 Preclinical Pharmacology

4.2.1 Binding to $GABA_A$ Subunits

The affinity of Lorediplon to the α_1 (α_1-enriched cortex tissue of the rat brain) and α_2 (α_2-enriched tissue of the rat GABA-A receptor) to GABA-A subunits was evaluated in vitro and compared to that of other Z-drugs. Lorediplon preferentially binds with high affinity to the α_1 subunit of the $GABA_A$ receptor and with much lower affinity to the α_2 subunit. The K_i for the α_1 and α_2 subunit is 2.1 nM and 20 nM, respectively, giving a ratio α_2/α_1 of 9.6 (Table 4). Therefore, Lorediplon can

Table 4 In vitro pharmacology of Lorediplon and related compounds

$GABA_A$ receptor subunit affinity (K_i nM)[a]		
	α_1	α_2
Lorediplon	2.1	20.0
Zaleplon	182.6	1,571.3
Zolpidem	38.2	656.9
Zopiclone	82.3	49.4
Potency and efficacy on most common $GABA_A$ receptor subtypes[b]		
	Potency (EC_{50} nM)	Efficacy (%)
$\alpha_1\beta_2\gamma_2$	8	64
$\alpha_2\beta_{2/3}\gamma_2$	23	62
$\alpha_3\beta_{2/3}\gamma_2$	31	73
$\alpha_5\beta_2\gamma_2$	36	44

[a]Enriched cortex tissue of rat brain

[b]$GABA_A$ receptor subtypes express in *Xenopus laevis* oocytes. Full concentration response profiles were determined at a GABA concentration giving rise to 5–20 % of the maximal GABA-evoked response for the GABA receptor subtypes studied which would serve as control. 0.5 μM diazepam which would elicit > 90 % of maximal response was used as reference. Potentiation was normalized to the reference diazepam potentiation in the same oocyte

be classified as highly selective for the α_1 subunit of the $GABA_A$ receptor. The affinity of Lorediplon for the α_1 subunit of the GABA receptor was the highest among the commercialized Z-drugs tested (Table 4). On the other hand, Lorediplon did not show any significant binding activity to a variety of other receptors representing the major drug target classes (data not shown). The in vitro potency as well as the efficacy of Lorediplon was studied in an assay measuring the function of the most common $GABA_A$ receptor subtypes (namely, $\alpha_1\beta_2\gamma_2$, $\alpha_2\beta_{2/3}\gamma_2$, $\alpha_3\beta_{2/3}\gamma_2$, $\alpha_5\beta_2\gamma_2$) expressed in *Xenopus laevis*. The results showed that Lorediplon is a potent positive allosteric modulator of Cl^- currents mediated by the $GABA_A$ sub-receptors studied and a fourfold potency selectivity for $\alpha_1\beta_2\gamma_2$ receptors over other subtypes was observed (Table 4). With respect to efficacy, Lorediplon is a partial allosteric modulator relative to diazepam at all receptor subtypes. Similar efficacy was observed at α_1, α_2, and α_3 containing receptor, while lower efficacy was observed at α_5 receptor (Table 4).

4.2.2 Sedation

As mentioned, the α_1 subunit of the $GABA_A$ receptor plays a key role in sedation. Therefore, the high affinity of Lorediplon for this subunit should translate into a sedative effect in vivo. The effect on Lorediplon on sedation was evaluated in mice by measuring the inhibition of the spontaneous locomotor activity (SMA) in the open field after oral administration. A single administration of Lorediplon induced a potent sedative effect superior to that of Zolpidem and Zaleplon (interval 5–60 min

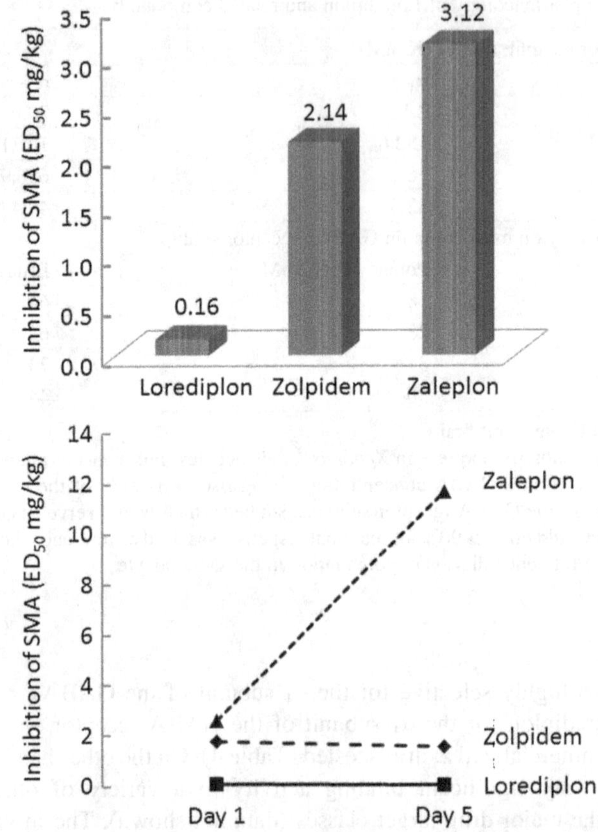

Fig. 2 Effect on inhibition of spontaneous motor activity in mice. Single (*upper panel*) or 5-day repeated (*lower panel*) oral administration. Group size 8–10 mice per treatment group. SMA = Spontaneous Motor Activity (traveled distance) calculated in the 5–60 min interval. ED_{50} calculated vs. the corresponding time-matched weekly control group

after dosing) (Fig. 2). The effect of Lorediplon was already evident in the 5–30 min interval and was maintained throughout the total record time. Contrary to Zaleplon and similarly to Zolpidem, no tolerance to this effect was observed after 5 days of repeated administration of Lorediplon (interval 5–60 min after dosing) (Fig. 2).

4.2.3 Inhibition of Muscular Tone

It is known that many sedative compounds also induce inhibition of muscular tone which, in the case of hypnotics, is an undesirable effect since it may cause dangerous falls in patients (Gunja 2013; Kolla et al. 2013). Therefore, the activity of Lorediplon after oral administration on muscular tone was evaluated by means of

Table 5 Effect of Lorediplon and related compounds on muscular tone, amnesia interaction with ethanol, and next-day hangover

Muscular tone[a]		
	Day 1 (ED_{50} mg/kg)	Day 5 (ED_{50} mg/kg)
Lorediplon	8.04	3.2
Amnesic effects[b]		
	MED (mg/kg)	Amnesic liability index
Lorediplon	10.0	62.5
Zaleplon	0.10	0.03
Zolpidem	30.0	14.02
Interaction with ethanol[c]		
	MED-30 min (mg/kg)	ETOH liability index
Lorediplon	1.00	0.16
Next-day hangover liability[d]		
	MED-24 h (mg/kg)	Hangover liability index
Lorediplon	>90	<0.001

All tests conducted in mice (8–12 animals per treatment group) after oral administration of the test compounds. MED = Minimal Effective Dose. Amnesic Liability Index = MED (mg/kg)/Spontaneous Motor Activity (SMA) ED_{50} (mg/kg). Hangover Liability Index = MED(mg/kg)/SMA ED_{50} (mg/kg). SMA values taken from Fig. 2

[a]Grip strength test

[b]Two-way active avoidance paradigm test

[c]Loss of the righting reflex test (30 min)

[d]Loss of the righting reflex test (24 h)

the grip strength test in mice 15 min after administration of the test compound. Results showed that the doses required to inhibit the muscular tone are much higher (ED_{50} 8.04 mg/kg) (Table 5) than those needed to induce sedation (ED_{50} 0.16 mg/kg) (Table 4), suggesting that falls due to inhibition of muscular tone are not likely to occur in patients receiving Lorediplon at hypnotic doses. Similarly to what observed for sedation, no tolerance was observed on inhibition of muscular tone after 5 days daily administration of the compound (Table 5).

4.2.4 Amnesic Effects

The amnesic potential of Lorediplon compared to that of Zolpidem and Zaleplon after oral administration in mice was evaluated by the two-way active avoidance paradigm. Animals received different doses of the drugs for 5 consecutive days and then were subjected to 100-trial active avoidance sessions in a shuttle box apparatus where the animals were trained to avoid a shock with the presentation of a conditions stimulus (light). The number of conditioned responses where the animal avoided the unconditional stimulus was recorded and expressed as the ratio of avoidance (conditioned changes/total changes) (Stienen et al. 2009). The Minimal Effective Dose (MED) that induced statistically significant impairment of avoidance was 10 mg/kg, 0.1 mg/kg, and 30 mg/kg for Lorediplon, Zaleplon, and

Zolpidem, respectively (Table 5). The Amnesia Liability Index, calculated by dividing the MED value for each compound obtained in this test by their ED_{50} obtained in the SMA test (Fig. 2), gave values of 62.5, 0.03, and 14.02 for Lorediplon, Zaleplon, and Zolpidem, respectively. Therefore of all the compounds tested, Lorediplon shows the widest margin between the doses required to induce sedation and those that cause amnesia (Table 5).

4.2.5 Ethanol Interaction and Next-Day Hangover Effects

The interaction between a hypnotic and ethanol is widely used as an indicator of possible next-day hangover effects of hypnotic drugs. In the case of Lorediplon, this interaction was evaluated by the loss of the righting reflex test in mice 30 min or 24 h after oral compound administration. The measured variables were the latency time (sleep onset latency) and narcosis time (sleep duration). No statistically significant differences were observed on sleep latency whereas narcosis time was dose dependently increased by Lorediplon. The MED that significantly increased the EtOH-induced narcosis after oral administration of Lorediplon was 1.0 mg/kg and >90.0 mg/kg at 30 min and 24 h post-oral drug administration, respectively (Table 5). The Lorediplon EtOH Liability Index and the Next-Day Hangover Liability Index, calculated by dividing the ED_{50} obtained in the SMA test (Fig. 2) by the MED value obtained in this test, were 0.16 and < 0.001, respectively, suggesting that Lorediplon is unlikely to show interaction with EtOH and/or next-day hangover effect.

4.2.6 Physical Dependence

The potential of Lorediplon to induce physical dependence was evaluated in NMRI mice that were treated orally twice a day for 4 consecutive days with vehicle, 3 mg/kg triazolam, or different doses of Lorediplon. Approximately 18 h after the last administration, all treatment groups received FG-7142 (40 mg/kg ip), a proconvulsant benzodiazepine receptor inverse agonist, and were individually placed in separate observation cages. The presence of Straub tail and tonic/clonic seizures in each animal was recorded for 60 min after FG-7142 administration.

The results showed that, under these experimental conditions, Lorediplon at dose up to 4.0 mg/kg, did not induce any sign of physical dependence (Table 6).

4.2.7 Electroencephalogram in Mice

The effect on EEG of Lorediplon as compared to Zolpidem was studied in mice. Animals were implanted with four cortical electrodes and three muscle electrodes to record the EEG and electromyogram (EMG) and to monitor sleep–wake cycle. EEG and EMG were recorded during lights-off (7:00 p.m.) continuously for 12 h.

Table 6 Effect of Lorediplon on physical dependence

Physical dependence		
Treatment	Straub tail (%)	Seizures (%)
Vehicle (negative control)	0	0
Triazolam 3 mg/kg (positive control)	100*	50*
Lorediplon 1.3 mg/kg	20	10
Lorediplon 4.0 mg/kg	10	0
Lorediplon 7.0 mg/kg	40*	10
Lorediplon 13.0 mg/kg	50*	30

Percentage of animals which shows Straub tail and seizures after precipitated withdrawal using FG-7142 (40 mg/kg ip). Lorediplon was administered orally twice a day for 4 consecutive days. 8–10 animals/group. $*p \leq 0.05$ vs. vehicle

The animals received different oral doses of Lorediplon (0.13, 0.40, and 1.2 mg/kg) or Zolpidem (2.0, 6.0, and 12.0 mg/kg) selected based on their inhibitory effect on locomotor activity assessed by open field test (Stienen et al. 2009).

Both Lorediplon and Zolpidem induced sleep and had significant effects on several sleep parameters (Figs. 3 and 4).

Lorediplon induced a decrease in sleep latency and wake (W) and an increase of slow-wave sleep (SWS), paradoxical sleep (PS), and total sleep (TS). All these effects were already noted at doses of 0.13 and 0.4 mg/kg and became statistically significant at a dose of 1.2 mg/kg. Increase of SWS was due to both increase of episode number and duration while increase of PS was mainly due to an increase of episode number. The effects on SWS, PS, and W induced by Lorediplon were maintained during the 6–9 h recording session. Contrary to other hypnotics (Gottesmann et al. 1998; Landolt et al. 2000) and Zolpidem (Anaclet et al. 2012), Lorediplon did not change the SWS/W power ratio and had a very well-balanced ratio PS/SWS and preserved the quality of SWS, all this indicating a good quality of sleep closer to that of physiological sleep (Anaclet et al. 2012).

Overall Lorediplon and Zolpidem had similar effects; however, several differences between the two compounds were noted:

- *Potency*: Lorediplon was more potent (about 10 fold) than Zolpidem and similar effects were noted with doses of 1.2 mg/kg of Lorediplon and 12 mg/kg of Zolpidem. This higher potency could be explained by the higher affinity of Lorediplon compared to Zolpidem for the α_1 subunit of $GABA_A$.
- *Sleep fragmentation*: The sleep induced by Lorediplon was characterized by smaller number of wake or SWS episodes compared to Zolpidem and the SWS episodes lasted longer than those induced by Zolpidem. All this suggests that the sleep induced by Lorediplon is less fragmented than that induced by Zolpidem, and since insomnia is also characterized by sleep fragmentation, this would be a desirable effect in the treatment of this condition (Sateia and Nowell 2004).
- *Middle-of-the-night awakenings*: Although Zolpidem induces sleep quickly, it is known that patients wake up frequently in the middle of the night and then have difficulties falling asleep again (Wallace 2005; Edinger and Means 2005). This

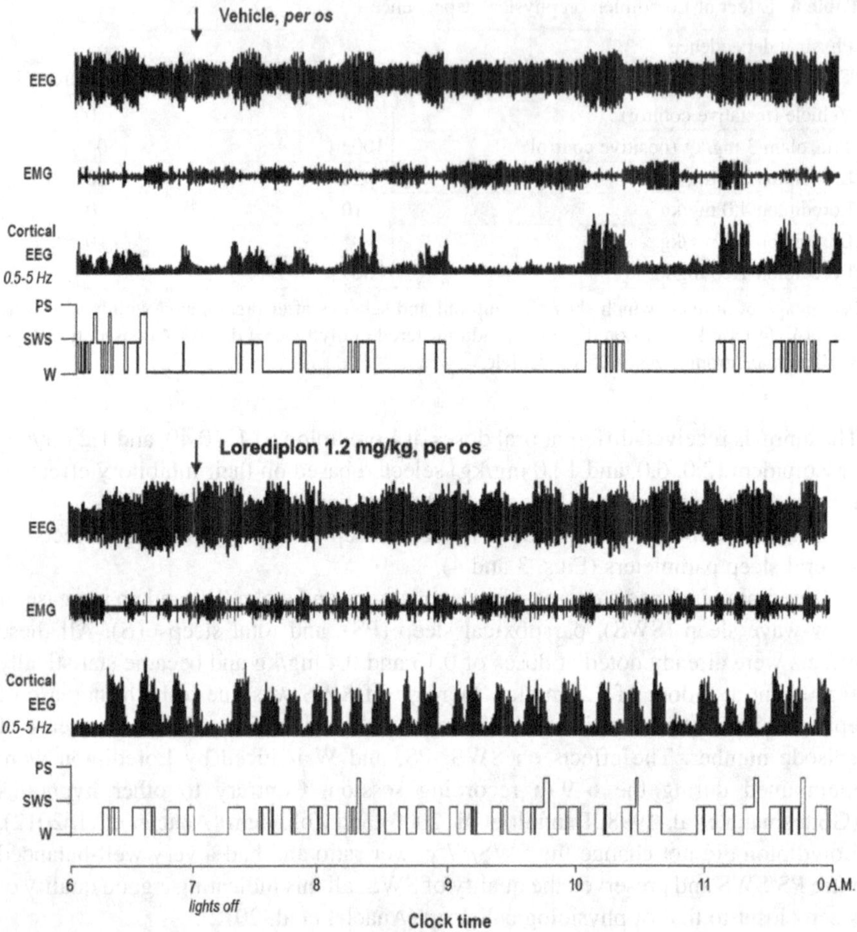

Fig. 3 Polygraphic recording in mice after oral administration of vehicle or Lorediplon. EEG = Electroencephalogram; EMG = Electromyogram; PS = Paradoxical Sleep; W = Wake (Reproduced with permission from Anaclet et al. 2012)

effect was also seen in mice after Zolpidem or Lorediplon administration; however, it was markedly less prominent after Lorediplon administration (Anaclet et al. 2012).

– *Quality of sleep*: Zolpidem decreased the cortical EEG power ratio between SWS and W. Lorediplon on the other hand did not change this ratio and no major changes of cortical EEG power slow activity during SWS were detected (Anaclet et al. 2012). This data indicated that the sleep induced by Lorediplon, contrary to that of Zolpidem, is of physiological quality.

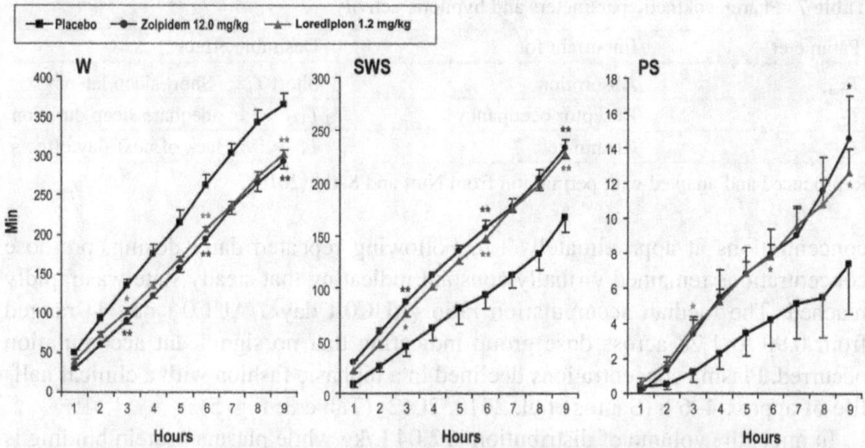

Fig. 4 Effect of oral administration of Lorediplon and Zolpidem on the cumulative amount of the sleep–wake state in mice. Compounds were administered just before lights-off at 7:00 p.m. Results are expressed as the mean cumulative time (mean ± SEM) spent in each behavioral state. W = Wake; SWS = Slow Wave Sleep; PS = Paradoxical Sleep. *$p < 0.05$; **$p < 0.01$, Mann–Whitney nonparametric test, $n = 9$) [Reproduced with permission from Anaclet et al. (2012)]

4.3 *Pharmacokinetics*

The pharmacokinetic profile of a hypnotic is very important for the treatment of insomnia since, among other things, it allows to predict which parameters of sleep, such as sleep induction or sleep duration, are likely to be affected by a given drug. A rapid absorption of a hypnotic is essential for a fast action as needed for example for patients with difficulties falling asleep while an adequate time occupancy of the receptor by the parent compound or its active metabolites as well as their elimination from the body will determine the maintenance of sleep and the presence of next-day effects (Table 7).

The pharmacokinetic profile of Lorediplon was evaluated in several animal studies and in two clinical trials.

In mice, Lorediplon is readily and quickly absorbed after single oral administration across a wide range of doses. The plasma exposure is proportional to the dose in the 0.5–200 mg/kg range. Less proportionality is observed in the 200–1,000 mg/kg dose range while doses higher than 1,000 mg/kg do not induce any further increase of systemic exposure. The mean bioavailability observed was 58.2 %. The maximal brain concentration was observed at 1 h after single oral administration and 24 h post-dose levels were below limit of quantification.

The pharmacokinetics of Lorediplon was also evaluated in two human studies. In the first study subjects received a single oral dose of Lorediplon ranging from 1 to 40 mg, while in the second study, oral doses up to 25 mg were administered daily for 7 consecutive days. Both studies demonstrated that Lorediplon is rapidly absorbed from the gastrointestinal tract and reaches maximum plasma

Table 7 Pharmacokinetic parameters and hypnotic activity

Parameter	Important for	Desirable effect
T_{max}	Absorption	Short T_{max}: Short sleep latency
$T_{1/2}$	Receptor occupancy	$T_{1/2} > 3$ h: adequate sleep duration
$T_{1/2}$	Clearance	$T_{1/2} < 7$ h: lack of next-day effects

Reproduced and adapted with permission from Nutt and Sathl (2010)

concentrations at approximately 2 h. Following repeated daily dosing, pre-dose concentrations remained virtually constant indicating that steady state was rapidly reached. The median accumulation ratio (AUC0-t day 7/AUC0-t day 1) ranged from 0.90 to 1.28 across dose group indicating that no significant accumulation occurred. Plasma concentrations declined in a biphasic fashion with a clinical half-life of approx 4–6 h (Santos et al. 2010, 2012) (Table 8; Fig. 5).

In mice, the volume of distribution is 2.04 L/kg while plasma protein binding is in the middle range with values of 82.8 % for human, 81.9 % for mouse, and 77.8 % for dog. Lorediplon is mainly metabolized through CYP3A4 and four major metabolites (monohydroxy Lorediplon, *N*-desmethyl Lorediplon, *N*-desacetyl Lorediplon, and *N*-desmethyl and *N*-desacetyl Lorediplon) were identified across a range of species. Lorediplon is highly unlikely to cause induction or inhibition of P_{450} system.

4.4 Activity in a Model of Insomnia in Humans

The activity of various sleep parameters of Lorediplon vs. Zolpidem was evaluated in a phase advance model of transient insomnia in healthy volunteers where subjects went to sleep several hours before their usual sleep time. This model produces a transient insomnia where subjects wake up in the middle of the night and is associated with alterations of several sleep parameters (Svetnik et al. 2010; Krystal et al. 2010). This model of insomnia, of which several variations based on how many hours before their normal sleep time subjects go to sleep exist, has been used extensively in the evaluation of hypnotics (Walsh et al. 1990, 2007; Svetnik et al. 2010).

Thirty-four Caucasian healthy male volunteers with normal polysomnography (PSG) and not placebo responders (WASO < 45 min) participated in a single-dose, randomized, double dummy, placebo, and positive controlled 5-way crossover study. Transient insomnia was induced by sending subjects to sleep 5 h before their usual sleep time. On day 1 subjects went to bed at their usual sleep time (11 p.m.) and slept for about 8 h (habituation night). On awakening, they filled out a post-sleep questionnaire and their degree of sleepiness/alertness was assessed by means of the psychomotor vigilance test (PVT), digit span test (DST), and visual analogue scale (VAS). At approximately 5:30 p.m., subjects received different oral doses of Lorediplon and 30 min later the phase advance regimen started. Lorediplon was

Table 8 Pharmacokinetic of Lorediplon in humans

Single ascending oral administration								
	1 mg	2 mg	5 mg	10 mg	20 mg	25 mg	30 mg	40 mg
T_{max} (h)	**1.67** *(0.67–2.33)*	**1.33** *(1.00–4.95)*	**1.83** *(1.00–4.00)*	**2.84** *(2.00–3.00)*	**2.33** *(1.00–3.50)*	**2.67** *(1.67–4.00)*	**2.17** *(1.67–3.50)*	**2.84** *(2.00–6.00)*
C_{max} (ng/mL)	**2.09** *(37)*	**3.14** *(32)*	**6.06** *(52)*	**13.9** *(30)*	**23.9** *(34)*	**34.7** *(61)*	**41.6** *(41)*	**43.9** *(49)*
AUC_{0-t} (h*ng/mL)	**10.55** *(55)*	**17.59** *(64)*	**40.56** *(74)*	**118.63** *(32)*	**198.71** *(42)*	**294.75** *(65)*	**378.33** *(40)*	**484.59** *(61)*
Terminal half-life (h)	**2.80** *(18)*	**3.69** *(50)*	**7.27** *(55)*	**14.42** *(68)*	**17.76** *(43)*	**15.84** *(38)*	**11.12** *(36)*	**17.54** *(66)*
C_{max} half-life (h)[a]	**4.32** *(22)*	**4.29** *(29)*	**4.41** *(23)*	**5.03** *(12)*	**4.35** *(35)*	**4.69** *(16)*	**6.67** *(75)*	**6.54** *(24)*
Multiple ascending oral administration								
	5 mg		10 mg		15 mg		25 mg	
	Day 1	Day 7	Day 1	Day 7	Day 1	Day 7	Day 1	Day 7
T_{max} (h)	**1.73** *(0.98–3.98)*	**1.75** *(0.97–3.52)*	**2.73** *(1.0–3.5)*	**3.47** *(0.67–3.5)*	**1.99** *(1.0–3.5)*	**1.52** *(1.03–3.48)*	**3.51** *(0.67–4.0)*	**2.5** *(0.67–4.0)*
C_{max} (ng/mL)	**6.23** *(67)*	**7.08** *(57)*	**13.9** *(26)*	**16** *(32)*	**16.1** *(53)*	**15.4** *(49)*	**27.7** *(55)*	**21.5** *(39)*
$AUC_{0-24\,h}$ (h*ng/mL)	**38.2** *(72)*	**52.6** *(81)*	**98.2** *(48)*	**101** *(52)*	**94.7** *(51)*	**96.2** *(42)*	**173** *(59)*	**150** *(61)*
Terminal half-life (h)		**10.3** *(32)*		**10.1** *(35)*		**9.95** *(30)*		**9.72** *(74)*
C_{max} half-life (h)	**3.25** *(31)*	**3.57** *(25)*	**3.17** *(36)*	**2.63** *(43)*	**3.15** *(41)*	**3.76** *(30)*	**2.61** *(43)*	**3.50** *(36)*

T_{max} values represent **median** (*min–max*). All other values are the **geometric mean** *(% Coefficient of variation)*. C_{max} half-life = Clinical half-life (i.e., time it takes for C_{max} to reduce by 50 %); AUC_{0-t} = Area under the curve from 0 to the last measurable point; AUC_{0-24} = Area under the curve from 0 to 24 h. In the single ascending oral administration study, the apparent increase of the terminal half-life across the dose level is most probably due to the fact that for lower doses a significant number of samples at the time point > 12 h fell below the limit of quantification. Dose-dependent elimination is unlikely, since the AUC increases proportionally with the dose

tested at 1, 5, or 10 mg, while Zolpidem was used as positive control and tested at a dose of 10 mg. PSG recording started around 6 p.m. and continued for 8 h until 2 a. m. At 7:00 a.m. on the morning of day 2 subjects repeated the post-sleep questionnaire and the PVT, DST, and VAS tests.

As expected, the phase advance model produced sleep disruption as evidenced by a WASO of 125.01 min in the placebo treatment group. The effects of Lorediplon and Zolpidem on the sleep parameters evaluated are reported on Table 9 and are summarized below:

1. Induction of Sleep

   Sleep onset. In this model of insomnia, subjects who received Lorediplon or Zolpidem did not show any significant change of sleep onset latency at any of the

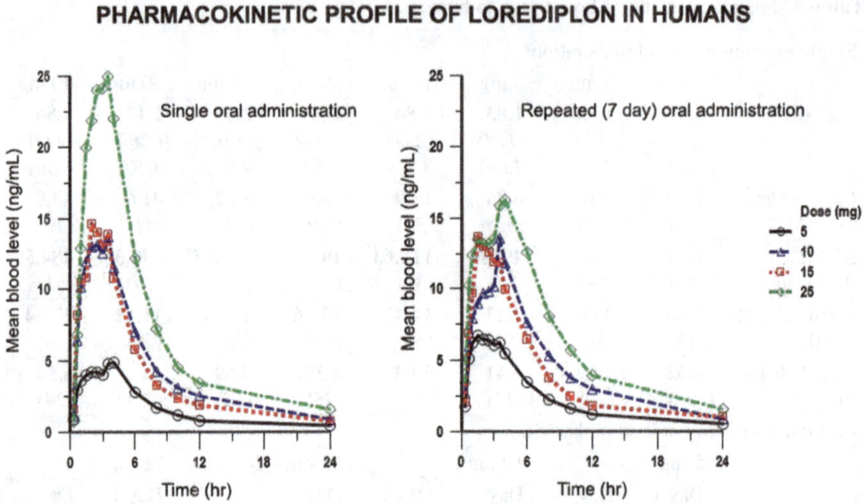

Fig. 5 Pharmacokinetic profile of Lorediplon in humans. Double-blind, placebo-controlled studies. Parallel groups. Sixty-four (single oral administration) or 32 (repeated oral administration) subjects divided into groups of 8 subjects. In each group 6 subjects received Lorediplon while 2 received placebo

doses tested (LPS, Table 9). The values of LPS observed after administration of Lorediplon or Zolpidem were not statistically different from the values observed in the placebo group (Table 9). This data however should be interpreted with caution since the 5 h phase advance model of insomnia may not be the best model to detect an effect on sleep onset latency for which other models such as 2–3 h phase advance model, first-night effect, and nap models are more appropriate. As a matter of fact, although in this study Zolpidem did not significantly change the LPS compared to the placebo group, it has been shown to reduce this parameter in other studies (Greenblatt and Roth 2012).

2. Maintenance of Sleep

   Wakefulness After Sleep Onset (WASO). Lorediplon dose dependently reversed the effect on WASO induced by the 5-h phase advance model. The effect reached statistical significance at a dose of 5 mg and peaked at a dose of 10 mg (Table 9 and Fig. 6). Zolpidem at the dose tested (10 mg) produced a similar effect although to a lesser degree than the same dose of Lorediplon (Table 9). Interestingly, when the effect on WASO was evaluated by quarters of the night, it was observed that the reduction of WASO induced by Lorediplon reached a peak during the 3rd quarter of the night, was sustained through the first three-quarters, and then returned to baseline in the fourth quarter (Fig. 7). Zolpidem 10 mg, on the other hand, showed a reduction of WASO mainly in the first two quarters (Fig. 7). Both in the second and particularly in the third quarter, the reduction of WASO induced by Lorediplon 10 mg was greater than the one induced by Zolpidem 10 mg (Fig. 7).

Table 9 Effect of Lorediplon and Zolpidem on sleep parameters in the 5-h phase advance model of insomnia

Parameter	Placebo	Lorediplon 1 mg	Lorediplon 5 mg	Lorediplon 10 mg	Zolpidem 10 mg
Sleep induction					
LPS (min)	−2.06 ± 4.82	−1.24 ± 5.71	−5.16 ± 4.27	−2.04 ± 5.22	−7.26 ± 4.29
Sleep maintenance					
WASO (min)	−11.38 ± 11.83	−13.37 ± 14.09	−52.24 ± 14.48**	−78.04 ± 14.99**	−60.2 ± 14.05**
W (number)	−0.59 ± 0.52	−0.62 ± 0.41	−1.03 ± 0.44	−1.68 ± 0.45**	−1.09 ± 0.52
Overall sleep					
TST (min)	11.24 ± 12.03	12.19 ± 14.14	51.71 ± 15.12**	73.78 ± 15.18**	61.37 ± 14.02**
Sleep efficiency index (%)	2.34 ± 2.51	2.54 ± 2.95	10.77 ± 3.15**	15.37 ± 3.16**	12.78 ± 2.92**
Stages of sleep					
Stage 1 (min)	1.43 ± 1.38	1.99 ± 1.48	2.1 ± 1.73	−0.26 ± 1.37	1.5 ± 2.11
Stage 2 (min)	−0.94 ± 7.98	−2.35 ± 9.33	21.28 ± 9.86*	47.88 ± 10.08**	30.12 ± 9.54**
Stage 3–4, SWS (min)	6.34 ± 3.06	11.93 ± 3.35	16.72 ± 4.01*	23.41 ± 3.40**	22.97 ± 3.90**
REM					
REM (min)	4.50 ± 5.33	1.22 ± 4.47	12.06 ± 5.81	3.32 ± 4.70	7.04 ± 4.10
REM latency (min)	−5.07 ± 14.66	−2.50 ± 13.02	−16.37 ± 12.57	0.15 ± 14.95	−15.13 ± 16.40

Polysomnographic endpoints (changes from baseline). Mean ± SEM (Standard Error Mean). LPS = Latency to persistent sleep; WASO = Wake After Sleep Onset; W = number of awakenings; TST = Total Sleep Time; REM = Rapid Eye Movement. *$p < 0.05$; **$p < 0.01$

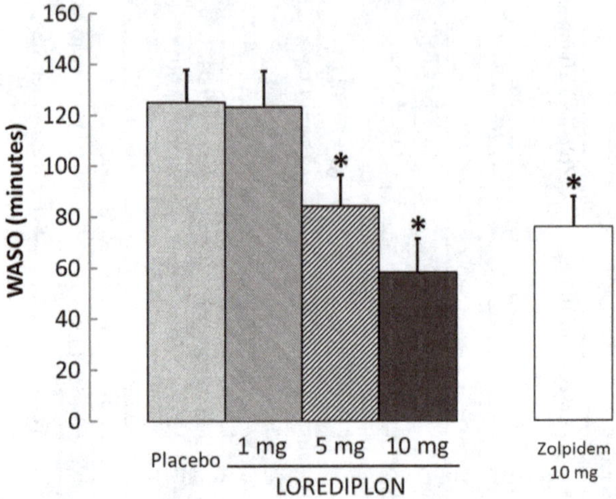

Fig. 6 Effect of Lorediplon on WASO in humans (phase advance model of insomnia). Drug was administered orally. Values represent Mean ± SEM (minutes). *$p < 0.01$ compared to placebo

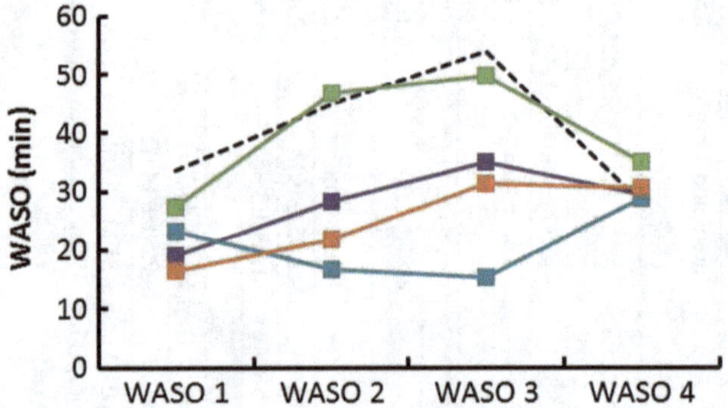

	Baseline	Placebo	Lorediplon 1 mg	Lorediplon 5 mg	Lorediplon 10 mg	Zolpidem 10 mg
WASO 1	37.27 ± 5.1	33.64 ± 4.2	27.28 ± 3.8	19.01 ± 2.2**	23.22 ± 4.3*	16.42 ± 2.5**
WASO 2	61.46 ± 8.8	45.05 ± 8.9	46.74 ± 8.8	28.24 ± 6.9	16.85 ± 5.9**	21.89 ± 6.0*
WASO 3	47.39 ± 8.8	53.96 ± 8.6	49.85 ± 9.5	34.92 ± 7.6	15.45 ± 5.7**	31.34 ± 7.4*
WASO 4	28.12 ± 6.6	27.98 ± 6.8	35.01 ± 6.7	29.45 ± 6.3	28.82 ± 6.2	30.65 ± 6.3

Fig. 7 Wake after sleep onset per 2 hourly quarters of the night (5-h phase advance model of insomnia). Thirty-four healthy human volunteers participated in the study. Values represent min ± SEM. *$p < 0.05$; **$p < 0.01$. Baseline data not represented in the graph [Reproduced and adapted from Horoszok et al. (2014)]

This behavior of zolpidem might explain why this drug produces so many middle-of-the-night awakening. The effect of Lorediplon seems to be longer lasting suggesting therefore that middle-of-the-night awakening should not be observed with this drug. Also, it should be noted that, after treatment with Lorediplon, the WASO returned to baseline in the 4th quarter of the night (Fig. 7). Since at this stage, the blood level of Lorediplon would be expected to have returned to baseline, this data might suggest that treatment with this drug is not likely to be associated with any next-day effects.

Number of awakenings (W). Lorediplon also dose dependently reduced the number of awakening which was statistically significant at a dose of 10 mg. The reduction of number of awakenings induced by Zolpidem 10 mg on the other hand was not statistically significant (Table 9).

3. Overall Sleep

   Total Sleep Time (TST) and Sleep Efficiency Index (SEI). An increase of TST and SEI was observed with all doses of Lorediplon. These changes were statistically significant at 5 and 10 mg dose. Zolpidem also significantly increased TST and SEI although to a lesser degree than Lorediplon (Table 9).

   The promoting effects on sleep observed with Lorediplon are mainly due to an increase of Nonrapid Eye Movement (NREM). On the other hand, Rapid Eye Movement (REM) sleep was not affected by Lorediplon (all doses) or Zolpidem 10 mg (Table 9).

4. Post-Sleep Questionaire

   A Post-Sleep Questionnaire (PSQ) was given to the subjects when they woke up the next morning. Except for sleep onset, all the other parameters were subjectively felt improved by the subjects after administration of Lorediplon 10 mg (Table 10). Although Zolpidem 10 mg was also felt to improve total sleep time and WASO, no significant effect was reported on number of awakenings, sleep onset, and sleep quality.

5. Next-Day Effects

   The presence of next-day residual drug effect was evaluated by specific tests (Digit forward score, Digit backward score, PVT, VAS) given to subjects after they woke up the day after drug administration. No statistically significant next-day residual effects were detected for Lorediplon (all doses) or Zolpidem 10 mg (data not shown).

4.5 Safety

In all clinical studies conducted so far, Lorediplon has shown an excellent safety profile in terms of both physical symptoms and signs and clinical laboratory testing. Under the conditions of the studies, the Maximum Tolerated Dose after single oral administration was considered to be 25 mg. The number of possible or probably drug-related adverse events detected in subjects exposed to Lorediplon is reported in Table 11 by dose and clinical study. Overall the number of adverse events

Table 10 Subjective assessment of sleep parameters after administration of Lorediplon or Zolpidem

Parameter	Placebo	Lorediplon 1 mg	Lorediplon 5 mg	Lorediplon 10 mg	Zolpidem 10 mg
Q1: Total Sleep Time (min)	122.35 ± 18.42	132.38 ± 21.50	152.45 ± 21.65*	184.85 ± 15.16*	171.94 ± 16.73*
Q2: Wake (number)	3.71 ± 0.55	2.90 ± 0.41	3.31 ± 0.44	2.47 ± 0.38*	2.89 ± 0.48
Q3: WASO (min)	109.74 ± 15.88	92.93 ± 19.72	89.09 ± 15.48*	74.50 ± 16.57*	81.78 ± 16.99*
Q4: Sleep onset (min)	−17.68 ± 3.44	−12.32 ± 3.38	−3.33 ± 3.92	−4.92 ± 3.27	−4.8 ± 3.76
Q5: Sleep quality	−23.91 ± 3.21	−20.00 ± 3.20	−18.53 ± 3.66	−11.9 ± 2.97*	−16.46 ± 3.34

Post-sleep questionnaire given after subjects woke the next morning after Lorediplon or Zolpidem administration. Numbers represent mean of changes from baseline ± Standard Error Mean (SEM). $*p < 0.01$

Questionnaire questions:

Q1: How long, in total, would you estimate you slept last night?

Q2: After you fell asleep for the first time last night, how many times did you wake up?

Q3: During those awakenings, how much time, in total, would you estimate that you spent awake?

Q4: How long, in minutes, would you estimate it took for you to fall asleep last night?

Q5: How would rate your overall quality of sleep last night?

Table 11 Possible or probably related adverse events after administration of Lorediplon or Zolpidem in humans

Dose	SAD ($n = 48$)	MAD ($n = 24$)	PAMI ($n = 35$)
1 mg	4 (3)	–	0 (0)
2 mg	1 (1)	–	–
5 mg	2 (2)	2 (1)	1 (1)
10 mg	2 (2)	18 (3)	0 (0)
15 mg	–	18 (4)	–
20 mg	2 (2)	–	–
25 mg	5 (4)	32 (5)	–
30 mg	17 (5)	–	–
40 mg	10 (5)	–	–
Placebo	*7 (6)*	*6 (1)*	*0 (0)*
Zolpidem 10 mg	–	–	*1 (1)*

Number of subjects exposed to placebo was 16 in the SAD study, 8 in MAD study, and 35 in the PAMI study. The number of subjected exposed to Zolpidem 10 mg was 35 in the PAMI study. Numbers represent the total number of possible or probably related adverse events by study and dose. Numbers in parentheses represent the number of subjects having experienced at least one possible or probably related adverse event by study and dose. SAD = Single Ascending Dose study; MAD = Multiple Ascending Dose study; PAMI = Phase Advance Model of Insomnia study

reported was small and dose and study design related. No serious adverse event was reported in any of the study and none of the adverse events detected lead to discontinuation of the study. The most frequent adverse event detected was somnolence whose intensity and duration increased with dose level. However, considering that Lorediplon is a hypnotic drug, somnolence could be expected.

4.6 Positioning and Advantages Compared to Other Hypnotics

The preclinical and clinical data on the use of Lorediplon in insomnia indicate that this molecule has a good hypnotic activity and safety profile and has several advantages over existing treatments that makes it an attractive potential new drug for treatment of insomnia.

First of all, Lorediplon belongs to a class of compounds the $GABA_A$ modulators whose pharmacology and safety are well known. Hypnotics that act by modulating the activity of this receptor have been around for quite some time and at least three generations of $GABA_A$ modulating compounds have been used as hypnotic with each new generation improving the profile of the previous one. As a matter of fact, the last generation of $GABA_A$ modulating compounds the so-called Z-drugs is today the most prescribed hypnotics. All this indicates that the medical community feels comfortable with this class of drugs and the introduction of a new molecule of the same pharmacologic class such as Lorediplon that improves the profile of the other members of the same class should be well received by both patients and prescribers.

Both preclinical and clinical data available indicate that Lorediplon has several advantages over Zolpidem the most prescribed drug for treatment of insomnia today. First of all, clinical studies have shown that the sleep induced by Lorediplon is of longer duration than that induced by Zolpidem while, at the same time, the two drugs have similar effects on sleep onset. Furthermore, all the indicators of sleep maintenance such as WASO, Total Sleep Time, Sleep Efficiency Index, and number of awakenings are significantly modified by Lorediplon to a greater extent than Zolpidem. Perhaps in this regard the most striking and convincing data is the effect of Lorediplon and Zolpidem on 2-h intervals WASO which clearly shows that, while the hypnotic effect of Zolpidem starts to decline after the second quarter, the effect induced by Lorediplon is maintained during the 3rd quarter and then, similarly to what observed with Zolpidem, returns to baseline at the end of the 4th quarter. This indicates that one of the most annoying effects observed in several patients after administration with Zolpidem (i.e., middle-of-the-night awakenings) is not likely to occur after administration of Lorediplon.

The objective beneficial effects of Lorediplon on sleep in humans were corroborated by the subjective perception of subjects treated with Lorediplon. Very importantly and contrary to what was reported after administration of Zolpidem 10 mg, subjects treated with 10 mg Lorediplon significantly rated their sleep as of better quality. This perception of improved sleep quality goes along with the EEG data collected in mice which indicates that the sleep induced by Lorediplon is more physiological than that induced by Zolpidem.

In order to better characterize the profile of a new hypnotic, the effects on muscular tone, memory, interaction with ethanol, next-day hangover effects, and development of physical dependence should also be evaluated. Promising results on the effect of Lorediplon on memory and next-day hangover effects were obtained in animals and confirmed clinically in the phase advance model of insomnia in humans. In addition, encouraging results on the effect of Lorediplon on inhibition of muscular tone, interaction with ethanol, and potential to induce physical dependence were obtained in animals, but so far they lack clinical confirmation. Although the overall picture on the effect of Lorediplon on this group of effect is encouraging, in order to draw more robust conclusions, these parameters should be evaluated in specifically designed clinical studies.

Conclusions

All the data collected so far on the pharmacology and safety profile of Lorediplon indicate that this molecule has several important advantages

(continued)

over other members of the same pharmacologic class and therefore has the potential to become the best in class drug for the treatment of insomnia. Lorediplon could therefore become part of a selected group of drugs available to the physician for the treatment of insomnia.

References

Anaclet C, Zhang M, Zhao C et al (2012) Effects of GF-015535-00, a novel α1 $GABA_A$ receptor ligand, on the sleep-wake cycle in mice with reference to Zolpidem. Sleep 35:103–111

Baur R, Tan KR, Lüscher BP et al (2012) Covalent modification of $GABA_A$ receptor isoforms by a diazepam analogue provides evidence for a novel benzodiazepines-binding site that prevents modulation by these drugs. J Neurochem 106:2353–2363

Belelli D, Lambert JJ (2005) Neurosteroids: endogenous regulators of the $GABA_A$ receptor. Nat Rev Neurosci 6:565–575

Benarroch EE (2012) GABA B receptors. Structure, functions and clinical implications. Neurology 78:5754

Bormann J (2000) The "ABC" of GABA receptors. Trends Pharmacol Sci 21:16–19

Browne SH, Kang J, Akk G et al (2001) Kinetic and pharmacological properties of GABA (A) receptors in single thalamic neurons and GABA(A) subunit expression. J Neurophysiol 86:2312–2322

Edinger JD, Means MK (2005) Overview of insomnia: definitions, epidemiology, differential diagnosis and assessment. In: Kryger MH, Roth T, Dement WC (eds) Principles and practice of sleep medicine, 4th edn. Elsevier Saunders, Philadelphia, pp 702–713

Fullerton DS (2006) The economic impact of insomnia in managed care: a clearer picture emerges. Am J Manag Care 12:S246

Gottesmann C (2002) GABA mechanisms and sleep. Neuroscience 111:231–239

Gottesmann C, Gandolfo G, Arnaud C et al (1998) The intermediate and paradoxical sleep in the rat: influence of three generations of hypnotics. Eur J Neurosci 10:409–414

Greenblatt DJ, Roth T (2012) Zolpidem for insomnia. Expert Opin Pharmacother 13:879–893

Gunja N (2013) The effects of Z-drugs on human performance and driving. J Med Toxicol 9:163–171

Hagmann WK (2008) The many roles for fluorine in medicinal chemistry. J Med Chem 51:4359–4369

Horoszok L, Baleeiro T, D'Aniello F et al (2014) A single-dose, randomized, doublé-blind, double dummy, placebo and positive-controlled, five-way cross-over study to assess the pharmacodynamic effects of lorediplon in a phase advance model of insomnia in healthy Caucasian adult male subjects. Hum Psychopharmacol Clin Exp 29:266–273

Istvan ES, Deisenhofer J (2001) Structural mechanism for statin inhibition of HMG-CoA reductase. Science 292:1160–1164

Kolla BP, Lovely JK, Mansukhani MP et al (2013) Zolpidem is independently associated with increased risk of inpatients falls. J Hosp Med 8:1–6

Kopp C, Rudolph U, Löw K et al (2004) Modulation of rhythmic brain activity by diazepam: GABAA receptor subtype and state specificity. Proc Natl Acad Sci USA 101:3674–3679

Krystal AD, Zammit GK, Wyatt JK et al (2010) The effect of vestibular stimulation in a four-hour sleep phase advance model of transient insomnia. J Clin Sleep Med 6:315–321

Landolt HP, Finelli LA, Roth C et al (2000) Zolpidem and sleep deprivation: different effect on EEG power spectra. J Sleep Res 9:175–183

López-Muñoz F, Ucha-Udabe R, Alamo C (2005) The history of barbiturates a century after their clinical introduction. Neuropsychiatry Dis Treat 1:329–343

Möhler H, Fritschy JM, Rudolph U (2002) A new benzodiazepine pharmacology. J Pharmacol Exp Ther 300:2–8

Möjler H (2007) Molecular regulation of cognitive functions and developmental plasticity: impact of GABAA receptors. J Neurochem 102:1–2

Nutt D (2006) GABAA receptors: subtypes, regional distribution and function. J Clin Med 2:S7–S11

Nutt DJ, Sathl SM (2010) Searching for perfect sleep: the continuing evolution of $GABA_A$ receptor modulators as hypnotics. J Psychopharmacol 24:1601–1612

Nys M, Kesters D, Ulens C (2013) Structural insights into Cys-loop receptor function and ligand recognition. Biochem Pharmacol 86:1042–1053

Olsen RW, Sieghart W (2008) International Union of pharmacology. LXX. Subtypes of γ-aminobutyric $acid_A$ receptors: classification on the basis of subunit composition, pharmacology, and function. Update. Pharmacol Rev 60:243–260

Perfilova VN, Tiurenkov IN (2010) GABA-B type receptors: structure and functions. Eksp Klin Farmakol 73:44–48

Perfilova VN, Tiurenkov IN (2011) GABA-C receptors: structure and functions. Eksp Klin Farmakol 74:45–49

Ramerstorfer J, Furtmüller R, Sarto-Jackson I, Varagic Z, Sieghart W, Ernst M (2011) The GABAA receptor $\alpha+\beta$ interface: a novel target for subtype selective drugs. J Neurosci 31:870–877

Roth BD (2002) The discovery and development of atorvastatin, a potent novel hypolipidemic agent. Prog Med Chem 40:1–22

Roth T (2005) Prevalence, associated risks and treatment patterns of insomnia. J Clin Psychiatry 66:10

Rudolph U, Knoflach F (2011) Beyond classical benzodiazepines: novel therapeutic potential of GABA-A receptor subtypes. Nat Rev Drug Discov 10:685–697

Rudolph U, Crestani F, Benke D et al (1999) Benzodiazepine actions mediated by specific gamma-aminobutyric acid (A) receptor subtypes. Nature 401:796–800

Ruth BD, Octwine DF, Hoefle ML et al (1990) Inhibitors of cholesterol biosynthesis. 1. Trans-6-(2-pyrrol-1-ylethyl)-4-hydroxypyran-2- ones, a novel series of HMG-CoA reductase inhibitors. 1. Effects of structural modifications at the 2- and 5-positions of the pyrrole nucleus. J Med Chem 33:21–31

Santos B, Lindauer R, Coll S et al (2010) Pharmacokinetics of a novel non-benzodiazepine hypnotic drug GF-015535 in healthy volunteers. In: 20th congress of the European sleep society. Lisbon, pp 14–18

Santos B, Lindauer A, Baleeiro T et al (2012) Pharmacokinetics following single and multiple administration of the novel non-benzodiazepine hypnotic drug Lorediplon in healthy volunteers. Sleep 2012, Boston

Sateia MJ, Nowell PD (2004) Insomnia. Lancet 364:1959–1973

Savič MM, Huang S, Furtmüller R et al (2008) Are GABAA receptors containing alpha5 subunits contributing to the sedative properties of benzodiazepine site agonists? Neuropsychopharmacology 33:332–339

Schenck CH, Mahowald MW, Sack RL (2003) Assessment and management of insomnia. JAMA 289:2475–2479

Sharman JL, Benson HE, Pawson AJ et al (2013) IUPHAR-DB: updated database content and new features. Nucleic Acids Res 41:D1083–8

Siegel E, Baur R, Rácz I et al (2011) The major central endocannabinoid directly acts at $GABA_A$ receptors. Proc Natl Acad Sci USA 108:18150–18155

Sieghart W, Ramerstorfer J, Sarto-Jackson I et al (2012) A novel $GABA_A$ receptor pharmacology: drugs interacting with the $\alpha^+\beta^-$ interface. Br J Pharmacol 166:476–485

Sigel E, Steinmann ME (2012) Structure, functions and modulation of $GABA_A$ receptors. J Biol Chem 287:40224–40231

Stienen PJ, Pérez C, Alcántara AM et al (2009) In-vivo pharmacological profile of GF-015535-00, a novel non-benzodiazepine sedative-hypnotic. In: Society for neuroscience annual meeting. Chicago

Svetnik V, Ferri R, Ray S et al (2010) Alteration in cyclic alternating pattern associated with phase advanced sleep are differentially modulated by gaboxadol and zolpidem. Sleep 33:1562–1570

Taylor DJ, Lichstein KL, Durrence HH (2003) Insomnia as a health risk factor. Behav Sleep Med 1:227–247

Tobler I, Kopp C, Deboer T et al (2001) Diazepam-induced changes in sleep: role of the α1 $GABA_A$ receptor subtype. Proc Natl Acad Sci USA 98:6464–6469

Van Rijnsoever C, Täuber M, Choulli MK et al (2004) Requirement of alpha 5-GABAA receptors for the development of tolerance to the sedative action of diazepam in mice. J Neurosci 24:6785–6790

Vedula SS, Bero L, Scherer RW, Dickersin K (2009) Outcome reporting in industry-sponsored trials of gabapentin for off-label use. N Engl J Med 361:1963–1971

Wafford KA, Eber B (2008) Emerging anti-insomnia drugs: tackling sleeplessness and the quality of wake time. Nat Rev Drug Discov 7:530–540

Wallace MB (2005) Hypnotic medications: mechanisms of action and pharmacologic effects. In: Kryger MH, Roth T, Dement WC (eds) Principles and practice of sleep medicine, 4th edn. Elsevier Saunders, Philadelphia, pp 444–451

Walsh JK, Schweitzer PK, Sugerman JL, Muehlbach MJ (1990) Transient insomnia associated with a 3-hour phase advance of sleep time and treatment with zolpidem. J Clin Psychopharmacol 10:184–189

Walsh JK, Deacon S, Dijk DJ (2007) The selective extrasynaptic GABAA agonist, gaboxadol, improves traditional hypnotic efficacy measures and enhances slow wave activity in a model of transient insomnia. Sleep 30:593–602

Whiting PJ (2003) GABA-A receptor subtypes in the brain: a paradigm for CNS drug discovery ? Drug Discov Today 8:445–450

Wick JY (2013) The history of benzodiazepines. Consult Pharm 28:538–548

Zolpidem Sublingual Formulations

Luc Staner

Abstract The present chapter discusses the pharmacology, efficacy, and safety of two recently developed sublingual formulation of zolpidem that specifically target sleep-onset insomnia (Edluar, a standard dose of zolpidem, SL-SD) and middle-of-the-night (MOTN) insomnia (Intermezzo, a low dose of zolpidem, SL-LD). These two SL formulations are bioequivalent to the standard immediate oral release (IOR) form of zolpidem and have a comparable elimination half-life but a somewhat shorter Tmax than the standard oral form. Due to a gender effect on zolpidem metabolism, half-doses are recommended in women (i.e., 5 mg for SL-SD and 1.75 mg for SL-LD). Efficacy has been shown for SL-SD in three double-dummy crossover polysomnographic studies that could demonstrate the superiority of the acute administration of 10 mg SL-SD zolpidem over the same dose of IOR zolpidem in healthy subjects using models of transient insomnia and in patients with DSM-IV primary insomnia. Both 1.75 and 3.5 mg of SL-LD zolpidem have been found effective in 2 large placebo-controlled studies performed in DSM-IV primary insomniacs having middle-of-the-night insomnia. Most common adverse events were somnolence, fatigue, headache, and dysgeusia for SL-SD zolpidem and headache, nausea, and fatigue for SL-LD zolpidem. More generally types of adverse events for SL zolpidem were consistent with the adverse event profile of IOR zolpidem. Because of its middle-of-the-night way of administration, concern regarding the next-morning safety of the SL-LD was addressed in a highway driving performance study. Results indicate that SL-LD zolpidem 3.5 mg taken 3 h before driving may impair driving performance, but that there is a minimal risk of impairing driving performance if the drug is taken ≥ 4 h before driving. SL-SD zolpidem has been approved for the short-term treatment of sleep-onset insomnia at a dose of 10 mg in non-elderly man and at the dose of 5 mg in women or in special population including elderly patients. SL-LD zolpidem has been approved for the treatment of insomnia when MOTN awakening is followed by difficulty in returning to sleep. Recommended dosages are 3.75 mg for non-elderly man and 1.75 mg in women and in special population including elderly. It has to be stressed

L. Staner (✉)
Unité d'Exploration des Rythmes Veille-Sommeil, Centre Hospitalier de Rouffach, Rouffach, France
e-mail: l.staner@ch-rouffach.fr

A. Guglietta (ed.), *Drug Treatment of Sleep Disorders*, Milestones in Drug Therapy,
DOI 10.1007/978-3-319-11514-6_7

that, due to safety issue related to next-morning residual effects, the dose should be taken following MOTN awakenings only if the patient has at least 4 h of sleep remaining.

1 Introduction

Insomnia is essentially characterized by a complaint of sleep dissatisfaction either in terms of duration or quality associated with difficulties to initiate or maintain sleep. General population studies indicate that these symptoms are extremely prevalent in adults [from 10 % to nearly 60 % according to the different ways insomnia is defined and assessed (Ohayon 2002)]. However, to be considered as a disorder, the sleep disturbance needs to be sustained and long lasting [for instance, at least three times a week for at least 3 months in the DSM-V classification system (American Psychiatric Association 2013)] and has to impact the daytime functioning. Prevalence is about 6–10 % when such criteria are used to define insomnia (American Psychiatric Association 2013), with higher rate in women and older adults particularly in those with chronic medical or psychiatric conditions. About 20 % of patients with insomnia estimate that their disorder seriously impacts their life (Staner et al. 2012). Indeed, poor sleep has been associated with emotional distress and recurrent health problem (Edinger et al. 2000). A prospective study showed that poor sleepers were less effective in their work, less likely to receive promotion, and more likely to be demoted and discharged (Johnson and Spinweber 1983). This is not surprising since, in non-insomniac subjects, a sleep loss of even 1–2 h a night might impair next-day alertness, concentration, attention, memory, mood, and pain threshold (Staner et al. 2012).

Effective treatments for insomnia include both pharmacological and non-pharmacological approaches. The use of non-pharmacological treatment such as sleep hygiene measures, cognitive therapies, relaxation, or sleep restriction is limited by lack of trained providers, understanding of the treatment method, and cost/third party reimbursement (Perlis et al. 2003). Indeed most patients initiate treatments with various nonprescription therapies of unknown risk and benefit such as OTC sedating antihistamines, herbal remedies, dietary supplements, or even rely on alcohol to relieve their sleep difficulties. Effective pharmacological treatment for insomnia should normalize sleep patterns without affecting daytime functioning. First generation hypnotics (such as barbiturates, carbamate, chloral hydrate, methaqualone) and long-acting benzodiazepines were effective in inducing sleep but put patients at risk for next-day residual effects and, for some of them, with additional risks of liver toxicity, dependence, and fatal overdose. The last 20 years have seen the development of safer hypnotic drugs with a better risk/benefit profile, such as cyclopyrrolones (zopiclone, eszopiclone), imidazopyridines (zolpidem), pyrazolopyrimidines (zaleplon, indiplon, lorediplon), melatonin receptor agonists (ramelteon, tasimelteon), orexin antagonists (suvorexant), and low-dose sedative antidepressant drugs (doxepin).

It has to be kept in mind that although sleep initiation and sleep maintenance disturbances most frequently coexist in patients with insomnia disorder, some individuals may complain specifically of either sleep-onset difficulties or of prolonged awakening after sleep onset. Although current diagnostic classification systems such as DSM-V or ICSD offer no operationalized diagnostic criteria for these two insomnia subtypes, the DSM-V still provides for illustrative purpose, quantitative criteria to define sleep initiation, and sleep maintenance difficulties. On this basis, initial (or sleep-onset) insomnia may be considered when sleep latencies are recurrently greater than 20–30 min despite appropriate bedtime hours/sleep habits and middle (or middle-of-the-night) insomnia when patients regularly experience wake times after sleep onset greater than 20–30 min despite adequate sleep conditions.

During these last years, interest in the development of hypnotics that could specifically target these forms of insomnia was growing. A "quick and short" mode of action was favored with transmucosal delivery formulations that disintegrate in the sublingual cavity allowing the drug to be rapidly absorbed and quickly available in the brain by bypassing the first-liver metabolization. With regard to the "short" side of the mode of action, hypnotic drugs such as zolpidem and zaleplon were used due to their short half-life. There are currently two FDA-approved sublingual forms of zolpidem, a standard dose (SL-SD) for the short-term treatment of sleep-onset insomnia (Edluar 5 and 10 mg by Meda Pharmaceuticals) and a low dose (SL-LD) for as-needed treatment of middle-of-the-night (MOTN) insomnia (Intermezzo 1.75 and 3.75 mg by Transcept Pharmaceuticals). This chapter discusses the pharmacology, the efficacy, and the safety of these two sublingual formulations in the treatment of these forms of insomnia.

2 Pharmacology

Zolpidem, a non-benzodiazepine hypnotic acting at GABAa launched in the eighties in France, rapidly became one of the most widely used drugs to treat insomnia, because of its favorable efficacy/balance profile compared to classical benzodiazepines. Due to a short half-life and a selective α_1GABAa receptor profile, the drug is a powerful sleep initiation agent with a lower incidence of side effects than benzodiazepines. A complete review of the mechanism of action of zolpidem can be found elsewhere (Staner et al. 2010a).

Sublingual formulations of zolpidem are designed to disintegrate in the sublingual cavity and to be rapidly absorbed after administration. The two zolpidem sublingual formulations (SL-SD and SL-LD) are bioequivalent to the standard immediate oral release (IOR) form of zolpidem (Ambien) (Edluar 2009; Intermezzo 2011). Tmax is somewhat shorter for the sublingual forms, the peak concentration of zolpidem occurring at a mean time of 96 min for IOR zolpidem 10 mg, whereas a median Tmax of 82 min (range: 30–180 min) was found for SL-SD zolpidem 10 mg and a mean Tmax range of 35–75 min for SL-LD zolpidem 3.75 mg. Elimination half-life is comparable between the three formulations [IOR zolpidem 10 mg: 2.53 h (range 1.4–3.8), SL-SD zolpidem 10 mg : 2.65 h (range 1.75–3.77) SL-LD zolpidem 3.75 mg: 2.5 h (range 1.4–3.6)]. As expected IOR zolpidem 10 mg has higher AUC and Cmax values than SL-LD zolpidem 3.75 mg (AUC: 589 ng.hr/mL for IOR zolpidem versus 295 ng.hr/mL in women and 197 ng.hr/mL in men for SL-LD zolpidem; Cmax: 121 ng/mL for IOR zolpidem versus 75 ng/mL in women, and 53 ng/mL in men for SL-LD zolpidem) (Edluar 2009; Intermezzo 2011; Swainston Harrison and Keating 2005).

The oral bioavailability of the drug is estimated at 65–70 % but food delays its absorption. It has been shown that food decreases AUC by 15 %, Cmax by 25 %, increases Tmax by 50 %, and does not influence half-life (Swainston Harrison and Keating 2005). Food induced lower bioavailabilities are also observed with sublingual forms, a finding that could relate to the effect of a meal on salivary pH and its consequence on zolpidem sublingual absorption. Thus, food decreases AUC by 20 % and 19 %, Cmax by 31 % and 42 %, and prolongs Tmax by 28 % and 54 % for Edluar and Intermezzo, respectively (Edluar 2009; Intermezzo 2011). Accordingly, sublingual zolpidem should not be administered with or immediately after a meal.

Zolpidem does not accumulate in the body after repeated administration due to its short half-life. There is a negligible direct excretion of the drug since about 90–95 % of zolpidem is bound to plasma proteins and its clearance is mainly a metabolic one. Cytochrome P450 extensively metabolizes the drug in three main inactive metabolites and only traces of the unchanged compound can be found in urine. The CYP3A4 is the principal isoform responsible for zolpidem metabolism, accounting for about 60 % of net cytochrome-mediated hepatic clearance. Consequently, drugs inducing CYP3A4 such as rifampicine decrease zolpidem AUC, Cmax, and half-life while the converse is observed with CYP3A4 inhibitors such as itraconazole. Detailed review of significant drug interactions can be found elsewhere (Hesse et al. 2003). Oral zolpidem dosage adjustment recommendations for special populations also apply to sublingual forms of the drug. Thus, initial dose should be adjusted to 5 mg for SL-SD zolpidem and 1.75 mg for SL-LD zolpidem in the elderly, in debilitated patients, or in patients with concomitant CNS depressant or with hepatic impairment.

The finding of a gender effect on SL-LD zolpidem metabolism (Greenblatt et al. 2013) (see Table 1 showing that plasma AUC and Cmax parameters are approximately 45 % higher in women than men) has led to the reanalysis of data

Table 1 Mean ± SD of gender specific pharmacokinetic parameter of low dose sublingual zolpidem (Intermezzo) (Intermezzo® 2011)

	Zolpidem 1.75 mg		Zolpidem 3.5 mg	
	Women ($n = 11$)	Men ($n = 13$)	Women ($n = 11$)	Men ($n = 13$)
AUC (ng h/mL)	151.36 ± 61.54	104.73 ± 35.04	295.6 ± 105.66	197.69 ± 72.43
Cmax (ng/mL)	37.47 ± 11.1	27.68 ± 7.5	77.13 ± 23.71	53.15 ± 14.29

from studies of standard oral zolpidem products. Results showed that 8 h after taking a single dose of 10 mg of IOR zolpidem, 15 % of women and 3 % of men still had blood zolpidem levels above the threshold of 50 ng per milliliter known to impair driving performance. It has been suggested that low plasma concentrations of free testosterone may contribute to lower CYP3A activity since exposure to testosterone induced CYP3A-mediated biotransformation (Farkas et al. 2013). These findings led the FDA to reconsider the recommended dose of zolpidem for women that has been reduced from 10 to 5 mg for immediate-release products such as IOR zolpidem and SL-SD zolpidem, and from 12.5 to 6.25 mg for controlled-release products (e.g., Ambien CR) (FDA Drug Safety Communication 2013). The recommended doses of SL-LD zolpidem did not change as the label already recommended the lower dosage (1.75 mg) for women.

3 Efficacy

The sleep-promoting effect of zolpidem has been demonstrated in healthy "good sleepers" and in elderly and non-elderly patients with chronic primary or secondary/comorbid insomnia. Zolpidem has been found to improve objective and subjective sleep initiation and maintenance parameters such as sleep-onset latency (SOL), latency to persistent sleep (LPS), wake after sleep onset (WASO), total sleep time (TST), and sleep efficiency (SE). Subjective sleep quality is generally improved but the effects on sleep architecture parameters are less consistent, with some placebo-controlled studies showing increases of stage 2 sleep or decreases of rapid eye movement (REM) sleep, others not. Extensive reviews of these efficacy studies of IOR or CR zolpidem have been published elsewhere (Staner et al. 2010a; Monti et al. 2008) and this section focuses on studies performed with the two sublingual formulations of zolpidem.

3.1 Sublingual Standard Dose of Zolpidem (Edluar)

Two sleep laboratory studies investigated the efficacy of two doses of SL-SD zolpidem (5 and 10 mg), a sublingual tablet designed to provide a more rapid relief of sleep initiation difficulties than IOR zolpidem in patients suffering from insomnia. These two double-dummy crossover trials were devoid of placebo arm since the main objective was to demonstrate the superiority of SL-SD zolpidem over IOR zolpidem in terms of sleep initiation parameters (i.e., SOL and LPS). The first study (Staner et al. 2009) was performed in healthy subjects (17 women and 4 men aged 26.7 $\pm$ 5.3 years) using a post-nap model of transient insomnia during which subjects had to perform a daytime nap in order to disrupt subsequent nighttime sleep. Subjects were recorded during 2 consecutive nights and on the day between during a 2-h nap. Treatment (either SL-SD zolpidem 5 mg, SL-SD zolpidem 10 mg, or IOR zolpidem 10 mg) was randomly administered before the second recording night to subjects demonstrating at least 30 min of sleep during the nap recording. Results showed that SL-SD zolpidem 10 mg significantly improved SOL (5.81 min, $p < 0.05$) and LPS (6.11 min, $p < 0.05$) compared to IOR zolpidem 10 mg (an improvement around 30 %). There was no difference in SOL and LPS between SL-SD zolpidem 5 mg and IOR zolpidem 10 mg. Sleep maintenance parameters (i.e., WASO, TST, and SE), sleep architecture parameters, and subjective sleep ratings were not significantly different among the three treatments.

The second efficacy study (Staner et al. 2010b) which compared SL-SD zolpidem 10 mg to IOR zolpidem 10 mg was performed in DSM-IV primary insomniacs (37 males aged 35.7 ± 11 years and 42 females aged 45.5 ± 9.5 years) that were recruited in 8 centers after a careful 2-night screening procedure aimed at excluding patients with sleep apnea, periodic leg movement, and including those with a mean LPS greater than 30 min. Other polysomnographic entry criteria were evidence on both nights of no LPS lower than 20 min, a TST lower than 6.5 h, and a WASO of at least 30 min. Results were comparable to those observed in the post-nap study with a 30 % improvement of SOL (8.63 min, $p < 0.01$) and LPS (10.28 min, $p < 0.001$) with SL-SD zolpidem 10 mg compared to IOR zolpidem 10 mg. Moreover, SL-SD zolpidem increases SE by 1.56 % ($p < 0.05$), compared to IOR zolpidem. Other sleep continuity parameters (TST and WASO) and subjective sleep were comparable between the sublingual and oral formulations (Fig. 1).

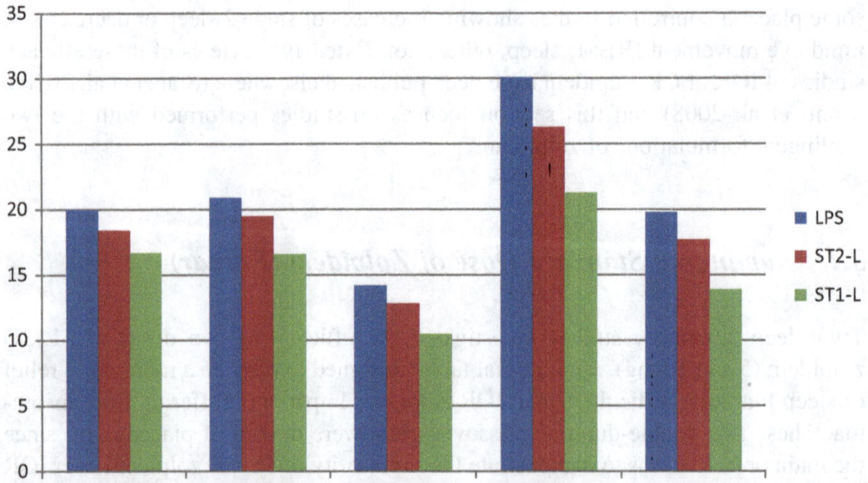

Fig. 1 Effects of immediate oral release of zolpidem 10 mg (IOR Zol 10), sublingual standard dose of zolpidem 5 mg (SLSD Zol 5) and 10 mg (SLSD Zol 10) on mean sleep EEG initiation parameters (in min), latency to persistent sleep (LPS), latency to stage 2 (ST2-L), and latency to stage 1 (ST1-L) in the post-nap (PN) study (Staner et al. 2009) and the primary insomnia (PI) study (Staner et al. 2010b)

These results were corroborated in another laboratory study (Valente et al. 2013) with healthy volunteers ($n=58$) that used a model of transient insomnia (sleep anticipation in 120 min). Compared to IOR zolpidem 10 mg, both SL-SD zolpidem 5 and SL-SD zolpidem 10 mg improved SOL and LPS. However, only the 10 mg SL-SD dose being superior to IOR zolpidem in terms of subjective sleep latency. On the whole the three studies agree with the idea that the sublingual form of zolpidem a more potent sleep inducer than the IOR formulation but is comparable in terms of sleep maintenance.

3.2 Sublingual Low Dose of Zolpidem (Intermezzo)

SL-LD zolpidem was developed to target MOTN insomnia and is currently the single FDA-approved hypnotic to treat this highly prevalent condition. Indeed, about one-third of the general population reported awakenings after sleep onset at least 3 nights per week [i.e., 35 % of 8.937 U.S. residents included in a telephone survey (Ohayon et al. 2010)]. About half of them (i.e., 15 % of the sample) had also difficulty resuming sleep and about a third of them (11 % of the total sample) had both difficulty resuming sleep after awakening and associated daytime impairment (Ohayon et al. 2010). SL-LD zolpidem clearly differs from many widely used hypnotics currently approved by the FDA for sleep onset. It also differs from those approved for nocturnal awakenings and/or prolonged wake time after nocturnal awakenings. Indeed these hypnotics are designed to be taken at bedtime in order to prevent possible sleep disruption while SL-LD zolpidem is approved for as-needed MOTN use after nocturnal awakening.

Two placebo-controlled studies investigating the efficacy of SL-LD zolpidem 3.5 and 1.75 mg are found in the literature. The first one is a 3-way crossover study conducted in 5 US sleep laboratories using a randomized, double-blind, placebo-controlled design that evaluated the efficacy and the safety of two doses of SL-LD zolpidem (1.75 and 3.5 mg) when taken during a scheduled MOTN awakening. Patients were included on basis of a DSM-IV primary insomnia diagnosis and a 4-week history of prolonged MOTN awakenings (i.e., at least 3 nights per week with a mean latency to fall back to sleep of more than 30 min post-awakening). The studied sample comprised of 58 females and 24 males with a mean age of 45.9 years. Treatment was dispended within 5 min after the scheduled MOTN awakening, 4 h after the initial lights out, and outcomes refer to the second 4 h polysomnographic recording. Results show that, compared to placebo, both doses of SL-LD zolpidem significantly decrease LPS and improve TST after the scheduled MOTN awakening (Fig. 2). Subjective sleep-onset latency and subjective TST were also improved with both doses (Roth et al. 2008a).

The second study (Roth et al. 2013) was performed in a sample of 295 - non-elderly outpatients (median age 43.1 years, 68.1 % females) with primary insomnia, a history of at least 3 months of MOTN awakenings occurring 3 or more times a week, and an average TST of less than 6.5 h. To qualify for

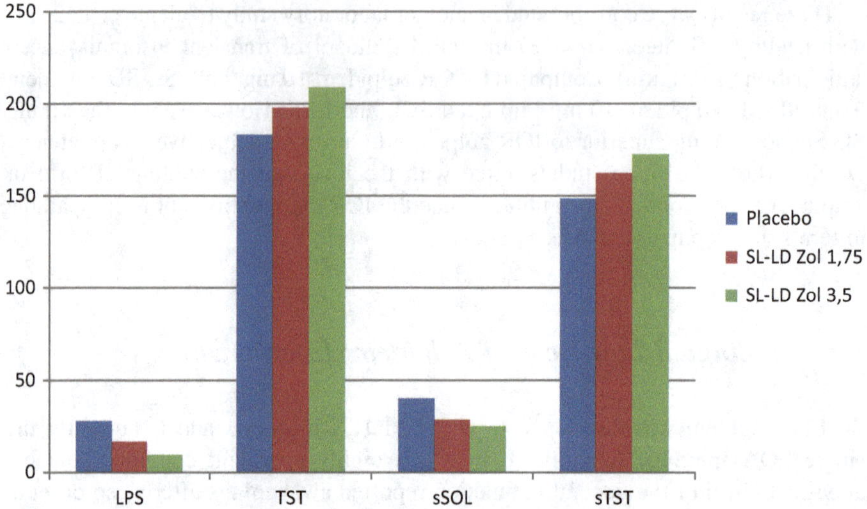

Fig. 2 Mean of main sleep EEG parameters (*LPS* latency to persistent sleep, *TST* total sleep time) and subjective sleep (*sSOL* subjective sleep-onset latency, *sTST* subjective total sleep time) after MOTN administration of either 1.75 mg SL-LD Zolpidem, 3.5 mg SL-LD Zolpidem, or placebo in the sleep laboratory study (Roth et al. 2008a)

randomization, evidences of at least 3 MOTN awakenings/week of at least 30 min with one of them lasting more than 60 min have to be documented during a 2-week placebo run in period. Patients were then randomized 1:1 to as-needed MOTN dosing with SL-LD zolpidem 3.5 mg or placebo for 28 nights. Across the 4-week period, significantly shorter SOL was reported in patients in the active treatment arm (38.2 min versus 56.4 min, $p < 0.001$). A significant improvement of TST after MOTN with as-needed SL-LD zolpidem 3.5 mg was only observed during the two first study weeks ($p < 0.05$), a finding that has to be considered in the light of lower baseline TST values in the placebo group ($p < 0.05$) and non-treatment specific effects (i.e., those related to protocol requirement consistent with good sleep hygiene practices). Interestingly, patients randomized to SL-LD zolpidem took the drug during the night on 62 % of study nights and did not experience an increase in drug utilization over the 4-week treatment period or rebound insomnia during nights they did not take study drug.

4 Safety

The most common adverse events (>1 % of treated patients) reported in the published trial were somnolence, fatigue, headache, and dysgeusia for SL-SD zolpidem (Staner et al. 2009, 2010b) and headache, nausea, and fatigue for SL-LD zolpidem (Roth et al. 2008a, 2013). More generally types of adverse events for sublingual zolpidem were consistent with the adverse event profile of oral zolpidem. Long-term safety profile for sublingual zolpidem is currently not available; for instance, it is not known whether the two sublingual zolpidem formulations have a different abuse and dependence potential profile than the oral formulation [zolpidem is classified as a Schedule IV controlled substance by federal regulation because of its abuse and dependence potential, especially in patients with previous history of substance abuse and/or concomitant psychiatric illness (Staner et al. 2010a)].

For SL-SD zolpidem 5 and 10 mg, next-day residual effects in terms of vigilance, psychomotor performance as assessed by the critical flicker frequency test and the choice reaction time test (CRT), and attention and concentration as assessed by the digit symbol substitution test (DSST) did not differ from IOR zolpidem 10 mg (Staner et al. 2009, 2010b). Residual effects of SL-SD zolpidem were first estimated on the basis of a daytime pharmacokinetic–pharmacodynamic study performed in healthy subjects ($n = 24$, range 21–44 years of age). This crossover randomized double-blind study investigated the effects of SL-SD zolpidem (1, 1.75, and 3.5 mg) against placebo administered at 8 am. Results show that psychomotor performance and attention measured by the CRT, the DSST, and the symbol copy test are impaired from 20 min post-dose up to 4 h post-dose with the 3.5 mg dose for certain parameters (Roth et al. 2008b). Because of its middle-of-the-night way of administration, these results led to some concern regarding the next-morning safety of the SL-LD zolpidem 3.5 mg.

In the pivotal laboratory study (Roth et al. 2008a) no statistical differences were observed in the active treatment (i.e., SL-LD zolpidem 1.75 and 3.5 mg) condition compared to placebo condition in terms of DSST and subjective sleepiness both assessed about 4.5 h post-dose. In the repeated-dose outpatients study (Roth et al. 2013), participants were instructed to take their treatment (SL-LD zolpidem 3.5 mg or placebo) if they have at least 4 h of time remaining in bed and to report their daytime sleepiness/alertness every morning whether or not study medication was taken during the night. Results showed that sleepiness/alertness significantly ($p < 0.01$) improved compared with placebo at every time point after nights during which study medication was taken. On non-dosing nights, no statistically significant differences were noted between drug and placebo for morning sleepiness/alertness.

Finally, a highway driving performance study (Vermeeren et al. 2014) investigated the effect of a single dose of SL-LD zolpidem 3.5 mg administered in the MOTN at 3 and 4 h before driving in 40 healthy volunteers (50 % females). Results indicate that SL-LD zolpidem 3.5 mg taken 3 h before driving may impair driving performance, but that there is a minimal risk of impairing driving performance if the drug is taken ≥ 4 h before driving. No gender differences were observed.

Conclusion

SL-SD zolpidem is a sublingual formulation of zolpidem that improves the onset of sleep compared to the classical oral formulation of zolpidem. The drug has been approved for the short-term treatment of sleep-onset insomnia at a dose of 10 mg in non-elderly man and at the dose of 5 mg in women or in special population (elderly, debilitated patients, patients with concomitant CNS depressant, or with hepatic impairment). Since the sublingual formulation is more potent than the oral formulation with the effect of SL-SD zolpidem 5 mg being similar to those of IOR zolpidem 10 mg, a substitution from the oral 10 mg form to the sublingual 5 mg form would lower drug exposure and possibly improve the safety/efficacy balance. Studies are however needed to test this hypothesis.

SL-LD zolpidem improves the onset of sleep compared to placebo and it is indicated for the treatment of insomnia when MOTN awakening is followed by difficulty returning to sleep. It is an innovative treatment because, in contrast to hypnotics currently labeled for sleep maintenance insomnia, the drug is prescribed on an "as-needed" basis and not prophylactically at bedtime. Recommended dosages are 3.75 mg for non-elderly man and 1.75 mg in women and in special population (elderly, debilitated patients, patients with concomitant CNS depressant, or with hepatic impairment). It has to be stressed that, due to safety issue related to next-morning residual effects, the dose should be taken following MOTN awakenings only if the patient has at least 4 h of sleep remaining.

References

American Psychiatric Association (2013) Diagnosis and statistical manual of mental disorders, 5th edn. American Psychiatric Association, Arlington

Edinger JD, Fins AI, Glenn DM, Sullivan RJ Jr, Bastian LA, Marsh GR, Dailey D, Hope TV, Young M, Shaw E, Vasilas D (2000) Insomnia and the eye of the beholder: are there clinical markers of objective sleep disturbances among adults with and without insomnia complaints? J Consult Clin Psychol 68:586–589

Edluar^{TM} (zolpidem tartrate) sublingual tablets, CIV (2009) Prescribing information. Meda Pharmaceuticals Inc., Somerset, NJ; May 2009. http://www.edluar.com/EDLUAR-PI.pdf

Farkas RH, Unger EF, Temple R (2013) Zolpidem and driving impairment. Identifying persons at risk. N Engl J Med 369:689–691

FDA Drug Safety Communication (2013) Risk of next-morning impairment after use of insomnia drugs; FDA requires lower recommended doses for certain drugs containing zolpidem (Ambien, Ambien CR, Edluar, and Zolpimist); Jan 2013. http://www.fda.gov/Drugs/DrugSafety/ucm334033.htm

Greenblatt DJ, Harmatz JS, Roth T, Singh NN, Moline ML, Harris SC, Kapil RP (2013) Comparison of pharmacokinetic profiles of zolpidem buffered sublingual tablet and zolpidem oral immediate-release tablet: results from a single-center, single-dose, randomized, open-label crossover study in healthy adults. Clin Ther 35:604–611

Hesse LM, von Moltke LL, Greenblatt DJ (2003) Clinically important drug interactions with zopiclone, zolpidem, and zaleplon. CNS Drugs 17:513–532

Intermezzo® (zolpidem tartrate) sublingual tablets, CIV (2011) Prescribing information. Transcept Pharmaceuticals Inc., Port Richmond, CA; Nov 2011. http://www.accessdata.fda.gov/drugsatfda_docs/label/2011/022328lbl.pdf

Johnson LC, Spinweber CL (1983) Good and poor sleepers differ in Navy performance. Mil Med 148:727–731

Monti JM, Pandi-Perumal SR, Langer SZ (2008) Zolpidem: its use in the treatment of sleep disorder. In: Pandi-Perumal SR, Verster JC, Monti JM, Lader M, Langer SZ (eds) Sleep disorders, diagnosis and therapeutics. Informa Healthcare, London, pp 295–323

Ohayon MM (2002) Epidemiology of insomnia: what we know and what we still need to learn. Sleep Med Rev 6:97–111

Ohayon MM, Krystal A, Roehrs TA, Roth T, Vitiello MV (2010) Using difficulty resuming sleep to define nocturnal awakenings. Sleep Med 11:236–241

Perlis ML, Smith MT, Cacialli DO, Nowakowski S, Orff H (2003) On the comparability of pharmacotherapy and behavior therapy for chronic insomnia: commentary and implications. J Psychosom Res 54:51–59

Roth T, Hull SG, Lankford A, Rosenberg R, Scharf MB (2008a) Low-dose sublingual zolpidem tartrate is associated with dose-related improvement in sleep onset and duration in insomnia characterized by middle-of-the-night (MOTN) awakenings. Sleep 31:1277–1284

Roth T, Mayleben D, Corser BC, Singh NN (2008b) Daytime pharmacodynamic and pharmacokinetic evaluation of low-dose sublingual transmucosal zolpidem hemitartrate. Hum Psychopharmacol 23:13–20

Roth T, Krystal A, Steinberg FJ, Singh NN, Moline M (2013) Novel sublingual low-dose zolpidem tablet reduces latency to sleep onset following spontaneous middle-of-the-night awakening in insomnia in a randomized, double-blind, placebo-controlled, outpatient study. Sleep 36:189–96

Staner L, Eriksson M, Cornette F, Santoro F, Muscat N, Luthinger R, Roth T (2009) Sublingual zolpidem is more effective than oral zolpidem in initiating early onset of sleep in the post-nap model of transient insomnia: a polysomnographic study. Sleep Med 10:616–620

Staner L, Cornette F, Otmani S, Nedelec JF, Danjou P (2010a) Zolpidem in the treatment of adult and elderly primary insomnia patients. In: Monti JM, Pandi-Perumal SR, Möhler H (eds) GABA and sleep: molecular, functional and clinical aspects. Springer, New York, pp 383–411

Staner C, Joly F, Jacquot N, Vlasova ID, Nethin M, Lundqvist T, Edenius C, Staner L (2010b) Sublingual zolpidem in early onset of sleep compared to oral zolpidem: polysomnographic study in patients with primary insomnia. Curr Med Res Opin 26:1423–1431

Staner L, Danjou P, Luthringer R (2012) A new sublingual formulation of zolpidem for the treatment of sleep-onset insomnia. Expert Rev Neurother 12:141–153

Swainston Harrison T, Keating GM (2005) Zolpidem: a review of its use in the management of insomnia. CNS Drugs 19:65–80

Valente KD, Hasan R, Tavares SM, Gattaz WF (2013) Lower doses of sublingual Zolpidem are more effective than oral Zolpidem to anticipate sleep onset in healthy volunteers. Sleep Med 14:20–23

Vermeeren A, Vuurman EF, Leufkens TR, Van Leeuwen CJ, Van Oers AC, Laska E, Rico S, Steinberg F, Roth T (2014) Residual effects of low-dose sublingual zolpidem on highway driving performance the morning after middle-of-the-night use. Sleep 37:489–496

Eszopiclone: Review and Clinical Applications in Chronic and Comorbid Insomnia

Jadwiga S. Najib

Abstract Eszopiclone, the active S(+)-enantiomer of zopiclone [(R,S)-zopiclone], was approved by the Food and Drug Administration (FDA) in 2004 for the treatment of insomnia in adult patients ≥18 years of age, and in 2005 became the first sedative hypnotic approved by the FDA for the long-term treatment of chronic insomnia. It is classified by the FDA as a Schedule IV drug under the Controlled Substances Act and is available in the United States as the brand Lunesta® in dosages of 1, 2, and 3 mg. Eszopiclone 2 and 3 mg consistently demonstrated significant improvements of the primary variable, latency to persistent sleep, assessed subjectively or objectively, in short- and long-term clinical trials. Additionally, improvements in sleep maintenance, sleep quality, and daytime functioning have also been observed. Eszopiclone 3 mg has been shown to improve insomnia associated with comorbid conditions. Tolerance with eszopiclone was not apparent in any of the sleep parameters for up to 12 months treatment duration and rebound insomnia, where documented, tends to be transient and limited to the first night of withdrawal. Eszopiclone is generally well tolerated and the most frequent adverse event is unpleasant bitter taste. Both eszopiclone and its racemic compound, zopiclone, are nonbenzodiazepine hypnotic agents that are derivatives of the cyclopyrrolone class, with eszopiclone exhibiting greater affinity than the R(−)-enantiomer for the $GABA_A$/benzodiazepine receptor complex; hence the sedative hypnotic effects of the racemate are primarily linked to the S(+)-enantiomer rather than the R(−)-enantiomer. The racemate zopiclone is available as 3.75, 5, and 7.5 mg tablets; the latter contains 3.75 mg of S(+)-zopiclone, which is more than the highest dose (3 mg) available in the United States. Eszopiclone is not marketed in the European Union, as it was too similar to the racemate to be considered a patentable product.

J.S. Najib (✉)
Division of Pharmacy Practice, Arnold & Marie Schwartz College of Pharmacy and Health Sciences, Long Island University, 5th Floor, 75 DeKalb Avenue, Brooklyn, NY 11201, USA

Departments of Pharmacy and Psychiatry, St. Luke's/Roosevelt Hospital Center, New York, NY, USA
e-mail: jadwiga.najib@liu.edu

A. Guglietta (ed.), *Drug Treatment of Sleep Disorders*, Milestones in Drug Therapy,
DOI 10.1007/978-3-319-11514-6_8

1 Pharmacology and Mechanism of Action

Eszopiclone is the S(+)-enantiomer of racemic zopiclone [(R,S-zopiclone)], and as enantiomers both hypnotics possess the same molecular and structural formula but have different three-dimensional configurations and are non-superimposable mirror images of each other. Both eszopiclone and its racemic compound, zopiclone, are non-benzodiazepine derivatives of the cyclopyrrolone class with no structural similarity to pyrazolopyrimidines, imidazopyridines, benzodiazepines, barbiturates, or other drugs with known hypnotic properties. Although eszopiclone's exact mechanism of action in enhancing and promoting sleep is unknown, its effects appear to be related to its interaction with GABA receptor complexes at binding sites located close or allosterically coupled to benzodiazepine receptors. Eszopiclone has demonstrated balanced selectivity for the α_1, α_2, α_3, and α_5 subtypes of the $GABA_A$ receptor (Najib 2006). This differs from other hypnotics such as zolpidem and zaleplon, which have primary selectivity for the α_1 subtype, and from triazolam, which has selectivity for all the $GABA_A$ receptor subtypes. Sedative effects from hypnotics are primarily associated with activity at the α_1 receptor, whereas anxiolytic effects are associated with activity at the α_2, α_3 receptors (Sieghart 2006). The racemate zopiclone contains 50:50 mixture of enantiomeric structures and binds with approximately equal affinity to $GABA_A$ receptors containing different α subunits. The S(+)-enantiomer (eszopiclone) has >50-fold higher affinity for the $GABA_A$/benzodiazepine receptor compared to the R(−)-enantiomer, and approximately twice that of the racemate. Hence, virtually all sedative and hypnotic activity of the racemate is attributable to S(+)-zopiclone, while R(−)-zopiclone is not only almost inactive but more toxic than racemic zopiclone. This is consistent with the recommended dosage of eszopiclone being approximately 50 % lower compared to its racemate (Greenblatt and Zammit 2012). It has been proposed that cyclopyrrolones, like eszopiclone, may exhibit more selectivity for certain subunits of the $GABA_A$ receptor than benzodiazepines, which could translate to a lower incidence of adverse events including rebound insomnia and withdrawal effects (Monti and Pandi-Perumal 2007).

2 Pharmacokinetics

The pharmacokinetic parameters in men and women, as well as for all races studied, appeared similar with no significant differences noted. Both zopiclone and eszopiclone are rapidly absorbed following administration and have similar time to reach peak plasma concentrations (t_{max}) reported at 1.5–2 h vs. 1 h, respectively, and elimination half-lives ($t_{1/2}$) of 5 h vs. 6 h, respectively. Steady state levels are attained 24–48 h after initiation of once-daily administration, and pharmacokinetic parameters did not differ appreciably between single dose and multiple daily dosing, indicating no accumulation of either the racemate or eszopiclone.

Following oral administration, eszopiclone is extensively metabolized by the cytochrome P450 (CYP) isoenzymes 3A4 and 2E1 via oxidation and demethylation to (S)-desmethyl-zopiclone and (S)-zopiclone-N-oxide, and less than 10 % of the drug is excreted in the urine as the parent drug. Both metabolites appear in plasma exposure at lower levels than the parent drug, suggesting that the clinical activity of eszopiclone is due mainly to the parent drug (Greenblatt and Zammit 2012).

Following administration of eszopiclone 3.5 mg compared to zopiclone 7.5 mg, the active metabolite, (S)-desmethyl-zopiclone, was noted to have a $t_{1/2}$ of 8.7 h vs. 12.3 h, with lower exposure (AUC) of 48.9 ng/ml h vs. 59.3 ng/ml^{-1} h, respectively. A possible explanation for the slower elimination has been the proposal of a stereoselective metabolism. Minimal hypnotic effects with little cognitive impairment are associated with eszopiclone levels $\leq$10 ng/ml. Therapeutic levels are achieved well within 30 min with eszopiclone 3.5 mg and decrease to 10 ng/ml in approximately 8 h. However, following administration of zopiclone 7.5 mg, levels of S(+)-zopiclone remain below the sleep-inducing threshold > 30 min and decrease to 10 ng/ml > 9 h after the dose (Brunello et al. 2009), suggesting that patients who take zopiclone may be at a higher risk for residual effects for at least 1 h longer in the morning following nighttime administration, compared to those who take eszopiclone.

3 Clinical Efficacy and Safety

The New Drug Application (NDA) for eszopiclone contained data from a total of 24 clinical trials and over 60 preclinical studies that assessed the safety and efficacy of eszopiclone for the treatment of insomnia in adults and elderly patients. The 24 clinical trials included 16 Phase I trials which gathered information on drug interaction and special population studies, 2 Phase II trials which assessed next-day residual effects in healthy subjects and in patients with chronic insomnia, and 6 Phase III trials. The primary study on which the Food and Drug Administration (FDA) in 2005 approved eszopiclone for long-term use was a 6-month trial in adults with primary insomnia (Krystal et al. 2003), hence making eszopiclone the only approved sedative/hypnotic at that time without short-term restriction. Since then, long-term use of eszopiclone has been evaluated in adult and elderly patients. Table 1 summarizes the key study design features and outcomes of pivotal trials evaluating the efficacy and safety of eszopiclone.

3.1 Transient Insomnia

Eszopiclone at doses 1, 2, 3, or 3.5 mg produced significant improvements over placebo on measures of latency to persistent sleep (LPS) and sleep maintenance in healthy subjects without sleep disturbances (Rosenberg et al. 2005). The "first-night

Table 1 Summary of clinical trials evaluating the efficacy and/or safety of eszopiclone

References	Population	Study design	Eszopiclone dose	Comparator	Assessment	Primary (1°) and key secondary (2°) endpoints
Rosenberg et al. (2005)	436 healthy normal adults with primary insomnia (25–50 years old)	R, PC, DB, PG, MC (single-dose first-night effect model)	1 mg 2 mg 3 mg 3.5 mg	Placebo	PSG, self-reports	1°: objective LPS Key 2°: objective sleep efficiency
Zammit et al. (2004)	308 patients with primary insomnia (21–64 years old)	R, PC, DB, PG MC (44 consecutive nights/2 nights of SB placebo)	2 mg 3 mg	Placebo	PSG, self-reports	1°: objective LPS Key 2°: objective sleep efficiency, objective WASO
Krystal et al. (2003)	788 patients with primary insomnia (21–69 years old)	R, PC, DB, MC (Monthly visits for 6 months)	3 mg	Placebo	Self-reports	1°: SL Key 2°: TST
Roth et al. (2005)	471 patients with primary insomnia (21–64 years old)	OL, 6-mo extension of Krystal et al. (2003) study (monthly visits for 6 months)	3 mg	Placebo	Self-reports	1°: SL Key 2°: TST
Walsh et al. (2007)	830 patients with primary insomnia (21–64 years old)	R, PC, DB, MC (6 months)	3 mg	Placebo	Self-reports	1°: SL Key 2°: TST
Erman et al. (2008)	65 patients with primary insomnia (21–64 years old)	R, PC, MC, 6-way CO (2 nights/3–7 day washout period)	1 mg 2 mg 2.5 mg 3 mg	Placebo, Zolpidem 10 mg	PSG, self-reports	1°: objective LPS Key 2°: objective sleep efficiency, objective WASO
Uchimura et al. (2012a)	72 patients with primary insomnia (21–64 years old)	R, PC, DB, MC, 5-way CO (2 nights/washout period of 5 days)	1 mg 2 mg 3 mg	Placebo, Zolpidem 10 mg	PSG, self-reports	1°: objective LPS, subjective SL Key 2°: objective and subjective TST, sleep efficiency

Uchimura et al. (2012b)	164 elderly patients (65–84 years old) 161 non-elderly patients (20–64 years old)	R, DB, PG, MC (24 weeks)	Low dose: 1 mg (elderly) 2 mg (non-elderly) High dose: 2 mg (elderly) 3 mg (non-elderly)	Placebo	Self-reports	1°: incidence of adverse events Key 2°: SL, TST
Scharf et al. (2005)	231 patients with primary insomnia (65–85 years old)	R, PC, DB, PG, MC (2 weeks)	1 mg 2 mg	Placebo	Self-reports	1°: SL Key 2°: TST
McCall et al. (2006)	264 patients with primary insomnia (65–85 years old)	R, PC, DB, PG (2 weeks)	2 mg	Placebo	PSG, self-reports	1°: SL, sleep efficiency Key 2°: TST
Ancoli-Israel et al. (2010)	388 elderly patients with primary insomnia (65–85 years old)	R, PC, DB, MC (12 weeks/then SB placebo for 2 weeks, additional 2 week evaluation period)	2 mg	Placebo	Self-reports	1°: TST Key 2°: SL, WASO

CO = crossover; DB = double blind; LPS = latency to persistent sleep; MC = multicenter; OL = open label; PC = placebo-controlled; PG = parallel group; PSG = polysomnography; R = randomized, SB = single blind, SL = sleep latency, TST = total sleep time, WASO = wake time after sleep onset effect model" dealt with insomnia encountered on the first assessment night in a sleep laboratory, acting as a model of transient insomnia. Efficacy was assessed via polysomnography (PSG) and a morning questionnaire. Compared to placebo, subjects receiving eszopiclone, regardless of dose, had significantly less wake time after sleep onset (WASO) and significantly greater sleep efficiency. Additionally, subjects treated with eszopiclone at all doses, except 1 mg, had significantly lower (better) objective LPS, and the number of awakenings (NAW) per night was significantly decreased with the 3 mg and 3.5 mg doses. Self-reported assessments were consistent with PSG results and included significantly less time to sleep onset, and fewer NAW for all doses of eszopiclone. Total sleep time (TST) was significantly greater in the 2, 3, and 3.5 mg groups compared to placebo.

3.2 Chronic Primary Insomnia in Non-elderly Patients

The efficacy of eszopiclone for the treatment of chronic primary insomnia in non-elderly adults with no other concurrent psychiatric disorders was evaluated in six trials (Krystal et al. 2003; Zammit et al. 2004; Roth et al. 2005; Walsh et al. 2007; Erman et al. 2008; Uchimura et al. 2012a), with a post hoc analysis by Krystal and colleagues (2012a) of their earlier trial. All patients met DSM-IV criteria for primary insomnia ($\leq$6.5 h of sleep per night, sleep onset latency $\geq$ 30 min for $\geq$ 1 month) and three studies also investigated rebound insomnia and tolerance (Zammit et al. 2004; Roth et al. 2005; Walsh et al. 2007). One study that assessed eszopiclone in both elderly and non-elderly patients with chronic insomnia who presented with and without comorbid psychiatric disorders is discussed in the section with eszopiclone use in elderly patients (Uchimura et al. 2012b).

Efficacy and safety of eszopiclone 2 and 3 mg were assessed in a double-blind trial for 44 consecutive nights, followed by 2 nights of single-blind placebo (Zammit et al. 2004). Patients who received eszopiclone, regardless of dose, fell asleep approximately 50 % faster, compared to those who received placebo, as noted by a reduction in LPS. Results from PSG findings were consistent with patient-reported sleep endpoints. Both eszopiclone doses significantly improved sleep efficiency, TST, sleep onset latency, sleep depth, and sleep quality compared to placebo. Improved sleep maintenance, reported as a reduction in WASO, was significant only for the 3 mg dose compared to placebo. No significant differences between NAW were noted between treatment and placebo groups. Psychomotor performance, evaluated by Digit Symbol Substitution Test (DSST) scores, approximately 1–1.5 h after awakening, was similar in both eszopiclone groups compared to placebo, and patients reported improvements in measure of daytime alertness and ability to function. Rebound insomnia was noted on first night after eszopiclone 2 mg was discontinued (sleep efficiency was reduced by 2.5 % and WASO increased by 7.0 min), but by the second night, both sleep efficiency and WASO had returned back to baseline ranges. Patients who received eszopiclone 3 mg maintained efficacy post discontinuation of dosing on Night 46 (sleep efficiency increased 3.7 % and WASO decreased by 8.5 min).

The first large, long-term study of nightly eszopiclone use was undertaken by Krystal and colleagues (2003) in adults with primary insomnia carried out at 70 sites. Efficacy was assessed weekly, and by the first week and throughout the study period, eszopiclone 3 mg relative to placebo significantly improved SL, WASO, TST, quality of sleep, and NAW. Additionally, subjective ratings of function, alertness, and physical well-being were more favorable with eszopiclone compared to placebo. No evidence of tolerance was noted. Most common adverse effects with eszopiclone compared to placebo, respectively, were unpleasant taste (26 %, 6 %), headache (20 %, 19 %), infection (16 %, 7 %), and nausea (11 %, 6 %).

In a 6-month, open-label extension phase of the same sample, Roth and colleagues (2005) evaluated eszopiclone 3 mg relative to placebo in patients at weekly intervals, thus further assessing the efficacy and safety of eszopiclone in adults with primary insomnia for a total of 12 months of continuous nightly treatment. During this 6-month open-label trial, eszopiclone maintained efficacy on all the sleep parameters in patients who had received eszopiclone during the previous double-blind trial. Additionally, patients previously treated with placebo reported improvements in SL, WASO, TST, quality of sleep NAW, and number of nights with awakenings at all monthly time points compared with their 6-month baseline. All patients treated with eszopiclone 3 mg during this period reported improvements in sleep and daytime functioning, without any evidence of tolerance or rebound insomnia. Most common adverse effects with eszopiclone were unpleasant taste (7 %), headache (4 %), somnolence (4 %), abnormal dreams (3 %), and dizziness (2.5 %).

One long-term trial in patients with primary insomnia demonstrated that eszopiclone 3 mg nightly treatment improved subjective SL and quality of life (Walsh et al. 2007). Significant improvements were noted in SL, WASO, TST, and NAW at months 1–6, and patient-reported sleep and daytime function as reported were significantly improved for all study months. At the end of 6 months, 50 % of subjects treated with eszopiclone reported no clinically significant insomnia, reported on the Insomnia Severity Index (ISI) score ≤ 7, compared to 19 % placebo-treated subjects. Improvements also were noted with eszopiclone in the Short-Form 36 Health Survey (SF-36) (domains: Physical Function, Vitality, and Social Functioning), and for all domains of the Work Limitations Questionnaire (WLQ). No significant evidence of rebound insomnia, no tolerance, and no withdrawal CNS effects were reported. Most common adverse effects with eszopiclone compared to placebo were unpleasant taste (20 %, 1 %), infection (17 %, 12 %), and headache (both 15 %).

In a post hoc analysis of the 6-month study of 2003, Krystal and colleagues (2012a) evaluated the relationship between baseline WASO severity and treatment outcome. At baseline, the proportion of patients in the eszopiclone and placebo groups for each WASO severity subgroups were similar and the distribution of baseline WASO was as follows: WASO $\leq$30 min (32.2 %), WASO > 30 min to $\leq$90 min (33.0 %), WASO >0 to $\leq$45 min (41.5 %), WASO >45 to $\leq$90 min

(23.7 %), and WASO > 90 min (22.6 %). In all WASO subgroups combined, treatment with eszopiclone 3 mg, compared with placebo, was associated with significant sleep maintenance efficacy at each month for the 6-month study period. However, greater baseline WASO severity was associated with increasingly greater drug–placebo differences on WASO. For example, compared to placebo, patients treated with eszopiclone 3 mg in the mild WASO subgroup (≤30 min) demonstrated a 20 % reduction from baseline in WASO, compared to patients in the more severe WASO subgroups (>0 to ≤45 min and >45 to ≤90 min) and the most severe WASO subgroup (>90 min), who showed a 64 % and 60 % reduction in WASO, respectively. Likewise, the eszopiclone/placebo difference in WASO at Month 1 in the mild WASO subgroup (≤30 min) vs. severe WASO subgroup (>90 min) was −8.8 and −41.2 min, respectively, while at Month 6, the eszopiclone/placebo difference in those same WASO subgroups was reported as −7.4 and −20.1 min, respectively. Higher baseline WASO severity was associated with an increase in placebo response at both Month 1 and Month 6, although the increase was less than that observed in the eszopiclone treatment group. Possible explanations for the increase in response to placebo noted in this post hoc analysis may be an increased tendency for subjective overestimation with increasing symptom severity, variation in self-reporting styles, greater tendency for regression to the mean, lack of objective PSG data, or other variables. Nonetheless, these findings of increased placebo response in patients with higher baseline severity differ from other CNS trials (e.g., migraine, anxiety, schizophrenia, and mood disorders including depression and bipolar), which consistently note lower placebo responses.

Erman et al. (2008) assessed efficacy and safety of a range of doses of eszopiclone relative to placebo in patients with primary insomnia. Patients received 2 nights treatment each with one of four eszopiclone doses (1, 2, 2.5, or 3 mg), a placebo, and zolpidem 10 mg, as an active control, after randomization to one of six treatment sequences. Objective efficacy was assessed via PSG and all active treatments were effective in reducing LPS and increasing sleep efficiency, relative to placebo. Eszopiclone 2.5 mg and 3 mg were significantly better for improved LPS and sleep efficiency than eszopiclone 1 mg. Additionally, objective sleep measures of WASO, NAW, and wake time during sleep were significantly improved with eszopiclone 3 mg, but not with other eszopiclone doses or zolpidem 10 mg. Significant differences were not noted on PSG measured outcomes between zolpidem 10 mg and eszopiclone 2 or 3 mg; however the study was not powered to detect differences between active treatments. Findings from this study support earlier PSG studies noting that higher doses of eszopiclone are more effective for sleep maintenance insomnia while the lower doses are effective for sleep onset insomnia (Rosenberg et al. 2005; Zammit et al. 2004). In this study, treatments were well tolerated and unpleasant taste was the only adverse event that was observed with a higher frequency in the eszopiclone 3 mg treatment group compared to zolpidem or placebo. Approximately twice the number of patients reported CNS adverse events with zolpidem 10 mg compared with eszopiclone 3 mg, or placebo, which included dizziness (10.9 %, 4.7 %, 4.8 %), somnolence (9.4 %, 4.7 %, 3.2 %), and hallucinations (4.7 %, 0 %, 0 %) respectively.

Uchimura and colleagues (2012a) conducted a multicenter, randomized, double-blind, five-way crossover study in non-elderly Japanese patients with primary insomnia. Patients received placebo, eszopiclone 1, 2, and 3 mg, or zolpidem 10 mg for two consecutive nights, with a washout period of 5 days. Zolpidem was utilized as an active reference to allow qualitative comparisons of eszopiclone, although quantitative comparisons of the active treatment groups were not possible. Patients met the DSM-IV-TR criteria for primary insomnia and additionally had a history $\geq$ 4weeks duration of SL $\geq$ 30 min for $\geq$ 3 days/week and TST $\leq$ 390 min for $\geq$ 3 days/week. Compared to placebo, all active treatments produced significant improvements in objective and subjective LPS and TST. A linear dose–response relationship was observed for eszopiclone, with greater efficacy obtained with increased dose. Treatment with eszopiclone 2 mg and 3 mg and zolpidem 10 mg significantly improved PSG assessments of WASO and NAW and subjective patient-reported measures of WASO, NAW, sleep quality, sleep depth, and daytime functioning compared to placebo. All treatment groups increased non-REM sleep and stage 2 sleep, but did not alter REM or slow-wave sleep. The most common adverse events with eszopiclone 1 mg, 2 mg, and 3 mg were dysgeusia (5.7 %, 8.7 %, 16.2 %, respectively) and somnolence (1.4 %, 4.3 %, 5.9 %, respectively). The most common adverse events with zolpidem 10 mg were somnolence and dizziness, reported as 4.3 % each.

3.3 Chronic Insomnia in Elderly Patients

Three studies evaluated eszopiclone in elderly patients with chronic insomnia (Scharf et al. 2005; McCall et al. 2006; Ancoli-Israel et al. 2010), and one study assessed eszopiclone in both elderly and non-elderly patients with chronic insomnia who presented with and without comorbid psychiatric disorders (Uchimura et al. 2012b). Three of the abovementioned studies also investigated rebound insomnia and tolerance (McCall et al. 2006; Ancoli-Israel et al. 2010; Uchimura et al. 2012b).

Two separate 2-week randomized, double-blind, placebo-controlled multicenter trials of eszopiclone were conducted in elderly patients with chronic insomnia who met the DSM-IV criteria for primary insomnia (Scharf et al. 2005; McCall et al. 2006). Eszopiclone 2 mg compared to placebo significantly improved subjective measures of SL, TST, and WASO in both trials. Scharf et al. (2005) noted that eszopiclone 1 mg was effective in inducing sleep, whereas eszopiclone 2 mg was effective at both inducing and maintaining sleep in elderly patients. McCall et al. (2006) also noted eszopiclone 2 mg demonstrated improvements in some quality of life measures, ISI total severity, and quality of sleep scores, along with improved objective LPS, WASO, and sleep efficiency. Eszopiclone was well tolerated during these short-term trials of eszopiclone use in elderly patients and the overall incidence of adverse events was similar across all treatment groups, with headache, unpleasant taste, somnolence, dizziness, and dyspepsia being the most commonly reported (≥5 %) adverse events in the eszopiclone groups.

Uchimura and colleagues (2012b) evaluated short-term efficacy (4 weeks) and long-term safety (24 weeks) of eszopiclone in elderly and non-elderly Japanese patients with chronic insomnia who presented with and without comorbid psychiatric disorders. Patients were randomized to receive either low-dose (1 mg for elderly subjects, 2 mg for non-elderly subjects) or high-dose (2 mg for elderly subjects or 3 mg for non-elderly subjects) eszopiclone. Patients had a diagnosis of primary insomnia as defined by DSM-IV-TR, or insomnia associated with a physical or psychiatric disorder. Additionally, patients had reported symptoms with SL ≥ 30 min on ≥3 nights/week and TST ≤ 390 min on ≥3 nights /week for ≥ 4 weeks. Comorbid psychiatric disorders were noted in 49.4 % of elderly patients and 49.7 % of non-elderly patients enrolled. The most common comorbid psychiatric disorders among patients with insomnia were major depressive disorder, generalized anxiety disorder, dysthymic disorder, and agoraphobia. The most common concomitant medications for patients with insomnia and comorbid psychiatric disorders included psychoneurotics, anxiolytics, peptic ulcer medications, analgesics, antihyperlipidemics, and antihypertensives. Measures of efficacy included self-reported improvement in both SL and TST at Week 4 compared to baseline in 88 % of elderly and 83 % of non-elderly patients, and improvements were similar in those with and without comorbid psychiatric disorders. At Week 1, eszopiclone 1 mg and 2 mg, relative to placebo, significantly decreased SL from 60 min to 20 or 30 min in elderly patients with and without psychiatric comorbidities, respectively, and effects were sustained through the end of the study period assessing efficacy (Week 4). Likewise, eszopiclone 2 and 3 mg significantly improved SL in non-elderly patients with and without psychiatric comorbidities throughout Week 1–Week 4. Additionally, eszopiclone significantly improved WASO, NAW, and daytime sleepiness and function from baseline to Week 4, in both elderly and non-elderly patients, irrespective of psychiatric comorbidity. Improvement in scores for Mental Component Summary and Mental Health Domain was observed in the Medical Outcomes SF-36 in elderly and non-elderly patients with insomnia and psychiatric comorbidities.

This study notes a lack of dose discrimination between eszopiclone 1 and 2 mg in elderly patients and between eszopiclone 2 and 3 mg in non-elderly patients, regarding sleep efficacy assessments at Weeks 1–4. These results differ from the previously reported US trials. The previously reported 2-week US study in elderly patients treated with 2 mg eszopiclone noted improved SL, TST, and WASO, and while the lower dose of 1 mg improved SL, it did not improve WASO and TST (Scharf et al. 2005). The 6-week US study in non-elderly patients treated with eszopiclone 2 and 3 mg noted improvements in SL and TST with both doses, but the lower 2 mg dose did not improve WASO (Zammit et al. 2004). Likewise, in another US study involving non-elderly patients, both eszopiclone 2 and 3 mg improved SL but only the 3 mg dose was associated with improvement in WASO (Erman et al. 2008). Other US trials have also noted efficacy for sleep induction and maintenance, with eszopiclone 2 and 3 mg doses in elderly patients and non-elderly patients, respectively, irrespective of the presence of psychiatric comorbidity (Ancoli-Israel et al. 2010; McCall et al. 2010; Fava et al. 2006; Pollack et al. 2008; Walsh et al. 2007).

Overall, the rate of adverse events reported with eszopiclone 1 mg and 2 mg in the abovementioned study (Uchimura et al. 2012b) with the elderly group was 81.5 % and 79.5 %, respectively. For elderly patients with psychiatric comorbidities receiving eszopiclone 1 mg and 2 mg, the rate of adverse events was 79.5 % and 81.0 %, respectively, while for those without psychiatric disorders, the rates were 83.3 % and 78.0 %. In non-elderly patients, the adverse events reported with eszopiclone 2 and 3 mg were 82.1 % and 87.0 %, respectively, and overall, no adverse events occurred at a substantially higher rate among patients with psychiatric comorbidities versus those without psychiatric disorders. The most frequently reported adverse events in elderly patients receiving eszopiclone 1 mg and 2 mg, compared to non-elderly patients receiving eszopiclone 2 mg and 3 mg, respectively, were dysgeusia (18.5 %, 27.7 % vs. 42.9 %, 57.1 %), nasopharyngitis (17.3 %, 21.7 % vs. 26.2 %, 18.2 %), headache (4.9 %, 6.0 % vs. 3.6 %, 1.3 %), and somnolence (4.9 %, 2.4 % vs. 3.6 %, 7.8 %). No evidence of rebound insomnia or dependency was observed upon discontinuation of eszopiclone following 24 weeks of treatment with the hypnotic in Japanese elderly and non-elderly patients with chronic insomnia, regardless of psychiatric comorbidity.

3.4 *Studies in Long-Term Use in Elderly*

The longest trial of eszopiclone in elderly patients with primary and comorbid insomnia was a Phase IV study that included patients with stable or chronic medical conditions (Ancoli-Israel et al. 2010). Eszopiclone improved mean TST over placebo by a mean of 63.24 min, and additionally, eszopiclone-treated patients compared to placebo reported a greater decrease in SL (mean decrease of 24.62 min vs. 19.92 min) and greater decrease in WASO (mean decrease of 36.4 min vs. 14.8 min), respectively. Improvements were noted during the first week of treatment and were maintained for the remainder of weekly assessments throughout the 12-week study period. Additionally, significantly greater improvements from baseline ISI total scores were observed in subjects treated with eszopiclone and, at week 12, the number of patients who reported "no insomnia" or "subthreshold insomnia" was 78 % in the eszopiclone group, compared to 61.1 % in the placebo group. Significant improvements in patients who received eszopiclone, relative to placebo, were noted in the vitality score of the SF-36 at Week 6 and Week 12, and in the general health scale at Week 12. Patient reports of NAW, sleep quality, and depth of sleep noted significant improvements at all time points for those who received eszopiclone compared to placebo.

The overall incidence of adverse events during the double-blind period for eszopiclone and placebo was 59.3 % and 50.5 %, respectively. The most common adverse events ($\geq$5 %) reported for eszopiclone and placebo, respectively, were headache (13.9 %, 12.4 %), unpleasant taste (12.4 %, 1.5 %), and nasopharyngitis (5.7 %, 6.2 %). Adverse events in older adults observed with eszopiclone compared to placebo, respectively, included dizziness (4.1 %, 1.5 %), falls (1.0 %, 0.5 %), hallucinations (0.5 %, 0 %), memory impairment (1.0 %, 0 %), attention disturbance (0.5 %, 0 %), nervousness (1.5 %, 0 %), and anxiety (2.1 %, 1.0 %).

Following the 12-week treatment period, patients entered a 4-week follow-up period, which included a single-blind placebo for 2 weeks to assess rebound and withdrawal effects, followed by another 2-week evaluation where no drug or placebo was administered. Following discontinuation of treatment, SL, TST, and WASO all statistically significantly improved compared to baseline measures, indicting no evidence of rebound insomnia upon discontinuation of eszopiclone. No withdrawal symptoms were noted upon drug discontinuation, nor any evidence of tolerance. Although objective measures were not utilized in this study, the authors claim that in clinical practice patients with insomnia do not receive overnight sleep recordings, and hence, when assessing the benefit of a sedative hypnotic, clinicians rely on patient's subjective reports of improved sleep and well-being.

4 Use of Eszopiclone in Comorbid Insomnia

Comorbid insomnia is defined as insomnia that occurs along with another medical or psychiatric condition. Eszopiclone has a substantial body of clinical evidence supporting its efficacy in insomnia patients with certain comorbid disorders. Key design features along with primary and secondary outcomes of trials utilizing eszopiclone in patients with insomnia comorbid with other psychiatric and medical conditions are outlined in Table 2. Additional ongoing trials are under way evaluating the use of eszopiclone in the treatment of patients with comorbid conditions such as fibromyalgia, low back pain, and attention-deficit hyperactivity disorder.

4.1 Perimenopausal and Postmenopausal Women

Two studies evaluated the effects of eszopiclone 3 mg and placebo in perimenopausal and postmenopausal women who also met criteria for insomnia (Soares et al. 2006; Joffe et al. 2010). The first trial by Soares and colleagues noted that women who received eszopiclone compared to placebo reported significant improvements in SL, WASO, TST, sleep quality, and next-day functioning. The frequency or severity of hot flashes did not differ between treatment groups. Menopause-related measures, including awakenings due to hot flashes, Greene Climacteric Scale scores, and vasomotor and physical domain scores of the menopause-specific QoL questionnaire (MenQoL), significantly improved at Week 4 in patients receiving eszopiclone. Additionally, eszopiclone produced

Table 2 Summary of clinical trials evaluating eszopiclone with insomnia comorbid with other conditions

References	Population	Study design	Eszopiclone dose	Comparator	Assessment	Primary (1°) and key secondary (2°) endpoints
Soares et al. (2006)	410 peri/postmenopausal women with insomnia (40–60 years old)	R, PC, DB, PG, MC (4 weeks/1 week SB placebo run-out period)	3 mg	Placebo	Self-reports	1°: SL Key 2°: MADRS, MenQOL, GCS
Joffe et al. (2010)	59 peri/postmenopausal women with insomnia (40–65 years old)	R, PC, DB, MC (CO of 4 weeks/2 week washout period)	3 mg	Placebo	Self-reports	1°: ISI Key 2°: MADRS, BAI, MenQOL
Fava et al. (2006)	545 patients with insomnia and comorbid MDD (21–64 years old)	R, PC, DB, PG, MC (8 weeks treatment (Fava et al. 2006)	Eszopiclone 3 mg/fluoxetine 20 mg with dose titration to 40 mg	Placebo	Fava et al.: Objective, self-reports	Fava et al.: 1°: WASO Key 2°: HAM-D
Krystal et al. (2007)		2 weeks SB placebo run-out period) (Krystal et al. 2007)			Krystal et al.: Objective, self-reports	Krystal et al.: 1°: SL, WASO, TST, measures of daytime function Key 2°: HAM-D
McCall et al. (2010)	60 patients with insomnia and comorbid MDD (aged 18–70 years old)	R, PC, DB (8 weeks)	Eszopiclone 3 mg/fluoxetine 20 mg with dose titration to 40 mg	Placebo	Objective, self-reports, PSG, actigraphy	1°: DLRF and RSO subscales of the Basis-32 Key 2°: Q-LES-Q, HAM-D
Pollack et al. (2008)	595 patients with insomnia and comorbid GAD (18–64 years old)	R, PC, DB, PG, MC (8 weeks/2 weeks SB placebo run-out period)	Eszopiclone 3 mg/escitalopram 10 mg	Placebo	Objective, self-reports	1°: SL Key 2°: TST, HAM-A
Pollack et al. (2011)	24 patients with PTSD (18–64 years old)	R, PC, DB (CO of 3 weeks/1 week washout period)	3 mg	Placebo	Objective, self-reports	1°: SPRINT scores, PSQI Key 2°: WASO, CAPS

(continued)

Table 2 (continued)

References	Population	Study design	Eszopiclone dose	Comparator	Assessment	Primary (1°) and key secondary (2°) endpoints
Rosenberg et al. (2007)	21 OSA patients (35–64 years old)	R, PC, DB, MC (2 nights CO/5–7 day washout)	3 mg	Placebo	PSG	1°: AHI score
Lettieri et al. (2008)	226 OSA patients (18–64 years old)	R, PC, DB (1 night)	3 mg	Placebo	PSG, self-reports	1°: rates of non-usable and poor quality PSGs Key 2°: SL, WASO
Lettieri et al. (2009a)	160 OSA patients (mean age 45.7 years)	R, PC, PG (14 nights)	3 mg	Placebo	Objective, self-reports	1°: CPAP adherence at 24 weeks Key 2°: rate of CPAP discontinuation
Lettieri et al. (2009b)	117 OSA patients (18–64 years old)	R, PC, DB (4–6 weeks)	3 mg	Placebo	PSG, self-reports	1°: CPAP compliance over 4–6 weeks Key 2°: Quality of CPAP titrations
Eckert et al. (2011)	23 OSA patients (19–62 years old)	R, PC, DB (1 night)	3 mg	Placebo	PSG,	1°: Effect on arousal threshold, AHI score
Roth et al. (2009)	153 patients with insomnia and RA (25–64 years old)	R, PC, DB, PG, MC (4 weeks/2 week run-out)	3 mg (plus stable treatment regimen for RA for ≥ 3 months)	Placebo	Self-reports,	1°: WASO Key 2°: ISI, SF-36, ASES
Menza et al. (2010)	30 patients with Parkinson's disease (35–85 years old)	R, PC, DB, PG, MC (6 weeks)	Patients < 65 years: 3 mg Patients ≥ 65 years: 2 mg	Placebo	Self-reports	1°: TST Key 2°: WASO

Attarian et al. (2011)	30 patients with RRMS (aged 25–64 years)	R, PC, DB (7 weeks)	2 mg titrated to 3 mg	Placebo	Objective, self-reports, actinography	1°: TST Key 2°: FDS, MFIS, ESS
Dimsdale et al. (2011)	45 patients with mucositis 2° to malignancies (20–75 years old)	R, PC, DB, PG (2 days)	>64 years old, or on concomitant CYP 3A4 inhibitor: 2 mg <64 years old: 3 mg	Placebo	Self-reports	1°: pain assessments, POMS-SF, Key 2°: SL

AHI = Apnea-Hypopnea Index; ASES = Arthritis Self- Efficacy Scale (on a scale 0–10, 10 being best function and least pain); BAI = Beck Anxiety Inventory score; Basis-32 = Behavior and Symptom Identification Scale, a self-report measure assessing mental health treatment outcomes; CAPS = Clinician-Administered PTSD Scale; CO = crossover; CPAP = continuous positive airway pressure; CYP 3A4 = cytochrome P 450 isoenzyme 3A4; DB = double blind; DLRF = Daily Living and Role Functioning; ESS = Epsworth Sleepiness Scale; FDS = Fatigue Descriptive Scale; GAD = Generalized Anxiety Disorder; GCS = Greene Climacteric Scale (21 questions related to climacteric symptomatology); HAM-A = Hamilton Anxiety Rating Scale; HAM-D = 17-item Hamilton Depression Rating Scale; ISI = Insomnia Severity Index; MADRS = Montgomery-Asberg Depression Rating Scale (9 questions related to mental status); MC = multicenter; MDD = Major Depressive Disorder; MenQOL = Menopause-specific quality of life questionnaire (29 questions related to symptoms of menopause); MFIS = Modified Fatigue Impact Scale; OSA = obstructive sleep apnea; PC = placebo-controlled; PG = parallel group; POMS-SF = Profile of Mood States Scale, Short Form; PSG = polysomnography; PSQI = Pittsburgh Sleep Quality Index (patient-administered scale assessing 7 sleep-related domains); PTSD = Posttraumatic stress disorder; Q-LES-Q = Quality of Life Enjoyment and Satisfaction Questionnaire- measures 16 aspects of quality of life (e.g., mood, social and family relationships, physical health, leisure and household activities, ability to engage in work or hobbies, overall sense of well-being, overall life satisfaction) and includes a global summary; R = randomized; RA = Rheumatoid Arthritis; RRMS = relapsing-remitting multiple sclerosis; RSO = Relationship to Self and Others; SB = single blind; SF-36 = Medical Outcomes Study 36-Item Short-Form Health Survey (to evaluate quality of life); SL = sleep latency; SPRINT = Short PTSD Rating Interview (10-item clinician-administered scale assessing core symptoms of PTSD); TST = total sleep time; WASO = wake time after sleep onset significant improvements in mood, assessed via the Montgomery Asberg Depression Rating Scale (MADRS), and in menopausal symptoms especially vasomotor, psychological, and physical subscales and family life/home domains of the Sheehan Disability Scale. Following a 1-week placebo run-out period after the 4-week treatment, no significant differences were noted between treatments with regard to changes from baseline in menopause symptoms. Most frequently reported adverse effect was unpleasant taste (17.6 % vs. 0.5 %). In the second study in peri/postmenopausal women, eszopiclone significantly reduced ISI scores by 8.7 ± 1.4 more points than placebo, and after 4 weeks, the ISI score was ≤ 7 in 87 % of women on eszopiclone compared to 34 % of those on placebo (Joffe et al. 2010). Significant improvements were noted in SL, TST, WASO, and sleep efficiency with eszopiclone. Additionally, significant improvements with eszopiclone compared to placebo were noted in nighttime hot flashes, depressive symptoms (assessed via MADRS), anxiety symptoms (assessed via Beck Anxiety Inventory), and quality of life (assessed via MenQoL). Daytime hot flashes and functional impairment did not differ between treatment groups. Eszopiclone was well tolerated in both trials and adverse events occurring in ≥ 5 % of the women in the first trial, compared to placebo, included unpleasant taste (17.6 % vs. 0.5 %), headache (14.6 % vs. 14.9 %), and back pain (5.0 % vs. 1.9 %). The only adverse event associated with eszopiclone occurring in ≥ 5 % of the participants in the second trial was metallic taste, which was reported at a rate of 25 %.

4.2 Major Depressive Disorder

The first large-scale clinical trial of adjunctive use of eszopiclone with a selective serotonin reuptake inhibitor (SSRI) co-administered to patients who met DSM-IV criteria for both major depressive disorder (MDD) and insomnia was assessed by Fava and colleagues (2006), while the hypnotic discontinuation and withdrawal-related adverse events from this study were assessed in a separate 2-week single-blind placebo run-out phase (Krystal et al. 2007). Participants received fluoxetine (initially 20 mg/day with dose titration to 40 mg/day) every morning for 10 weeks and were randomized to concomitantly receive either eszopiclone 3 mg or placebo at night for 8 weeks, followed by a 2-week, single-blind placebo run-out period. Participants in the eszopiclone and fluoxetine group reported significant subjective improvements in SL, WASO, TST, depth of sleep, and sleep quality. By week 8, there was a 95-min reduction in SL and 2.5-h increase in TST. Additionally, significant increases were reported in daytime alertness, ability to function, and ability to think clearly and concentrate, although physical sense of well-being did not significantly improve. Reduced 17-item Hamilton Rating Scale for Depression (HAM-D) scores at Weeks 4 and 8 were observed in the eszopiclone co-therapy group, even after removing the three insomnia-related items. Patients who were more severely depressed at baseline experienced the most significant changes in depressive symptoms at Weeks 4 and 8. Significantly fewer patients receiving eszopiclone compared to placebo had to be titrated to the higher dose of fluoxetine at Week 4, noting an antidepressant effect. Improved CGI-I and CGI-S scores were noted at all time points. Patients who received eszopiclone compared to placebo had significantly more responders (patients with ≥50 % reduction in HAM-D, 59 % vs. 48 %) and remitters (patients with HAM-D ≤7, 42 % vs. 33 %) at Week 8. Treatment with eszopiclone was well tolerated and the most prevalent adverse events observed in patients treated with eszopiclone compared to placebo were unpleasant taste (22.7 % vs. 0.7 %), headache (16.7 % vs. 14.6 %), nausea (13.0 % vs. 12.8 %), dry mouth (9.3 % vs. 8.8 %), somnolence (8.6 % vs. 7.3 %), and dizziness (8.5 % vs. 3.3 %) (Fava et al. 2006). The incidence rates of CNS and potentially CNS-related adverse events were similar between eszopiclone/fluoxetine (9.8 %) and placebo/fluoxetine (8.8 %) treatment groups. There was no evidence of withdrawal, rebound depression or rebound insomnia. Following discontinuation with eszopiclone, patients maintained improvements in SL, WASO, and TST during the 2-week placebo run-out phase. Likewise, improvements in CGI-I and HAM-D scores noted at Week 8 were maintained at Week 10, and significantly higher depression response and remission rates were noted in the co-therapy group at Week 10 (Krystal et al. 2007).

The first clinical trial to evaluate eszopiclone's effect on objective measures of sleep in insomniac patients with major depressive disorder by assessing the health-related quality of life (HRQOL) was undertaken by McCall and colleagues (2010). A 1-week open-label trial with fluoxetine, followed by 8 weeks of fluoxetine combined with either eszopiclone 3 mg or placebo nightly, was administered to depressed insomniac outpatients, mostly female (67 %), with either SL > 30 min and sleep efficiency < 85 % ≥ 4 nights/week, or Research Diagnostic Criteria insomnia criteria ≥ 4nights/week. At baseline, patients had ISI total scores that reflected moderate severity of insomnia, moderate to severe depression as assessed via Patient Health Questionnaire (PHQ9) Total scores and 24-item HAM-D, and "poor" baseline Quality of Life Enjoyment and Satisfaction Questionnaire (Q-LES-Q) scores. The primary outcome as measured by the Behavior Symptom and Identification Scale (Basis-32), a self-reported measure to assess mental health treatment outcomes, assessed the subscales of Daily Living and Role Functioning (DLRF) and Relationship to Self and Others (RSO), which at baseline both domains were in the "moderate difficulty" range. At the end of treatment, patients who received eszopiclone had significantly better DLRF scores compared to those who received placebo. Women had lower (better) scores than men for DLRF, RSO, and Q-LES-Q with an effect size of 0.62, 0.44, and 0.38 respectively, indicating a moderate effect for DLRF and RSO, and a small effect for Q-LES-Q. Self-reported sleep as assessed via sleep diaries also noted significant benefits with eszopiclone over placebo in SL, NAW, and TST. Significant effects favoring eszopiclone for CGI-severity, CGI-Improvement, ISI-total score, ISI-Improvement, and HAM-D were noted. Additionally, PSG comparing treatment groups at end of treatment noted significant improvements in WASO, TST, and sleep efficiency with eszopiclone. The fact that patient-reported improvement in sleep was accompanied by PSG improvements is noteworthy as many depressed patients will note improvement in sleep even though it may not correlate with PSG improvement. The findings that eszopiclone improved overall depression symptoms as reflected in HAM-D total score and percent responders are consistent with the previously reported study by Fava and colleagues (2006).

McCall and McCall (2012) did a post hoc analysis of this 2010 study and examined changes in sleep parameters in the laboratory and home before and after laboratory monitoring in depressed insomniacs. Mean actigraphy values demonstrated significantly improved sleep time and sleep efficiency and decreased wake time in the laboratory compared to home setting. Differences between actigraphy and PSG were not significant. Sleep diaries, however, noted a slight worsening of sleep in the laboratory compared to home environment, with significantly more NAW in laboratory compared to the week at home before and after the laboratory night. The differences between objective and subjective sleep measurements seen in depressed insomniacs may be influenced by where the monitoring is taking place and how it is measured. Although many individuals may sleep worse during their first night in a sleep laboratory (first-night effect), this study noted that actigraphic sleep was better in the laboratory than at home (reverse first-night effect). Although actigraphic monitoring is controversial in some populations, including insomniacs, it is considered a valid method to characterize sleep patterns in insomnia by the American Academy of Sleep Medicine. Discrepancies between objective and subjective methods emphasize the importance of using multiple methods for assessing sleep continuity in patients with a diagnosis of sleep disorders.

4.3 *Generalized Anxiety Disorder*

The efficacy of eszopiclone administered with escitalopram was evaluated in patients with insomnia and comorbid generalized anxiety disorder (GAD) who were randomized to receive escitalopram 10 mg plus eszopiclone 3 mg or placebo for 8 weeks, followed by a 2-week single-blind placebo run-out period (Pollack et al. 2008). Those who received eszopiclone with the SSRI had significantly greater improvements in the Hamilton Anxiety Scale (HAM-A) scores at each week and through Weeks 4 through 10 with the insomnia item removed. Patients receiving eszopiclone and escitalopram also reported significant improvement in SL, TST, WASO, NAW, daytime alertness, ability to function, and ability to concentrate and had no clinically meaningful insomnia (ISI ≤ 7) at week 8 compared to those who received placebo. CGI-I scores at each week significantly improved with eszopiclone co-therapy although CGI-S scores did not significantly differ between treatment groups after Week 1. At Week 8, significantly more patients who received eszopiclone achieved improvements in response rate (patients with ≥ 50 % reduction in HAM-A), and although remission rates also improved, they did not reach significance. No evidence of tolerance, rebound insomnia, or withdrawal effects was reported and treatment differences in anxiety measures were maintained during the placebo washout period. Overall discontinuation rates due to adverse events were similar in both groups, although higher discontinuation rates were noted in the escitalopram plus placebo group during the first 2 weeks of the double-blind period. Overall rates of treatment-emergent adverse events for eszopiclone and placebo groups were 77.6 % and 67.9 %, respectively, with unpleasant taste, headache, dry mouth, and somnolence reported more frequently with eszopiclone.

Fava et al. (2011a) did a post hoc analysis of data pooled from the 2 previously reported 8-week trials (Fava et al. 2006; Pollack et al. 2008) and assessed the effect of eszopiclone and a selective serotonin reuptake inhibitor in patients with insomnia and comorbid anxious depression. In the combined dataset, 30.5 % of patients had insomnia and comorbid anxious depression. At week 8, patients in the eszopiclone compared to placebo co-therapy groups had significantly greater improvements from baseline in ISI total scores, along with greater reductions in HAM-D scores when insomnia items were excluded or included, but not for anxiety/somatization. Response rates were significantly higher in anxious depressed patients in the eszopiclone co-therapy group compared to placebo co-therapy at week 8; however, when insomnia items were removed the difference between groups was not significant. Remission rates did not significantly differ between treatment groups. Results showed that the benefits previously reported with eszopiclone in patients with insomnia and comorbid MDD or GAD were also observed in patients with insomnia and comorbid anxious depression. Discontinuation of eszopiclone therapy did not result in withdrawal or rebound insomnia.

4.4 Posttraumatic Stress Disorder

Treatment with eszopiclone in patients with posttraumatic stress disorder (PTSD) and associated insomnia was evaluated in a double-blind, 3-week crossover trial with eszopiclone 3 mg administered at bedtime compared to placebo, followed by a washout week (Pollack et al. 2011). The mean duration of PTSD experienced by subjects was approximately 20 years, and over half the patients had a concurrent mood or anxiety disorder, and over a third reported a lifetime history of alcohol or substance abuse or dependence. Compared to placebo, eszopiclone 3 mg was associated with greater improvement in sleep as observed in a significant reduction in the Pittsburgh Sleep Quality Index (PSQI) score with significant improvements in PSQI subscales assessing subjective sleep quality, SL, sleep disturbances, and daytime dysfunction. Additionally, relative to placebo, eszopiclone 3 mg was associated with significant improvement in scores on the Short PTSD Rating Interview and the Clinician-Administered PTSD Scale. Although not significant, there was an increase in TST. Adverse events included dysgeusia (32 %), sedation (16 %), and headaches (12 %).

4.5 Chronic Obstructive Sleep Apnea and Pulmonary Disease

Eszopiclone 3 mg has been evaluated in patients with chronic obstructive sleep apnea (OSA) in several trials from 1 to 2 nights to 4 to 6 weeks (Rosenberg et al. 2007; Lettieri et al. 2008, 2009a, b; Eckert et al. 2011). Using PSG measures, a small crossover study of two consecutive nights demonstrated that eszopiclone's mean change from baseline in Apnea-Hypopnea Index (AHI) score did not differ significantly from that of placebo in patients with mild to moderate OSA (AHI $\geq$ 10 to $\leq$ 40). Significant differences favoring eszopiclone were noted in spontaneous arousals, sleep efficiency, TST, WASO, and wake time during sleep. Most common adverse event with eszopiclone compared to placebo was bitter taste (28.6 % vs. 9.5 %, respectively) (Rosenberg et al. 2007). Another larger study of one night's duration noted pretreatment with eszopiclone compared to placebo resulted in improved CPAP titrations with less residual events and fewer incomplete titrations, resulting in decreased need to repeat CPAP titration studies. The rate of poor quality studies was significantly lower in eszopiclone compared to placebo-treated patients (26.5 % vs. 46.0 %). Additionally, significant PSG improvement in sleep onset latency, sleep efficiency, WASO, and prolonged sleep time was noted (Lettieri et al. 2008). Another study noted eszopiclone during the first 2 weeks of CPAP improved CPAP adherence in adults with newly diagnosed mild to severe OSA (mean AHI of 37) for the first 6 months of therapy (Lettieri et al. 2009a). Use of CPAP resulted in subjective improvements in sleepiness, fatigue, and quality of life, with greater improvements in the eszopiclone group, most likely reflecting increased use of CPAP. The mean duration of regular use of CPAP (>4 h/night for > 70 % of nights) was 17.6 weeks for the eszopiclone group vs. 13.3 weeks for the placebo group. Another study (Lettieri et al. 2009b) also noted that one-time use of eszopiclone 3 mg as pretreatment during CPAP titration significantly improved compliance with CPCP over 4–6 weeks in patients with moderate to severe OSA (mean AHI of 29.2), as recorded objectively from a downloadable "smart card" adherence-monitoring device. CPA was used on a higher percentage of nights in patients who received eszopiclone (75.9 %) compared to those on placebo (60 %). Additionally, the percentage of regular use of CPAP (>4 h/night) was significantly greater among the eszopiclone group (46.9 % vs. 29.9 %, respectively). Good compliance, defined as use of CPAP > 4 h/night for > 70 % of nights, was significantly greater in those premedicated with eszopiclone (53.1 %) compared to placebo (27.1 %). Eszopiclone compared to placebo significantly improved mean sleep efficiency, TST, and WASO. A single night physiological study noted also that eszopiclone 3 mg significantly increased stage 2 respiratory arousal threshold and sleep duration, improved sleep quality, and lowered the AHI in OSA patients with moderate to severe OSA (mean AHI of 31), without respiratory event prolongation or worsening hypoxemia (Eckert et al. 2011). Eszopiclone reduced the total AHI by 23 $\pm$ 9 % and the greatest reduction in AHI occurred in patients with a low arousal threshold. A shift from lighter stage 1 sleep to a more consolidated stage

2 sleep is associated with reduced OSA severity. Eszopiclone was noted to increase arousal threshold by ~ 30 % (more negative) from stage 1 to stage 2 sleep, and this may allow for respiratory stimuli to accumulate and allow sufficient upper airway muscle dilation resulting in airway patency. The reduction in AHI with eszopiclone was noted in non-REM sleep with no change in REM AHI, consistent with the proposed mechanism of reduced apnea severity with the non-benzodiazepine sedatives, and it is known that upper airway dilator muscles are responsive during non-REM sleep in the presence of respiratory stimulation, but less so during REM sleep.

4.6 Rheumatoid Arthritis

In a pilot study in patients with insomnia associated with rheumatoid arthritis, eszopiclone 3 mg nightly for 4 weeks significantly improved SL, WASO, TST, sleep quality, and depth of sleep for every week relative to placebo (Roth et al. 2009). At week 4, significantly more patients on eszopiclone had no clinically meaningful insomnia (ISI score ≤ 7) and had greater improvements in ability to function averaged over the 4-week treatment period compared to those on placebo. Additionally, eszopiclone was significantly better than placebo on some pain measures including improvements in the Arthritis Self-Efficacy Scale, in the activities domain of the Health Assessment Questionnaire-Disability Index, and on the SF-36 questionnaire. Following a 2-week placebo run-out phase, there was some evidence of rebound insomnia on the first night following discontinuation with eszopiclone with respect to TST, but not for SL or WASO. Overall incidence of adverse events was 67.5 % with eszopiclone and 60.5 % with placebo. The most frequently reported adverse events with eszopiclone compared to placebo were unpleasant taste (27.3 % vs. 0 %), increased symptoms of rheumatoid arthritis (18.2 % vs. 9.2 %), headache (10.4 % vs. 7.9 %), pharyngitis (10.4 % vs. 2.6 %), asthenia, and viral infection (each 6.5 % vs. 2.6 %).

4.7 Parkinson's Disease

Eszopiclone was compared to placebo in patients with Parkinson's disease and insomnia in a 6-week, 5-site, double-blind, randomized controlled trial (Menza et al. 2010). Patients had been ill with Parkinson's disease an average of 4.5 years, and had an average Hoehn–Yahr stage of 1.6. Patients < 65 years of age received eszopiclone 3 mg, and patients ≥ 65 years of age received eszopiclone 2 mg. Most patients were taking more than one class of medications for Parkinson's disease, which included levodopa, dopamine agonists, catechol-O-methyltransferase inhibitors, and monoamine oxidase inhibitors. At the end of 6 weeks, there were no significant differences noted in TST, SL, and WASO between patients who received eszopiclone and those on placebo. However, significant improvements in secondary endpoints in NAW, quality of sleep, and in physician-rated CGI improvement were noted in patients who received eszopiclone. No significant differences between groups were noted in measures of daytime alertness, ability to function, or sleep impairment. Additionally, no significant differences were noted in quality of life (QoL), motor functioning, fatigue severity, as well as caregiver QoL and depression. Although no difference between groups was observed in the Parkinson's disease QoL scale or in measure of daytime functioning, it may be due to the low power of the study or that patients with Parkinson's disease have other impairments which may overwhelm QoL daytime functioning assessments. Eszopiclone use was not accompanied by any change in Parkinson's disease motor symptoms, and it was well tolerated. Incidence of adverse events with eszopiclone was 33 % compared to 27 % with placebo. Two of the 15 patients (13 %) in the eszopiclone group had an adverse event assessed as related to study drug: two had sedation and one of those also reported dizziness. None of the patients reported unpleasant taste.

4.8 *Multiple Sclerosis*

A significant relationship between fatigue and sleep disturbance has been established in patients with relapsing-remitting multiple sclerosis (RRMS) (Attarian et al. 2004), and thus the effect of eszopiclone on sleep disturbances and daytime fatigue and functioning was evaluated in ambulatory patients with RRMS (Attarian et al. 2011). Patients were excluded if they were receiving other sedatives, stimulants, amantadine, or modafinil, and all patients had a Fatigue Descriptive Scale (FDS) score of ≥ 5, indicating fatigue. Patients underwent a diagnostic phase for 2 weeks and were randomized to receive eszopiclone 2 mg or placebo by Day 14 and if they reported continued sleep problems, their dose was increased to 3 mg of either eszopiclone or placebo, for a minimum of 4 weeks at the maximum tolerated dose. Patient-maintained sleep logs and actigraphs revealed that eszopiclone compared to placebo was superior in increasing objective TST. There was no statistically significant correlation between increased TST and improved fatigue, and the hypnotic did not improve SL, SE, or WASO.

4.9 *Cancer Pain*

Pain is known to disrupt sleep and opioids themselves have been shown to disrupt sleep by significantly increasing light sleep and decreasing deep sleep, which may not only further increase next-day fatigue, but may also lower the threshold for pain stimuli necessitating continuous or higher doses of opioids, putting in place a vicious cycle of pain and insomnia. To assess how the quality of sleep influences daytime pain which in turn affects sleep at night, eszopiclone was evaluated in patients with malignancies who developed mucositis severe enough to require treatment with patient-controlled analgesia delivering either morphine or dilaudid (Dimsdale et al. 2011). Patients were randomized to eszopiclone or placebo for two consecutive days. Every 6 h after dosing and while awake, patients assessed pain with a 10-cm visual analog scale, and fatigue with the five-item Profile of Mood States Scale, Short-Form Fatigue-Inertia Scale. Eszopiclone usage was associated not only with significant improvements in sleep, but also reductions in self-reported pain throughout the entire day. Patients who received eszopiclone had significantly lower mean pain scores at all time points (morning, afternoon, evening), increased TST, fewer NAW, better self-reported sleep quality, and depth. Self-reports of fatigue or opioid usage did not differ significantly between treatment groups. The effects of eszopiclone on pain did not appear to be due to the sedative effects of the hypnotic, as lower pain ratings were not apparent only on the morning ratings, but were consistently lower throughout the day, an average reduction of pain reports by ~30 %. Additionally, fatigue scores, although not significantly different between treatment groups, and always a concern in cancer patients who already present with high fatigue levels, were reported at a lower rate with eszopiclone. Eszopiclone was well tolerated and drowsiness/confusion were each noted in both treatment groups with the same frequency (13 %).

5 Dosage and Administration/Drug Interactions

The lowest effective dose of eszopiclone should be used in patients to minimize any potential next day impairment associated with eszopiclone. The recommended initial dose is 1 mg, immediately before bedtime, with at least 7 to 8 hours remaining before the planned time of awakening. The dose may be increased if clinically indicated to a maximum of 3 mg, however for geriatric or debilitated patients, the dose should not exceed 2 mg. No dosage adjustments are required in patients with renal impairment. In patients with severe hepatic impairment, however, the starting dose of eszopiclone should be 1 mg. Eszopiclone became the first sedative hypnotic approved in the United States for long-term use in chronic insomnia, and subsequently, other hypnotics have received approvals with no restriction on the duration of treatment.

Eszopiclone is weakly bound to plasma proteins (52–59 %), and the large free fraction suggests that the drug should not be affected by drug interactions secondary to competitive plasma protein binding. However, the recommended starting dose of eszopiclone with potent CYP3A4 inhibitors (e.g., ketoconazole, itraconazole, clarithromycin, ritonavir, nelfinavir) is 1 mg, not to exceed 2 mg. Caution should be exercised when eszopiclone is administered with other drugs known to cause central nervous system depression, as this may lead to additive sedative and psychomotor effects. If eszopiclone is administered with or immediately following a high-fat/heavy meal, absorption may be delayed leading to a reduced effect on SL (Sunovion 2014).

6 Safety and Tolerability

Eszopiclone was generally well tolerated in non-elderly and elderly patients in both short-term and longer-term studies. The tolerability profile of eszopiclone during the longer-term trials was similar to that reported during the short-term trials. Pooled data from long-term studies in non-elderly patients noted the incidence of the following adverse events with eszopiclone 3 mg compared to placebo: unpleasant taste (22.8 % vs. 2.9 %), somnolence (8.3 % vs. 2.9 %), pain/back pain (3.5 % vs. 1.4 %), nausea/vomiting (5.7 % vs. 3.1 %), and somnolence (8.3 % vs. 2.9 %). Adverse event profile in the very elderly (≥75 years old) was similar to the elderly (<75 years old), with headache, asthenia, somnolence, and unpleasant taste being the most prevalent. In the ≥ 75 years old age group, asthenia was higher, while unpleasant taste was lower.

Unpleasant and bitter taste was the most commonly reported adverse event reported with eszopiclone across the dose range from 1 mg (8 %) to 3 mg (34 %). A double-blind study noted no meaningful relationship between the frequency or intensity of the taste disturbance associated with eszopiclone and a person's age, body mass index, or phenylthiocarbamide taste sensitivity status. However, the predominant dysgeusic sensations of bitter and metallic taste were more intense and longer lasting in females than males, stronger in morning than in evening hours, and positively correlated with eszopiclone's plasma and saliva levels (Doty et al. 2009).

Since $GABA_A$ receptors containing the various α subunits are located in various regions of the brain, the clinical effects of binding to these receptors may be determined by their location. Binding to $GABA_A$ receptors in the cerebellum may lead to problems with balance and coordination, whereas binding to $GABA_A$ receptors in the hippocampus is associated with memory impairment, and binding to $GABA_A$ receptors in the amygdala produces an anxiolytic effect (Nutt and Stahl 2010; Richey and Krystal 2011).

Reports of impairments in memory, cognitive, or psychomotor functions were infrequent, and trials utilizing various doses of eszopiclone did not depict evidence of a dose relationship associated with these events. Memory impairment was noted in 1.3 % vs. 0 % of patients treated with eszopiclone 3 mg compared to placebo in a 6-month trial, while the incidence of confusion was same between treatment groups (0.5 %) (Krystal et al. 2003). In elderly patients, spontaneous reporting rates of confusion and memory impairment with eszopiclone 2 mg or placebo were 0 % vs. 0.8 % and 1.5 % vs. 0 %, respectively (McCall et al. 2006). Another trial with elderly patients noted confusion and memory impairment with eszopiclone 1 mg, 2 mg, and placebo at 0 %, 2.5 %, and 0 % and 1.4 %, 0 %, and 0 %, respectively (Scharf et al. 2005). A 6-week trial reported the incidence of confusion and memory impairment with eszopiclone 2 mg, 3 mg, and placebo as 0 %, 3 %, and 0 % and 1 %, 1 %, and 0 %, respectively (Zammit et al. 2004).

Two randomized, double-blind, placebo-controlled crossover trials in healthy volunteers and patients with primary insomnia ($n = 32$, both) showed that eszopiclone 3 mg did not impair next-day driving ability or measures of cognitive or psychomotor function (Boyle et al. 2008). Next-day feelings of sedation did not increase in patients, but increased in volunteers. No objective impairments were noted in either group. Next-morning DSST scores, administered to objectively determine residual next-day psychomotor drug effects, demonstrated no clinically relevant impairment and no dose response pattern was evident, and all subjects reported they were not more or less sleepier than those who received placebo. Likewise, next-day effects of eszopiclone in healthy normal subjects and insomnia patients following a single nighttime dose demonstrated no significant effect of eszopiclone 3 mg on DSST at 9.5 h post dose. This is in contrast to the racemate, zopiclone 7.5 mg, that demonstrated significant impairment on DSST at both 9 and 10 h post dose (EMEA 2009). A double-blind study of 91 healthy adults (aged 25 to 40 years) assessed the effects of eszopiclone 3 mg on next-morning psychomotor and memory impairment (Boyle et al. 2012). Impairment was most severe at 7.5 hours, but still present and potentially clinically meaningful at 11.5 hours after dosing. Although subjective perception of sedation and coordination associated with eszopiclone 3 mg did not differ significantly from placebo, subjects were objectively impaired, and had an impairing effect almost as large as observed with zopiclone 7.5 mg (Boyle et al. 2012).

Tolerance with eszopiclone was not apparent in any of the sleep parameters for up to 12 months treatment duration. Rebound insomnia, where documented, tended to be transient and limited, and resolution typically was observed within one to two nights without intervention. Duration of treatment did not increase the incidence of patients experiencing rebound insomnia following eszopiclone discontinuation.

Eszopiclone is classified as a schedule IV medication, and doses of 6 mg–12 mg produce euphoria similar to the effects of diazepam 20 mg in patients with a history of benzodiazepine abuse (EMEA 2009). Withdrawal was evaluated using the Benzodiazepine Withdrawal Symptom Questionnaire (BWSQ) along with new or worsening adverse events following abrupt discontinuation of treatment in six trials ranging from 4 weeks to 6 months of double-blind treatment. The incidence of withdrawal adverse events and the total score for the BWSQ did not differ significantly between eszopiclone and placebo. Post-marketing data in the United States shows that tolerance, withdrawal syndrome, and dependence have occurred at 0.18 %, 0.19 %, and 0.11 %, respectively (EMEA 2009).

7 Hypnotics and Infections and Cancer

Recent epidemiologic studies and meta-analyses have raised concern about the risks associated with the use of hypnotics, specifically reports of infections and possibility of cancer (Joya et al. 2009; Kripke 2008; Kripke et al. 2012; Stebbing et al 2005). Review of the Approval History and Documents have noted that eszopiclone, along with zaleplon and ramelteon, was each reported to be clastogenic, that is, they cause breaks in chromosomes, and FDA files have revealed that those three and zolpidem were associated with cancers in rodents (Kripke

2008). Most infections reported with eszopiclone in trials were mild and viral and did not lead to dropouts. Data from trials of four hypnotics combined (eszopiclone, ramelteon, zaleplon, and zolpidem) noted that participants randomized to hypnotics had a higher rate of incident non-melanoma skin cancers and a significantly higher rate of incident skin cancers with possible malignant tumors, although the trails were not sufficient in size to detect if each hypnotic by itself caused cancer (Kripke 2008). Research is needed to determine if use of hypnotics may increase infections or cancer, and till then, data from post-marketing observational trials are the mainstay in detecting drug effects that are infrequent or have a delayed presentation.

8 Place in Therapy

Short- and long-term trials in adult patients demonstrated that eszopiclone 1–3 mg consistently improved measures of sleep latency, maintenance, and duration of sleep. Likewise data from trials demonstrated that in elderly patients, eszopiclone 1 mg was effective in inducing sleep, whereas eszopiclone 2 mg was effective at both inducing and maintaining sleep. Additionally, eszopiclone has a substantial body of clinical evidence supporting its efficacy in insomnia patients with certain comorbid disorders. When administered to patients with comorbid conditions receiving standard therapy, eszopiclone has shown greater improvements in those comorbidities according to certain parameters than those receiving placebo with standard therapy. Overall studies have also demonstrated that eszopiclone 3 mg did not worsen the severity of sleep-disordered breathing.

Data from 5 large, multicenter, randomized, placebo-controlled trials of adult outpatients of at least 1-month duration were analyzed by Krystal and colleagues (2012b) for effect size comparisons in patients with primary insomnia, and insomnia with medical and psychiatric comorbidity. As early as Week 1, effect sizes ranged from 0.40 to 0.69 (small to medium) and were maintained at 0.26–0.63 at Week 4 for SL, WASO, and TST in all five trials. Effect sizes on all measures of SL, WASO, TST, and ISI were consistently highest in the primary insomnia group, and tended to be lowest for patients with comorbid GAD and MDD. It is also too early to tell if the improvement noted across the insomnia subgroups is unique to eszopiclone or common to all hypnotics that enhance $GABA_A$ receptor-mediated inhibition.

Eszopiclone, apart from its known hypnotic effects, may have direct beneficial effects on comorbid disease symptomatology by its activity across a relatively broad range of $GABA_A$ receptor subunits. Eszopiclone appears to have a greater impact on the GABA α_2 and α_3 receptors, which may explain the antidepressant, anxiolytic, and possibly anti-pain effects associated with this hypnotic (Nutt and Stahl 2010). Likewise, its effect on the GABA α_3 subunit, which may be relevant for anxiolytic activity, and GABA α_5 subunits, which may be relevant in the extinction of learned fear in the hippocampus, may have a role in PTSD management. Since benzodiazepines are commonly administered as anxiolytics, eszopiclone's efficacy for comorbid anxiety and other affective disorders may be attributable to its broader range across $GABA_A$ receptor subunits compared to other sedative hypnotics. Sedative effects are associated with activity at α_1 subunit of the $GABA_A$ receptor, and zolpidem has a relatively higher affinity for α_1 compared to α_3, and studies with extended release zolpidem did not demonstrate anxiolytic and antidepressant benefits as were noted with eszopiclone (Fava et al. 2011b; Hanson et al. 2008).

Eszopiclone has been studied in other conditions where its role as a hypnotic may be beneficial. Treatment of chronic insomnia in patients with bipolar disorder is a therapeutic priority since it can cause or exacerbate manic-related symptoms. A preliminary study in patients with chronic insomnia and bipolar disorder noted treatment success rates with zolpidem, zolpidem CR, eszopiclone, zaleplon, and ramelteon were 60 %, 58 %, 46 %, 36 %, and 15 %, respectively, in this patient population (Schaffer et al. 2011).

Recently, reduction in sleep spindles observed in schizophrenics has been hypothesized to impair sleep-dependent memory consolidation. Sleep spindles correlate with various cognitive measures including memory consolidation, learning potential, and intellect. Reduced spindle activity has been reported to predict more severe positive symptoms of schizophrenia. Eszopiclone is known to increase the duration of stage 2 sleep and to act on GABA neurons in the thalamic reticular nucleus where spindles are generated. A small ($n = 21$) randomized placebo controlled trial noted significantly increased sleep spindle number and density during stage 2 sleep following 2 nights of eszopiclone 3 mg compared to placebo which correlated with enhanced though nonsignificant overnight motor sequence task improvement (Wamsley et al. 2013). Previous studies noted that triazolam led to overnight deterioration of motor sequence tasks, although spindles increased, and zolpidem did not differ from placebo in its effects on memory. Additionally, previous studies have reported that benzodiazepine-induced spindles differ in frequency, amplitude, and duration while eszopiclone did not alter those characteristics nor the topographical distribution of sleep spindles. It is unknown if these finding are specific to eszopiclone, but larger studies may be needed to detect if there is a significant effect on memory.

The European Medicines Agency (EMEA) in 2009 with advice from the Clinical Neurosciences Scientific Advisory Group (CHMP) confirmed their initial opinion from the previous year and cited no relevant and clinically meaningful differences between eszopiclone and its racemate, zopiclone, and denied granting eszopiclone a "new active substance" status. Hence, eszopiclone is not marketed in the European Union, as it was too similar to the racemate to be considered a patentable product. Racemic zopiclone itself has been available since 1987 outside of the United States, and has been widely utilized as a sedative/hypnotic with over 20 million patient-years of experience in many countries. Zopiclone is marketed as Imovane®, Zopinox®, or Zimovane® in Canada, Australia, Sweden, Norway, Finland, Russia, and the United Kingdom, and as Amoban® in Japan (Najib 2010).

Conclusion

Eszopiclone 2 and 3 mg has demonstrated statistically significant improvement of the primary variable, SL, measured objectively or subjectively in both short- and long-term trials. Secondary efficacy measures, including sleep maintenance, sleep quality, and daytime functioning, have also shown improvement with eszopiclone. Various trials for insomnia associated with comorbid conditions have noted improvement with eszopiclone 3 mg on sleep measures with no evidence of negative effects on the comorbid disease state. Long-term use with eszopiclone has not been associated with evidence of tolerance.

References

Ancoli-Israel S, Krystal AD, McCall WV, Schaefer K, Wilson A, Claus R, Rubens R, Roth T (2010) A 12-week, randomized, double-blind, placebo-controlled study evaluating the effect of eszopiclone 2 mg on sleep/wake function in older adults with primary and comorbid insomnia. Sleep 33(2):225–234

Attarian HP, Brown KM, Duntley SP, Carter JD, Cross AH (2004) The relationship of sleep disturbances and fatigue in multiple sclerosis. Arch Neurol 61(4):525–528

Attarian H, Applebee G, Applebee A, Wang B, Clark M, McCormick B, Salzman E, Schuman C (2011) Effect of eszopiclone on sleep disturbances and daytime fatigue in multiple sclerosis patients. Int J MS Care 13(2):84–90

Boyle J, Trick L, Johnsen S, Roach J, Rubens R (2008) Next-day cognition, psychomotor function, and driving-related skills following nighttime administration of eszopiclone. Hum Psychopharmacol 23(5):385–397

Boyle J, Groeger JA, Paska W, Cooper JA, Rockett C, Jones S, Gandhi P, Scott J, Atzori G, Dijk DJ (2012) A method to assess the dissipation of the [corrected] residual effects of [corrected] hypnotics: eszopiclone versus zopiclone. J Clin Psychopharmacol 32(5):704–709

Brunello N, Bettica P, Maier G, Nutt D (2009) Pharmacokinetics of (S)-zopiclone and (S)-desmethylzopiclone following dosing with zopiclone and eszopiclone. World Psychiatry [Int Congre World Psychiatr Assoc (April 1–4, Florence), 8 (Suppl 1):Abst P02.149

Dimsdale JE, Ball ED, Carrier E, Wallace M, Holman P, Mulroney C, Shaikh F, Natarajan L (2011) Effect of eszopiclone on sleep, fatigue, and pain in patients with mucositis associated with hematologic malignancies. Support Care Cancer 19(12):2015–2020

Doty RL, Treem J, Tourbier I, Mirza N (2009) A double-blind study of the influences of eszopiclone on dysgeusia and taste function. Pharmacol Biochem Behav 94(2):312–318

Eckert DJ, Owens RL, Kehlmann GB, Wellman A, Rahangdale S, Yim-Yeh S, White DP, Malhotra A (2011) Eszopiclone increases the respiratory arousal threshold and lowers the apnoea/hypopnoea index in obstructive sleep apnoea patients with a low arousal threshold. Clin Sci (Lond) 120(12):505–514

Erman MK, Zammit G, Rubens R, Schaefer K, Wessel T, Amato D, Caron J, Walsh JK (2008) A polysomnographic placebo-controlled evaluation of the efficacy and safety of eszopiclone relative to placebo and zolpidem in the treatment of primary insomnia. J Clin Sleep Med 4 (3):229–234

European Medicines Agency (2009) Withdrawal assessment for Lunivia. International non-proprietary name: Eszopiclone procedure No. EMEA/H/C/000895. Available at: http://www.ema.europa.eu/docs/en_GB/document_library/Application_withdrawal_assessment_report/2010/01/WC500061611.pdf. Accessed 22 Feb 2014

Fava M, McCall WV, Krystal A, Wessel T, Rubens R, Caron J, Amato D, Roth T (2006) Eszopiclone co-administered with fluoxetine in patients with insomnia coexisting with major depressive disorder. Biol Psychiatry 59(11):1052–1060

Fava M, Schaefer K, Huang H, Wilson A, Iosifescu DV, Mischoulon D, Wessel TC (2011a) A post hoc analysis of the effect of nightly administration of eszopiclone and a selective serotonin reuptake inhibitor in patients with insomnia and anxious depression. J Clin Psychiatry 72 (4):473–479

Fava M, Asnis GM, Shrivastava RK, Lydiard B, Bastani B, Sheehan DV, Roth T (2011b) Improved insomnia symptoms and sleep-related next-day functioning in patients with comorbid major depressive disorder and insomnia following concomitant zolpidem extended-release 12.5 mg and escitalopram treatment: a randomized controlled trial. J Clin Psychiatry 72 (7):914–928

Greenblatt DJ, Zammit GK (2012) Pharmacokinetic evaluation of eszopiclone: clinical and therapeutic implications. Expert Opin Drug Metab Toxicol 8(12):1609–1618

Hanson SM, Morlock EV, Satyshur KA, Czajkowski C (2008) Structural requirements for eszopiclone and zolpidem binding to the gamma-aminobutyric acid type-A (GABAA) receptor are different. J Med Chem 51(22):7243–7252

Joffe H, Petrillo L, Viguera A, Koukopoulos A, Silver-Heilman K, Farrell A, Yu G, Silver M, Cohen LS (2010) Eszopiclone improves insomnia and depressive and anxious symptoms in perimenopausal and postmenopausal women with hot flashes: a randomized, double-blinded, placebo-controlled crossover trial. Am J Obstet Gynecol 202(2):171.e1–171.e11

Joya FL, Kripke DF, Loving RT, Dawson A, Kline LE (2009) Meta-analyses of hypnotics and infections: eszopiclone, ramelteon, zaleplon, and zolpidem. J Clin Sleep Med 5(4):377–383

Kripke DF (2008) Possibility that certain hypnotics might cause cancer in skin. J Sleep Res 17 (3):245–250

Kripke DF, Langer RD, Kline LE (2012) Hypnotics' association with mortality or cancer: a matched cohort study. BMJ Open 2(1):e000850. doi:10.1136/bmjopen-2012-000850 (Print 2012)

Krystal AD, Walsh JK, Laska E, Caron J, Amato DA, Wessel TC, Roth T (2003) Sustained efficacy of eszopiclone over 6 months of nightly treatment: results of a randomized, double-blind, placebo-controlled study in adults with chronic insomnia. Sleep 26(7):793–799

Krystal A, Fava M, Rubens R, Wessel T, Caron J, Wilson P, Roth T, McCall WV (2007) Evaluation of eszopiclone discontinuation after cotherapy with fluoxetine for insomnia with coexisting depression. J Clin Sleep Med 3(1):48–55

Krystal AD, Huang H, Zummo J, Grinnell T, Marshall RD (2012a) A WASO sub-group analysis of a 6-month study of eszopiclone 3 mg. Sleep Med 13(6):691–696

Krystal AD, McCall WV, Fava M, Joffe H, Soares CN, Huang H, Grinell T, Zummo J, Spalding W, Marshall R (2012b) Eszopiclone treatment for insomnia: effect size comparisons in patients with primary insomnia and insomnia with medical and psychiatric comorbidity. Prim Care Companion CNS Disord. 14(4). pii:PCC.11m01296

Lettieri CJ, Quast TN, Eliasson AH, Andrada T (2008) Eszopiclone improves overnight polysomnography and continuous positive airway pressure titration: a prospective, randomized, placebo-controlled trial. Sleep 31(9):1310–1316

Lettieri CJ, Shah AA, Holley AB, Kelly WF, Chang AS, Roop SA (2009a) Effects of a short course of eszopiclone on continuous positive airway pressure adherence. Ann Intern Med 151 (10):696–702

Lettieri CJ, Collen JF, Eliasson AH, Quast TM (2009b) Sedative use during continuous positive airway pressure titration improves subsequent compliance: a randomized, double-blind, placebo-controlled trial. Chest 136(5):1263–1268

McCall C, McCall WV (2012) Objective vs. subjective measurements of sleep in depressed insomniacs: first night effect or reverse first night effect? J Clin Sleep Med 8(1):59–65

McCall WV, Erman M, Krystal AD, Rosenberg R, Scharf M, Zammit GK, Wessel T (2006) A polysomnography study of eszopiclone in elderly patients with insomnia. Curr Med Res Opin 22(9):1633–1642

McCall WV, Blocker JN, D'Agostino R Jr, Kimball J, Boggs N, Lasater B, Haskett R, Krystal A, McDonald WM, Rosenquist PB (2010) Treatment of insomnia in depressed insomniacs: effects on health-related quality of life, objective and self-reported sleep, and depression. J Clin Sleep Med 6(4):322–329

Menza M, Dobkin RD, Marin H, Gara M, Bienfait K, Dicke A, Comella CL, Cantor C, Hyer L (2010) Treatment of insomnia in Parkinson's disease: a controlled trial of eszopiclone and placebo. Mov Disord 25(11):1708–1714

Monti JM, Pandi-Perumal SR (2007) Eszopiclone: its use in the treatment of insomnia. Neuropsychiatr Dis Treat 3(4):441–453

Najib J (2006) Eszopiclone, a nonbenzodiazepine sedative-hypnotic agent for the treatment of transient and chronic insomnia. Clin Ther 28(4):491–516

Najib JS (2010) Efficacy and safety of zopiclone and eszopiclone in the treatment of primary and comorbid insomnia. In: Monti JM, Pandi-Perumal SR, Mohler H (eds) GABA and sleep: molecular, functional and clinical aspects. Springer, Basel, pp 413–452

Nutt DJ, Stahl SM (2010) Searching for perfect sleep: the continuing evolution of GABAA receptor modulators as hypnotics. J Psychopharmacol 24(11):1601–1612

Pollack M, Kinrys G, Krystal A, McCall WV, Roth T, Schaefer K, Rubens R, Roach J, Huang H, Krishnan R (2008) Eszopiclone coadministered with escitalopram in patients with insomnia and comorbid generalized anxiety disorder. Arch Gen Psychiatry 65(5):551–562

Pollack MH, Hoge EA, Worthington JJ, Moshier SJ, Wechsler RS, Brandes M, Simon NM (2011) Eszopiclone for the treatment of posttraumatic stress disorder and associated insomnia: a randomized, double-blind, placebo-controlled trial. J Clin Psychiatry 72(7):892–897

Richey SM, Krystal AD (2011) Pharmacological advances in the treatment of insomnia. Curr Pharm Des 17(15):1471–1475

Rosenberg R, Caron J, Roth T, Amato D (2005) An assessment of the efficacy and safety of eszopiclone in the treatment of transient insomnia in healthy adults. Sleep Med 6(1):15–22

Rosenberg R, Roach JM, Scharf M, Amato DA (2007) A pilot study evaluating acute use of eszopiclone in patients with mild to moderate obstructive sleep apnea syndrome. Sleep Med 8 (5):464–470

Roth T, Walsh JK, Krystal A et al (2005) An evaluation of the efficacy and safety of eszopiclone over 12 months in patients with chronic primary insomnia. Sleep Med 6:487–495

Roth T, Price JM, Amato DA, Rubens RP, Roach JM, Schnitzer TJ (2009) The effect of eszopiclone in patients with insomnia and coexisting rheumatoid arthritis: a pilot study. Prim Care Companion J Clin Psychiatry 11(6):292–301

Schaffer CB, Schaffer LC, Miller AR, Hang E, Nordahl TE (2011) Efficacy and safety of nonbenzodiazepine hypnotics for chronic insomnia in patients with bipolar disorder. J Affect Disord 128(3):305–308

Scharf M, Erman M, Rosenberg R, Seiden D, McCall WV, Amato D, Wessel TC (2005) A 2-week efficacy and safety study of eszopiclone in elderly patients with primary insomnia. Sleep 28 (6):720–727

Sieghart W (2006) Structure, pharmacology, and function of $GABA_A$ receptor subtypes. Adv Pharmacol 54:231–263

Soares CN, Joffe H, Rubens R, Caron J, Roth T, Cohen L (2006) Eszopiclone in patients with insomnia during perimenopause and early postmenopause: a randomized controlled trial. Obstet Gynecol 108(6):1402–1410

Stebbing J, Waters L, Davies L, Mandalia S, Nelson M, Gazzard B, Bower M (2005) Incidence of cancer in individuals receiving chronic zopiclone or eszopiclone requires prospective study. J Clin Oncol 23(31):8134–8136

Sunovion (2014) Lunesta prescribing information, Marlborough, MA, May 2014. Available at: http://www.lunesta.com/pdf/PostedApprovedLabelingText.pdf. Accessed 14 Oct 2014

Uchimura N, Kamijo A, Kuwahara H, Uchiyama M, Shimizu T, Chiba S, Inoue Y (2012a) A randomized placebo-controlled polysomnographic study of eszopiclone in Japanese patients with primary insomnia. Sleep Med 13(10):1247–1253

Uchimura N, Kamijo A, Takase T (2012b) Effects of eszopiclone on safety, subjective measures of efficacy, and quality of life in elderly and nonelderly Japanese patients with chronic insomnia, both with and without comorbid psychiatric disorders: a 24-week, randomized, double-blind study. Ann Gen Psychiatry 11(1):15–29

Walsh JK, Krystal AD, Amato DA, Rubens R, Caron J, Wessel TC, Schaefer K, Roach J, Wallenstein G, Roth T (2007) Nightly treatment of primary insomnia with eszopiclone for six months: effect on sleep, quality of life, and work limitations. Sleep 30(8):959–968

Wamsley EJ, Shinn AK, Tucker MA, Ono KE, McKinley SK, Ely AV, Goff DC, Stickgold R, Manoach DS (2013) The effects of eszopiclone on sleep spindles and memory consolidation in schizophrenia: a randomized placebo-controlled trial. Sleep 36(9):1369–1376

Zammit GK, McNabb LJ, Caron J, Amato DA, Roth T (2004) Efficacy and safety of eszopiclone across 6-weeks of treatment for primary insomnia. Curr Med Res Opin 20(12):1979–1991

Orexin/Hypocretin Antagonists in Insomnia: From Bench to Clinic

Alberto K. De la Herrán-Arita, Ana C. Equihua-Benítez, and René Drucker-Colín

Abstract Insomnia is a common clinical condition portrayed by difficulty in initiating or maintaining sleep, or non-restorative sleep with impairment of daytime functioning, such as irritability or fatigue during wakefulness. This ailment is one of the most rampant health concerns; however, it represents an everyday struggle to clinicians because of its many potential causes, unfamiliarity with behavioral treatments, and concerns about pharmacologic treatments. The etiology and pathophysiology of insomnia involve genetic, environmental, behavioral, and physiological factors culminating in hyperarousal.

Current pharmacological treatment for insomnia exists in the form of benzodiazepine receptor agonist drugs (GABA-A receptor). Nonetheless, the use of these hypnotic medications must be carefully monitored for adverse effects and concerns persist regarding their safety and limited efficacy.

The recent advances made in elucidating the processes of sleep/wake regulation have altered the way that insomnia is approached. Current studies have highlighted new targets for drug discovery. One of the most promising ones is the orexin (hypocretin) system. Orexin neuropeptides regulate transitions between wakefulness and sleep by promoting arousal through activation of cholinergic/monoaminergic neural pathways. This has led to a swift development of a novel class of drugs that antagonize the physiological effects of orexin. These pharmacological agents may lead to new therapies for insomnia without the side effect profile of benzodiazepines (e.g., impaired cognition, disturbed arousal, and motor balance difficulties); however, antagonizing the orexin system may produce an entirely new plethora of side effects.

Despite the impending side effect profile of orexin antagonists, these drugs will inevitably supplement or replace conventional BZD receptor agonists for treating insomnia.

A.K. De la Herrán-Arita
Center for Sleep Sciences and Medicine, Stanford School of Medicine, Palo Alto, CA 94304, USA

A.C. Equihua-Benítez • R. Drucker-Colín (✉)
Departamento de Neurociencias, Instituto de Fisiología Celular, Universidad Nacional Autónoma de México, Circuito exterior S/N, Apdo., Postal 70–600, 04510 México, D.F., Mexico
e-mail: drucker@unam.mx

A. Guglietta (ed.), *Drug Treatment of Sleep Disorders*, Milestones in Drug Therapy,
DOI 10.1007/978-3-319-11514-6_9

In this chapter, we will appraise the role of orexin antagonists for the treatment of insomnia.

1 Introduction

Pharmacological treatment is often characterized by ambiguities regarding diagnosis, selection of adequate treatment, and assessment of outcomes for individual patients. Likewise, multiple efficacious treatments are often available for a particular condition. This is indeed true for insomnia, with multiple behavioral and pharmacologic treatments demonstrating short-term and long-term efficacy data. However, what is considerably more challenging is trying to determine which treatment is best suited for which particular patient, based on the characteristics of their disorder, and the availability of treatments.

During the past half century, the pharmaceutical industry has invested a great deal of effort in the approval and usage of the very popular central nervous system (CNS) agents that enhance signaling of γ-aminobutyric acid (GABA), an inhibitory neurotransmitter (Zisapel 2012). These endeavors have succeeded in ascertaining the popularity of the GABA compounds as sleep aids; nevertheless, there are several issues that still raise many questions as to their use, namely, the fact that all of these compounds are classified as controlled substances and the label change that was required by the Food and Drug Administration (FDA).

Over the past several years, a novel mechanism for modulating sleep that has gained significant popularity is the orexin/hypocretin system, due to its highly conserved nature and its ability to regulate arousal and wakefulness (Fig. 1).

Orexins (also known as hypocretins) were first described in 1999 (de Lecea et al. 1998; Peyron et al. 1998; Sakurai et al. 1998) and, shortly, their absence was associated with the development of the sleep disorder narcolepsy (Chemelli et al. 1999; Lin et al. 1999; Thannickal et al. 2000). Since then, orexins have been intensely studied for their role in the sleep–wake cycle (SWC) primarily as wake-promoting neurotransmitters.

Orexin-producing neurons are found in the lateral hypothalamus (LH), where they synthesize two excitatory neuropeptides called orexin A and B (alternatively known as hypocretin 1 and 2) cleaved from a common protein precursor called prepro-orexin (prepro-hypocretin) (Sakurai et al. 1998). Orexinergic neurons extensively innervate the CNS, specifically areas known for their role in promoting arousal like the locus coeruleus (LC), the tuberomamillary nucleus (TMN), the basal forebrain (BF), the cerebral cortex, and the dorsal raphe (DR) (Peyron et al. 1998). Orexins exert their actions through their interaction with two G protein-coupled receptors called OX1R and OX2R (hcrtR1 and hcrtR2, respectively). These receptors have different affinities for the orexin peptides, while orexin A binds both receptors, orexin B selectively binds to OX2R (Sakurai et al. 1998).

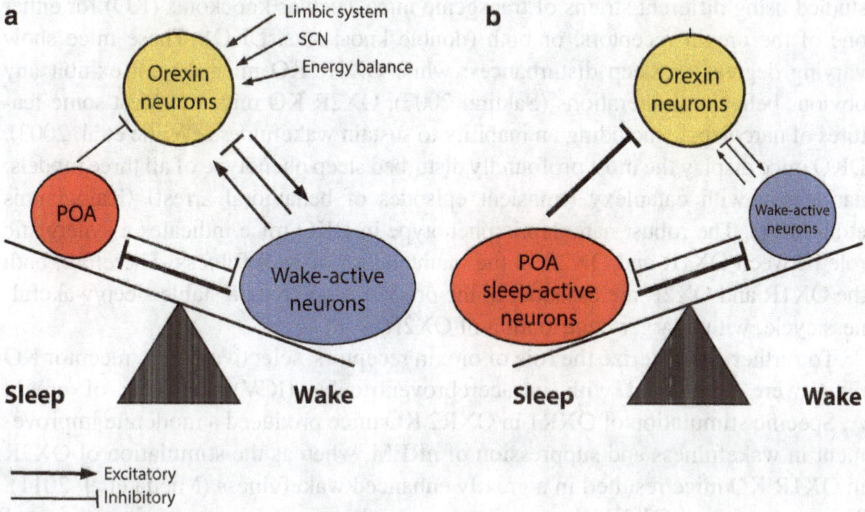

Fig. 1 Orexinergic control of sleep and wakefulness. (**a**) Awake state. Orexin neurons send excitatory projections to wake-active neurons, which send inhibitory feedback projections to orexin neurons. This system might maintain the activity of wake-active neurons. A small decrease in the activity of wake-active neurons results in decreased inhibitory influence to orexin neurons. Orexin neurons, therefore, are disinhibited and increase their excitatory influence on wake-active neurons to maintain their activity. These wake-active neurons send inhibitory projections to the preoptic area (POA) sleep center and excitatory projections to the thalamus and cerebral cortex. (**b**) Sleep state. GABAergic neurons in the sleep center are activated and send inhibitory projections to wake-active neurons and orexin neurons to maintain a sleep state. The figures represent the functional interaction between orexin neurons, wake-active centers, and sleep-active centers during various states of sleep and wakefulness. *Arrows* show excitatory and *lines* show inhibitory input. The *thickness of arrows* and *lines* represent relative strength of input. *Circle sizes* represent relative activities of each group of neurons. Modified from Lu and Zee, Clin Chest Med 31 (2010) 309–318

Orexin-excited neurons become more excitable through an inhibition of potassium channels, including GIRK (G protein-regulated inward rectifier) channels; in addition, signaling through the orexin receptors can induce a rapid and sustained rise in intracellular calcium through voltage-gated calcium channels, through transient receptor potential channels, or from intracellular stores (Mieda et al. 2013). In addition to these postsynaptic effects, orexins can also act presynaptically on nerve terminals to induce release of GABA or glutamate, thus generating more complicated effects on downstream neurons (Mieda et al. 2013). Through these many mechanisms, orexins are thought to generally excite neurons that promote many aspects of arousal.

Furthermore, orexin receptors are differentially located throughout the CNS; the LC mainly expresses OX1R, the TMN exclusively expresses OX2R, while the DR, BF, and cerebral cortex express both receptors (Marcus et al. 2001; Mieda et al. 2013). The divergent distribution and affinities of orexin receptors suggest they play different functions in the conservation of wakefulness. This has been studied using different strains of transgenic mice, such as knockouts (KO) for either one of the orexin receptors, or both (double knockouts, DKO). These mice show varying degrees of sleep disturbances; while OX1R KO mice do not exhibit any obvious behavioral alterations (Sakurai 2007), OX2R KO mice manifest some features of narcolepsy, including an inability to sustain wakefulness (Willie et al. 2003). DKO mice display the most profoundly disturbed sleep phenotype of all three models: narcolepsy with cataplexy (transient episodes of behavioral arrest) (Kalogiannis et al. 2011). The robust narcoleptic phenotype in DKO mice indicates a synergistic role between OX1R and OX2R in the maintenance of wakefulness. Therefore, both the OX1R and OX2R are essential in the process of keeping a stable sleep/wakefulness cycle, with a larger contribution of OX2R.

To further characterize the role of orexin receptors, selective orexin receptor KO mice were stimulated with intracerebroventricular (ICV) infusions of orexin A. Specific stimulation of OXR1 in OXR2 KO mice produced a moderate improvement in wakefulness and suppression of nREM, whereas the stimulation of OX2R in OX1R KO mice resulted in a greatly enhanced wakefulness (Mieda et al. 2011). This suggests that OX1R plays an important role in suppressing the instigation of nREM sleep, while OX2R has a major role in promoting wakefulness.

In another direction, overexpression of components of the orexinergic system also disrupts the SWC. For example in the zebrafish, overexpression of orexinergic neurons has been shown to induce an insomnia-like phenotype (Prober et al. 2006). Mice that overexpress prepro-orexin display sleep abnormalities, which include fragmentation of nREM sleep, reduced REM sleep, and increased motor activity during REM sleep, suggesting an inability to maintain sleep states.

The excitatory effects of the orexin peptides on these arousal centers are hypothesized to stabilize and maintain wakefulness (Fig. 1) and if we take into consideration that the activation of the orexinergic system promotes wakefulness and that its disruption brings about sleep disturbances, orexin antagonists could offer a very effective therapeutic alternative for insomnia.

Given that orexins essentially promote arousal, orexin antagonists have the potential to selectively promote sleep and cause fewer side effects. Dependence and abuse should be less of a concern, as animal studies have shown that orexin antagonists actually reduce drug seeking. Imbalance and falls should not be a problem, as there is no evidence that the orexin system affects balance or gait directly. Consequences of an overdose should not be too concerning, as orexin antagonists should not significantly depress respiration or affect blood pressure.

Potential side effects of orexin antagonists, including disinhibition of REM sleep, are discussed below, but overall, many researchers anticipate that these drugs should promote sleep without many of the side effects encountered with current medications.

2 Orexin Receptor Antagonism: A Novel Approach for Treating Insomnia

The newest molecules in the pipeline for the treatment of insomnia are orexin antagonists. Drugs targeting insomnia ideally promote sleep throughout the night, maintain normal sleep architecture, and are devoid of residual effects associated with morning sedation. These features of an ideal compound are not only dependent upon pharmacokinetics, receptor binding kinetics, potency, and pharmacodynamic activity, but also upon a compound's mechanism of action. There are many orexin antagonists currently being studied for the treatment of insomnia and they fall into one of two categories: single orexin receptor antagonists (SORAs) and dual orexin receptor antagonists (DORAs).

2.1 Single Orexin Receptor Antagonists (SORAs)

Evidence from experiments conducted in transgenic models of orexin receptor KO mice suggests that SORAs targeting OX1R will not promote sleep as effectively as those aimed at OX2R; nonetheless, OX1R antagonism could serve as a complementary treatment for insomnia or other sleep-related disorders.

2.1.1 OX1R

Of the available SORAs, SB-334867 was the first drug designed to selectively antagonize OX1R. This SORA is able to counteract the suppression of REM sleep after ICV infusion of OXA in rats (Smart et al. 2001; Smith et al. 2003); however, it does not decrease wakefulness, or increase the amount of time spent in sleep, nor does it reduce sleep latency by itself at any given dose. Also, SB-334867 at 3 and 30 mg/kg increased cumulative nREM during the first 4 and 6 h following administration (Morairty et al. 2012). SB-334867 is classified as a selective OX1R antagonist, but unspecific binding to adenosine and serotonin receptors has been reported; it also affects monoamine and norepinephrine transporters at high concentrations (Lebold et al. 2013).

Although the effect of SB-334867 on sleep induction was poor, this molecule has proven to be useful for the treatment of other conditions, such as substance abuse, withdrawal, obesity, and panic disorder (White et al. 2005; Johnson et al. 2010; Jupp et al. 2011; Smith and Aston-Jones 2012).

Other selective OX1R antagonists include SB-408124, SB-674042, and the newest AK-335827. So far, neither SB-408124 nor AK-335827 has been found to promote sleep (Dugovic et al. 2009; Steiner et al. 2013). In the case of SB-408124 however, insufficient brain penetration was observed and this could account in part for the absence of observable effects (Morairty et al. 2012).

Table 1 Summary of orexin receptor antagonists

Compound	Affinity (K_i, nM)			Potential applications
	OX1R	OX2R		
SORAs			Selectivity	
SB-334867	28	1704	OX1R	Withdrawal, substance abuse, obesity, panic disorder
SB-408124	22	1405		
SB-674042	1.1	129		
ACT-335827	6	417 (IC_{50})		
TCS-OX2-29	–	7.4 (pKi)	OX2R	Sleep promotion
JNJ-OX2-29	1644	6		
EMPA	900	1.1		
Antagonist 26	6.34	7.23 (pKi)		
DORAs			FDA phase	
Almorexant	13	8	III (discontinued)	Treatment of insomnia
SB-649868	0.3	0.4	II (completed)	
Suvorexant	0.6	0.4	III (pending approval)	
MK-6096	2.5	0.3	–	–
DORA 30	18	7 (IC_{50})	–	Sleep promotion

Modified from Equihua-Benítez et al., Front Pharmacol, 2013; 4:163

There are few studies characterizing the effect of these antagonists; nonetheless, there is some evidence that they can be useful in the treatment of substance abuse and withdrawal, and have potential for treating obesity and panic disorder (Table 1). For example, it has been shown that subcutaneous administration of SB-408124 lowers the release of dopamine in the nucleus accumbens (Dugovic et al. 2009), and orally administered AK-335827 has anxiolytic effects (Steiner et al. 2013).

It is interesting that despite the lack of sleep-promoting effects of OX1R SORAs on their own, these compounds have the capacity to thwart the sleep-inhibiting effects of ICV orexin infusion (Smith et al. 2003). Strikingly, they can also reduce the sleep-promoting effects of other antagonists, as observed under the co-administration of OX1R and OX2R antagonists, which has a milder sleep-promoting effect than when the OX2R antagonist is administered by itself (Dugovic et al. 2009). This could be due to the high concentrations used in these experiments (30 mg/kg) and the unspecific binding that follows.

2.1.2 OX2R

Type 2 orexin receptors are selectively expressed in both the paraventricular nucleus (PVN) and the TMN. The PVN is part of the hypothalamic-pituitary-adrenal axis (HPA), and the overactivation of the HPA has been proposed to be involved in the etiology of primary insomnia. Withholding the orexinergic stimuli to the HPA could help prevent the development of a vicious cycle of overactivation that could lead to chronic insomnia. Additionally, the TMN, a histaminergic nucleus, has a major role in the arousal effect observed after orexinergic stimulation (Huang et al. 2001). Inhibition of the TMN with orexinergic antagonists could facilitate the induction of sleep by allowing the sleep-promoting nuclei to prevail.

OX2R antagonists are less common than the other classes. Among the few available molecules that have been studied in the context of sleep promotion are EMPA, TCS-OX2-29, and JNJ-10397049. These antagonists have been more successful at diminishing wakefulness than OX1R antagonists (Table 1).

EMPA is the least effective sleep-promoting OX2R SORA ever studied. While intraperitoneal administration of EMPA (100 mg/kg) has been shown to selectively increase cumulative nREM sleep during the first 4 and 6 h after administration, these increases are not accompanied by any significant increase in REM sleep or reduction in latencies for either sleep stage (Morairty et al. 2012). On the other hand, rats that received an ICV infusion of TCS-OX2-29 (40 nmol) increased their total sleep time by 7 % in comparison to controls that received saline infusions. Interestingly, this effect was secondary to a selective increase in REM sleep (Kummangal et al. 2013).

Intraperitoneal administration (5, 25, or 50 mg/kg) of JNJ-10397049 6 h into the dark phase produced a robust increase in total sleep time, traced to increases in both REM and nREM sleep (Gozzi et al. 2011). Similar results have been observed with subcutaneous injections (Dugovic et al. 2009). Starting at doses of 3 mg/kg, administration of JNJ-10397049 2 h into the light phase significantly decreased the latency to nREM sleep while increasing the length of each bout. At higher concentrations (30 mg/kg), this drug also induced a decrease in REM sleep latency without noticeable changes in its duration. Overall, 3 mg/kg of JNJ-10397049 increased total sleep time by 42 % while keeping the proportion of nREM/REM sleep observed in vehicle-treated animals.

Furthermore, microdialysis assays showed that this compound reduces histamine release in the LH. As mentioned earlier, release of histamine in the TMN is fundamental for the wake-promoting effects of OXA ICV infusions (Dugovic et al. 2009; Huang et al. 2001).

Animal studies support the notion that OX2R antagonists are helpful as sleep-inducing agents (Table 1). Further research is needed to determine the degree of sleep generation achieved by these compounds in different species, including humans. It is possible that the sleep-promoting effect of selectively antagonizing OX2R is less pronounced than the one observed with DORAs, but it may also be more specific, which would be worth investigating.

2.2 *Dual Orexin Receptor Antagonists (DORAs)*

It had been long suspected that antagonizing both orexin receptors would elicit the most powerful sleep-promoting effects; therefore, many of the studies around orexin antagonists have focused on DORAs. So far, evidence has proven this to be the case, to the point that DORAs are the only orexin antagonists currently undergoing clinical trials in the hope that the FDA will approve them for the treatment of insomnia.

2.2.1 Almorexant

ACT-078573 (almorexant) is the most widely studied DORA and one of the first to enter phase III clinical trials (NCT00608985). Almorexant is a reversible, selective dual OX1 and OX2 receptor antagonist.

In wild-type mice, the administration of almorexant 15 min before lights-out reduced the amount of time spent awake, while increasing the length of nREM and REM sleep bouts in a dose-dependent manner (Mang et al. 2012). Notably, the proportion of REM sleep observed after almorexant administration during the dark phase was in the range of that observed during the light phase with vehicle treatment.

Further studies in KO mice determined that the sleep-inducing effect of almorexant was related to the stimulation of OX2R and not OX1R. This conclusion was reached after the authors did not observe any changes in the amount of sleep in OX2R KO, but did for OX1R KO mice. Interaction with sites other than orexin receptors that could account for the changes in sleep times was discarded when no changes were observed in the SWC of DKO mice.

When administered in healthy humans, almorexant increased deep sleep and REM sleep (unlike GABA receptor modulators which decrease REM sleep). In a Phase II study of 161 patients with primary insomnia, almorexant 400 mg significantly reduced LPS (mean treatment effect—18 min; $p = 0.02$) and WASO (mean treatment effect—54 min; $p < 0.001$) compared to placebo. Dose-related increases in TST were found in all (400, 200, and 100 mg) almorexant groups compared with placebo. The most commonly reported adverse effects in the almorexant-treated patients were dizziness, nausea, fatigue, headache, and dry mouth. Almorexant had no effect on subjective WASO at any dose, but objective WASO decreased in a dose-dependent fashion. The higher almorexant doses (400 and 200 mg) were associated with some residual effects on next-day performance as evidenced by increased mean reaction times and lower scores on the digit span test. At doses up to 1,000 mg, there were no reports of cataplexy or narcolepsy in almorexant-treated patients (Hoever et al. 2012).

Actelion's first Phase III study of almorexant (RESTORA 1) was a 2-week randomized, double-blind trial in 707 patients with primary insomnia. Patients were randomized to almorexant 100 mg, almorexant 200 mg, placebo, or zolpidem

Table 2 Dual orexin receptor antagonists (DORAs)

Compound	ACT-078573 (Almorexant)	MK-4305 (Suvorexant)	SB-649868
Company	Actelion & GlaxoSmithKline	Merck & Co.	GlaxoSmithKline
Clinical Trial	Phase III, discontinued	Phase III completed, NDA submitted Nov 2012, denied FDA approval July 2013	Phase II completed
Indications	Insomnia	Insomnia	Insomnia
$T_{1/2}$ (hours)	21	12	3–6
Metabolism	Hepatic CYP3A4 (S/I)	Hepatic CYP3A4 (S)	Hepatic CYP3A4 (S/I)

S substrate, *I* inhibitor, *NA* not available

10 mg (active comparator). The primary endpoint was change from baseline to Day 1 and 2 in WASO. Both almorexant groups showed statistically significant improvements in wake after sleep onset with median reductions of 29 and 40 min in the 100 and 200 mg groups, respectively, versus 11 min in the placebo group (Renzulli et al. 2011).

Doses of and above 200 mg elicited decreased alertness, with increased reports of fatigue, drowsiness, sleepiness, and sleep efficiency, measured as an increase in SWS and REM sleep (Brisbare-Roch et al. 2007). In patients with primary insomnia, it proved to be effective for boosting sleep, increasing total sleep time, and reducing both REM sleep latency and the frequency of awakening (Hoever et al. 2012). This effect was dose dependent, with the most notorious effect on sleep architecture achieved at doses of 400 mg; doses of 100 and 200 mg had modest effects on sleep, with fewer adverse effects (e.g., headache, dizziness, blurred vision).

Although almorexant was generally well tolerated in this study, safety signals led to further investigations of the clinical safety of almorexant. These expanded studies led to the discontinuation of almorexant development for undisclosed human tolerability issues (Tables 1 and 2). Currently, almorexant is in a new phase of clinical trials in order to evaluate its effect on cognitive performance (NCT01243060).

2.2.2 SB-649868

SB-649868 is a potent orally active DORA manufactured by the same pharmaceutical company as almorexant. There is also evidence for the effectiveness of SB-649868 in promoting sleep, in both animal studies and human trials.

When administered to rats, it elicited an increase in total sleep time (related to increases of both nREM and REM sleep) and reduced sleep latencies at doses of 10 and 30 mg. Moreover, the effect of SB-649868 on motor coordination was null, given that the rotarod model of coordination failed to reveal any motor impairment in rats treated with this compound, even when the orexin antagonist was administered concurrently with ethanol (Di Fabio et al. 2011). Compared to almorexant, the in vivo efficacy of this compound is excellent; thus it has been moved on to clinical trials.

The administration of SB-649868 to healthy volunteers who participated in a noise-disturbed sleep study showed that this compound is effective at inducing somnolence and fatigue at 10 and 30 mg doses (Bettica et al. 2012). Furthermore, patients diagnosed with PI reported that SB-649868 significantly improved the quality of sleep (10, 30, and 60 mg) while objectively increasing total sleep time, reducing sleep latency, and suppressing nighttime awakenings. During this study, the most common complaints were headaches, dry mouth, and nasopharyngitis; the number of complaints increased in a dose-dependent manner. Phase II clinical trials of SB-649868 have been completed (NCT00426816) (Table 1).

2.2.3 Suvorexant

Another promising DORA is the potent suvorexant (MK-4305, Merk), a compound variation from the diazepam series.

Animal studies have shown that suvorexant reduces active wake time by increasing nREM and REM sleep in rats, dogs, and monkeys. In all cases, these effects were achieved at much lower doses (10 mg) than with almorexant (Winrow et al. 2011).

It has been determined that its median peak plasma concentrations occur approximately 2 h after administration and are not affected by food. It has a volume of distribution of 105.9 L and is highly protein bound (99.5 %). It is primarily metabolized by the cytochrome P450 (CYP3A4) enzyme system, with some contribution from CYP2C19 into M9, an inactive metabolite. Finally, it is eliminated primarily via inactive metabolites in the feces; there is no renal elimination. The half-life is approximately 12.2 h on average (range 8–19 h) (Table 2). Steady-state plasma concentrations occur in about 3 days with daily administration (Sun et al. 2013).

This molecule is currently under evaluation for approval by the FDA. Merck's application to the FDA for approval included 32 studies that enrolled more than 900 subjects (healthy and those with insomnia).

Healthy participants received one of three doses of suvorexant (10, 50, or 100 mg) or placebo and were evaluated in two 8-h (polysomnography) PSG recording sessions in a general laboratory setting. Different parameters were analyzed in order to assess the efficacy of suvorexant administration; these included: sleep onset, or latency to persistent sleep (LPS); sleep maintenance, or waking after sleep onset (WASO); sleep efficiency (SE); and total sleep time (TST).

The lowest dose (10 mg) reduced the number of awakenings after sleep onset, and at higher doses (50 mg) it reduced sleep latency, while increasing SE and TST. High doses (50 and 100 mg) elicit undesirable side effects such as an increase in reaction time, difficulty waking up, and reduced alertness following awakening; in addition, it led to mild complaints of headaches and somnolence (Sun et al. 2013).

With the purpose of assessing the effectiveness of orexin receptor antagonism as a novel approach for treating insomnia, men and women, 18–64 years of age with primary insomnia, based on Diagnostic and Statistical Manual of Mental Disorders, 4th edition, text revision (DSM-IV-TR) criteria, were enrolled in this clinical study. They were given 10- and 30-mg tablets, along with their corresponding placebos, in order to achieve doses of 10, 20, 40, and 80 mg per dose.

Suvorexant reduced sleep latency and increased the time patients spent asleep after a single administration without reducing the number of awakenings after sleep onset. The increase in total sleep time was mostly attributable to an increase in REM sleep (Herring et al. 2012). The most frequent adverse effects were somnolence, headaches, dizziness, and abnormal dreams, all of which occurred in a dose-dependent manner. In addition, there were no next-day residual effects, no rebound insomnia, complex sleep-related behaviors, or withdrawal effects after 4 weeks. Instead, during this study there were a few reports of sleep paralysis ($n = 59$, at 40 mg), and at high doses (80 mg), excessive daytime sleepiness ($n = 61$) and hypnagogic hallucinations ($n = 61$) (Herring et al. 2012). These are symptoms of narcolepsy, and should be carefully monitored due to the close association between narcolepsy and the orexinergic system.

In general, suvorexant was well tolerated and, because the most consistently effective dosages were 30 and 40 mg, the pharmaceutical company manufacturing suvorexant submitted a dose range of 15–40 mg for FDA approval. However, the FDA requested a lower starting dose of 10 mg for the general population and a 5 mg dose for those taking concomitant CYP3A4 inhibitors (Table 2). Patients should avoid taking other CYP3A medications while they are taking suvorexant. Potent CYP3A inhibitors increase plasma concentrations, placing patients well above the desired therapeutic threshold. Suvorexant is a mild inhibitor of CYP3A, but when administered with CYP3A substrates, including oral contraceptives and warfarin, it had minimal effects. As of August 2014, the FDA approved suvorexant (Belsomra, Merk) in four different strengths -5, 10, 15, and 20 mg.

An advantage of suvorexant over previous insomnia therapies is the low potential for addiction or dependence.

2.3 *Advantages of DORAs over Other Pharmacological Alternatives*

Based on these preclinical evaluations of structurally divergent DORAs, suvorexant exhibits a pharmacokinetic profile predicted to be advantageous for the treatment of insomnia. Analysis of the time course of suvorexant plasma concentrations determined in an early Phase I clinical trial indicates that concentrations sufficient for efficacy in humans are restricted to normal sleep periods while efficacy is maintained for a range of doses. Given their high OX2R occupancy threshold for efficacy, DORAs require a plasma concentration sufficient to maintain occupancy in order to promote sleep throughout the resting period in humans (Table 2). For suvorexant, this is achieved with a plasma concentration T1/2 exceeding 8 h.

Although insomnia is a common disorder, its underlying mechanisms are still not fully understood. DORAs have been rationally designed specifically to promote sleep by blocking wakefulness and thus alleviate the symptoms of insomnia while minimizing potential off-target activity that occurs with widespread signaling via GABA-A receptor modulators. By using antagonist compounds like suvorexant to block the arousal-promoting effects of orexin signaling in the brain, the oscillation in endogenous orexin pathway activation (highest signaling during active periods and diminished signaling during normal sleep times) is modified to mimic more closely the expected physiological state in normal sleep and wakefulness (Fig. 1).

One potential advantage of DORAs over classic insomnia treatments, such as benzodiazepines, is the possibility of inducing a more physiological sleep (Lanoir and Killam 1968; Borbély et al. 1985; Gaillard et al. 2009). For instance, while DORAs enhance REM sleep, benzodiazepines have proven to suppress this sleep stage. In addition, orexin antagonists appear to have a better side effect profile, with mild complaints of headaches and dizziness being the most common. The only exception appears to be almorexant, given the surprising suspension of clinical trials. Although the reasons for halting clinical trials have not been disclosed to the public, it is conceivable that the high doses required to achieve therapeutic effects could also cause more severe adverse effects, not observed in other drugs that require doses ten times smaller.

2.4 *Considerations for Administering Orexin Antagonists*

One of the most important questions when characterizing an orexin antagonist is whether or not it elicits narcoleptic symptoms. Thus far, orexin antagonists have not been observed to cause cataplexy in animal models or in human patients. Up until now, reports of human patients complaining of sleep paralysis or hypnagogic hallucinations have been scarce, only occurring with high doses of suvorexant. As clinical trials progress, medical practitioners should still be on the alert for symptoms of narcolepsy.

Also, because orexin antagonists have a novel mechanism of action, they have the potential to improve insomnia in patients who have found other agents ineffective. Clinical studies now under way should better define the benefits and limitations of orexin receptor antagonists.

Thus far, little information is publicly available on the adverse effects of orexin antagonists, but on the basis of the available evidence and the predicted effects of orexin blockade, one may anticipate that orexin antagonists will have a better adverse event profile than many currently available hypnotics. Orexin antagonists may have their own unique set of concerns, and additional clinical data are needed. However, unlike benzodiazepines, they should have little potential for abuse or unsteady gait, and unlike sedating antidepressants, they are unlikely to cause autonomic side effects such as orthostasis.

Morning or daytime sleepiness may be a concern for drugs that continue to block orexin signaling upon awakening. Possibly, this sleepiness would present differently from how it presents with benzodiazepines because people with narcolepsy often feel alert upon awakening and then sleepy later in the day. Thus, in clinical trials, it will be important to monitor sleepiness during the entire day, not just in the morning.

Unlike other hypnotics, orexin antagonists may cause some dysregulation of REM sleep as is encountered in narcolepsy. Hypnagogic hallucinations and sleep paralysis could occur around the onset of sleep or upon awakening, although they have not been reported. In general, these symptoms are disturbing but fleeting, and they should be manageable with patient education and dose reduction. The potential for cataplexy is a bigger concern because a sudden fall could produce injury, yet people with narcolepsy rarely have cataplexy during their sleep period, and no cataplexy was observed with almorexant in rats, dogs, and humans despite high levels of receptor blockade. However, these studies may have overlooked a tendency for cataplexy because it is usually triggered by strong, positive emotions (e.g., laughing at a great joke) and it is hard to elicit in the lab. In addition, animals and subjects were allowed to sleep in most studies, which could have masked any tendency toward cataplexy. Most likely, cataplexy will be a concern only in unusual circumstances inconsistent with intended clinical use, such as if a patient is wide awake and socializing after taking a high dose of an orexin antagonist.

Considering that orexins promote wakefulness and suppress REM sleep, one might find it odd and surprising that some people with narcolepsy can have moderately fragmented sleep, sleepwalking during nREM sleep, and movements during REM sleep known as REM sleep behavior disorder. Possibly, these symptoms are a consequence of chronic orexin deficiency or injury to neurons other than those producing orexins. As these symptoms are hard to explain with current models, predicting whether they may occur with orexin antagonists is difficult. However, in dogs, almorexant increased twitching of distal parts of the limbs during sleep, and clinical studies should monitor for movements or other disruptions of sleep.

Orexin signaling seems to enhance activity in the mesolimbic pathways that regulate reward and motivation, and reduced activity in this system could worsen mood or motivation. People with narcolepsy may have a higher prevalence of depression, although it is unknown if this is a direct consequence of reduced orexin signaling or a response to the challenge of having a chronic illness. Alternatively, better sleep in patients with depression could improve mood. As many patients with insomnia have depression, clinicians should watch for any changes in mood.

Benzodiazepines can depress respiration and worsen obstructive sleep apnea or severe lung disease, and overdose can be fatal, especially if used in combination with alcohol or other sedatives (Gaillard et al. 2009). As orexin knockout mice have relatively normal baseline ventilation (Kalogiannis et al. 2011), it seems unlikely that orexin antagonists would significantly reduce respiratory drive. However, they may reduce the response to hypercarbia. High levels of CO2 increase respiratory rate and tidal volume, and orexin antagonists can blunt this response, especially during wakefulness. Thus it may be wise to closely evaluate orexin antagonists in patients with hypercarbia, such as individuals with severe chronic obstructive pulmonary disease (COPD) or respiratory muscle weakness.

Whether orexin affects appetite or metabolism in humans remains unclear, but people and mice with narcolepsy tend to be slightly overweight despite apparently eating less than normal. Thus orexin deficiency may lower metabolic rate, and orexin antagonists could promote mild weight gain if administered chronically. In practice, orexin antagonists will be mainly given at night, and normal orexin signaling during the day should offset any reductions in metabolism or hunger at night.

Sedating antidepressants can produce unsteady gait, dizziness, and falls, but these are unlikely to be concerns with orexin antagonists as the cerebellum and vestibular nuclei essentially lack orexin fibers and receptors (Peyron et al. 1998). Orexin knockout mice run at a normal speed, and rats treated with almorexant balance well on a rotating rod. Studies of almorexant in humans have shown only small increases in body sway 1–3 h after dosing, so it seems unlikely that orexin antagonists will substantially increase the risk of falls.

Orexin antagonists should probably be avoided in patients with narcolepsy because they could worsen some of the patients' symptoms. These compounds might also have a higher risk of producing narcolepsy-like effects during the day in other disorders in which the orexin-producing neurons are injured, such as Parkinson's disease and severe traumatic brain injury. Thus in those patients, clinicians might consider initiating treatment with low doses.

Some researchers have questioned whether there is value in producing behavioral effects similar to narcolepsy, but overall, it appears that orexin antagonists should promote sleep with fewer and less harmful side effects than many current hypnotics. Ongoing clinical trials are watching closely for any symptoms related to dysregulation of REM sleep, and watching for fragmented sleep, movements during sleep, and worsening mood will also be important. Daytime sedation should not be much of a concern for compounds with favorable kinetics that permit normal orexin signaling during the day.

In addition, orexin enhances activity in mesolimbic pathways regulating reward and motivation; reduced activity in the orexin system could theoretically worsen mood or motivation. This would dictate cautious use in patients with underlying mood disorders especially since there were reports of suicidal ideation in high-dose suvorexant studies. Orexin receptor antagonists may offer yet another viable option for the pharmacological management of insomnia. Longer-term studies and head-to-head comparisons with existing hypnotics will be crucial to determine the benefit–risk ratio of these agents.

Conclusion

Although benzodiazepines and non-benzodiazepines are effective for insomnia, their adverse effect profiles and recommended limitations on long-term use may prompt patients and clinicians to seek other options. Patients who experience both sleep onset and sleep maintenance insomnia may be particularly challenging to treat. The recent discovery of orexins and their receptors has led to the development of new therapy targets. Evidence suggests that some of these medications offer a sustained benefit for patients with symptoms of chronic insomnia (Suvorexant in particular, which has a unique clinical profile).

We know of no evidence that people with insomnia have abnormally high orexin tone, but under most conditions, any reduction in orexin signaling should make it easier to fall asleep and to return to sleep.

Since orexin activity is highest during active wakefulness, not during sleep periods, these agents may be ineffective for certain types of insomnia. These drugs might be especially effective in shift workers or individuals with jet lag who are trying to sleep during their biological active period when orexin tone is high. They may also be helpful in the many insomnia patients who have high sympathetic tone because high sympathetic tone delays the onset of sleep and sympathetic activation may promote arousal by exciting the orexin neurons. These agents may also be chosen preferentially in the elderly to avoid gait disturbances and confusion and in those with substance abuse histories to avoid dependence and abuse. Benzodiazepines can cause imbalance and confusion, and orexin antagonists may be a good choice for some older patients because they may be less likely to cause these side effects. On the other hand, drugs that block orexin signaling may be less effective in people suffering from insomnia caused by anxiety or pain; there is no evidence that orexin antagonists reduce anxiety or raise sensory thresholds as benzodiazepines do.

The better understanding of the biology of the orexin/hypocretin system has promoted drug discovery programs in several pharmaceutical companies, resulting in a series of patents and compounds with different selectivity and in vitro characteristics. Targeting the orexin receptor is now entering a translational pipeline, and we believe that orexin antagonists as a treatment modality in insomnia management will be the gold standard in the near future, especially for DORAs.

References

Bettica P, Squassante L, Zamuner S, Nucci G, Danker-Hopfe H, Ratti E (2012) The orexin antagonist SB-649868 promotes and maintains sleep in men with primary insomnia. Sleep 35:1097–1104

Borbély AA, Mattmann P, Loepfe M, Strauch I, Lehmann D (1985) Effect of benzodiazepine hypnotics on all-night sleep EEG spectra. Hum Neurobiol 4:189–194

Brisbare-Roch C, Dingemanse J, Koberstein R, Hoever P, Aissaoui H, Flores S et al (2007) Promotion of sleep by targeting the orexin system in rats, dogs and humans. Nat Med 13: 150–155

Chemelli R, Willie J, Sinton CM (1999) Narcolepsy in orexin knockout mice: molecular genetics of sleep regulation. Cell 98:437–451

de Lecea L, Kilduff T, Peyron C, Gao X, Foye P, Danielson P et al (1998) The hypocretins: hypothalamus-specific peptides with neuroexcitatory activity. Proc Natl Acad Sci USA 95: 322–327

Di Fabio R, Pellacani A, Faedo S, Roth A, Piccoli L, Gerrard P et al (2011) Discovery process and pharmacological characterization of a novel dual orexin 1 and orexin 2 receptor antagonist useful for treatment of sleep disorders. Bioorg Med Chem Lett 21:5562–5567

Dugovic C, Shelton JE, Aluisio LE, Fraser IC, Jiang X, Sutton SW et al (2009) Blockade of orexin-1 receptors attenuates orexin-2 receptor antagonism-induced sleep promotion in the rat. J Pharmacol Exp Ther 330:142–151

Equihua AC, De la Herrán-Arita AK, Drucker-Colín R (2013) Orexin receptor antagonists as therapeutic agents for insomnia. Front Pharmacol 4:163

Gaillard JM, Schulz P, Tissot R (2009) Effects of three Benzodiazepines (Nitrazepam, Flunitrazepam and Bromazepam) on sleep of normal subjects, studied with an automatic sleep scoring system. Pharmacopsychiatry 6:207–217

Gozzi A, Turrini G, Piccoli L, Massagrande M, Amantini D, Antolini M et al (2011) Functional magnetic resonance imaging reveals different neural substrates for the effects of orexin-1 and orexin-2 receptor antagonists. PLoS ONE 6:e16406

Herring WJ, Snyder E, Budd K, Hutzelmann J, Snavely D, Liu K et al (2012) Orexin receptor antagonism for treatment of insomnia: a randomized clinical trial of suvorexant. Neurology 79: 2265–2274

Hoever P, Dorffner G, Beneš H, Penzel T, Danker-Hopfe H, Barbanoj MJ et al (2012) Orexin receptor antagonism, a new sleep-enabling paradigm: a proof-of-concept clinical trial. Clin Pharmacol Ther 91:975–985

Huang ZL, Qu WM, Li WD, Mochizuki T, Eguchi N, Watanabe T et al (2001) Arousal effect of orexin A depends on activation of the histaminergic system. Proc Natl Acad Sci USA 98: 9965–9970

Johnson PL, Truitt W, Fitz SD, Minick PE, Dietrich A, Sanghani S et al (2010) A key role for orexin in panic anxiety. Nat Med 16:111–115

Jupp B, Krivdic B, Krstew E, Lawrence AJ (2011) The orexin receptor antagonist SB-334867 dissociates the motivational properties of alcohol and sucrose in rats. Brain Res 1391:54–59

Kalogiannis M, Hsu E, Willie JT, Chemelli RM, Kisanuki YY, Yanagisawa M et al (2011) Cholinergic modulation of narcoleptic attacks in double orexin receptor knockout mice. PLoS ONE 6:e18697

Kummangal BA, Kumar D, Mallick HN (2013) Intracerebroventricular injection of orexin-2 receptor antagonist promotes REM sleep. Behav Brain Res 237:59–62

Lanoir J, Killam EK (1968) Alteration in the sleep-wakefulness patterns by benzodiazepines in the cat. Electroencephalogr Clin Neurophysiol 25:530–542

Lebold TP, Bonaventure P, Shireman BT (2013) Selective orexin receptor antagonists. Bioorg Med Chem Lett 23:4761–4769

Lin L, Faraco J, Li R, Kadotani H, Rogers W, Lin X et al (1999) The sleep disorder canine narcolepsy is caused by a mutation in the hypocretin (orexin) receptor 2 gene. Cell 98:365–376

Lu BS, Zee PC (2010) Neurobiology of sleep. Clin Chest Med 31:309–318

Mang GM, Dürst T, Bürki H, Imobersteg S, Abramowski D, Schuepbach E et al (2012) The dual orexin receptor antagonist almorexant induces sleep and decreases orexin-induced locomotion by blocking orexin 2 receptors. Sleep 35:1625–1635

Marcus JN, Aschkenasi CJ, Lee CE, Chemelli RM, Saper CB, Yanagisawa M et al (2001) Differential expression of orexin receptors 1 and 2 in the rat brain. J Comp Neurol 435:6–25
Mieda M, Hasegawa E, Kisanuki YY, Sinton CM, Yanagisawa M, Sakurai T (2011) Differential roles of orexin receptor-1 and -2 in the regulation of non-REM and REM sleep. J Neurosci 31:6518–6526
Mieda M, Tsujino N, Sakurai T (2013) Differential roles of orexin receptors in the regulation of sleep/wakefulness. Front Endocrinol (Lausanne) 4:57
Morairty SR, Revel FG, Malherbe P, Moreau J-L, Valladao D, Wettstein JG et al (2012) Dual hypocretin receptor antagonism is more effective for sleep promotion than antagonism of either receptor alone. PLoS ONE 7:e39131
Peyron C, Tighe DK, van den Pol A, de Lecea L, Heller HC, Sutcliffe JG et al (1998) Neurons containing hypocretin (orexin) project to multiple neuronal systems. J Neurosci 18: 9996–10015
Prober DA, Rihel J, Onah AA, Sung R-J, Schier AF (2006) Hypocretin/orexin overexpression induces an insomnia-like phenotype in zebrafish. J Neurosci 26:13400–13410
Renzulli C, Nash M, Wright M et al (2011) Disposition and metabolism of [14C]SB-649868, an orexin receptor antagonist, in humans. Drug Metab Dispos 39:215–227
Sakurai T (2007) The neural circuit of orexin (hypocretin): maintaining sleep and wakefulness. Nat Rev Neurosci 8:171–181
Sakurai T, Amemiya A, Ishii M, Matsuzaki I, Chemelli RM, Tanaka H et al (1998) Orexins and orexin receptors: a family of hypothalamic neuropeptides and G protein-coupled receptors that regulate feeding behavior. Cell 92:573–585
Smart D, Sabido-David C, Brough SJ, Jewitt F, Johns A, Porter RA et al (2001) SB-334867-A: the first selective orexin-1 receptor antagonist. Br J Pharmacol 132:1179–1182
Smith RJ, Aston-Jones G (2012) Orexin/hypocretin 1 receptor antagonist reduces heroin self-administration and cue-induced heroin seeking. Eur J Neurosci 35:798–804
Smith MI, Piper DC, Duxon MS, Upton N (2003) Evidence implicating a role for orexin-1 receptor modulation of paradoxical sleep in the rat. Neurosci Lett 341:256–258
Steiner MA, Gatfield J, Brisbare-Roch C, Dietrich H, Treiber A, Jenck F et al (2013) Discovery and characterization of ACT-335827, an orally available, brain penetrant orexin receptor type 1 selective antagonist. ChemMedChem 8:898–903
Sun H, Kennedy WP, Wilbraham D, Lewis N, Calder N, Li X et al (2013) Effects of suvorexant, an orexin receptor antagonist, on sleep parameters as measured by polysomnography in healthy men. Sleep 36:259–267
Thannickal TC, Moore RY, Nienhuis R, Ramanathan L, Gulyani S, Aldrich MS et al (2000) Reduced number of hypocretin neurons in human narcolepsy. Neuron 27:469–474
White CL, Ishii Y, Mendoza T, Upton N, Stasi LP, Bray GA et al (2005) Effect of a selective OX1R antagonist on food intake and body weight in two strains of rats that differ in susceptibility to dietary-induced obesity. Peptides 26:2331–2338
Willie JT, Chemelli RM, Sinton CM, Tokita S, Williams SC, Kisanuki YY et al (2003) Distinct narcolepsy syndromes in orexin receptor-2 and orexin null mice. Neuron 38:715–730
Winrow CJ, Gotter AL, Cox CD, Doran SM, Tannenbaum PL, Breslin MJ et al (2011) Promotion of sleep by suvorexant-a novel dual orexin receptor antagonist. J Neurogenet 25:52–61
Zisapel N (2012) Drugs for insomnia. Expert Opin Emerg Drugs 17(3):299–317

Part III
Drugs to Treat Excessive Daytime Sleepiness

Modafinil and Armodafinil

Karel Šonka, Peter Šóš, and Marek Susta

1 Introduction

1.1 Excessive Daytime Sleepiness

Wakefulness of adequately long duration and quality is in every respect essential for good quality of life. Inadequately low wakefulness named excessive daytime sleepiness (EDS) is defined as a reduced ability to maintain continuous wakefulness during the day. EDS may take the form of lapses into sleep or as periods of somnolence leading to sleep onset in favorable circumstances and longer total duration of sleep within 24 h. EDS lowers the quality of life and complicates or disallows many regular activities.

Treatment of EDS starts with the identification of its cause and correction of all behavior bugs including avoidance of inappropriate drugs and substances. In case of a primary form of EDS a symptomatic therapy is the only option. The objective of any treatment is to eliminate EDS and to produce the best possible normal function for patients at workplace, school, at home, and society in general.

Historically, the most widely used and the most effective compounds to treat EDS were the amphetamine-like CNS stimulants (methamphetamine, dexamphetamine, and methylphenidate). Some of these drugs were withdrawn from markets in certain countries because of the risk of misuse and dependence. However, the use of

K. Šonka (✉)
Department of Neurology, Charles University, Prague, Czech Republic
e-mail: ksonka@lf1.cuni.cz

P. Šóš
Prague Psychiatric Center, Prague, Czech Republic

M. Susta
Department of Public Health, St. Elisabeth University, Bratislava, Slovakia

Department of Neurology, Charles University, Prague, Czech Republic

A. Guglietta (ed.), *Drug Treatment of Sleep Disorders*, Milestones in Drug Therapy,
DOI 10.1007/978-3-319-11514-6_10

the modafinil has considerably increased over the last years because of its favorable side-effect profile and several large controlled trials. Nowadays modafinil is the most frequently used drug for the treatment of EDS in narcolepsy.

1.2 History of Modafinil and Armodafinil

The story of modafinil started in 1974 in France. Two chemists—Assous and Gombert from Laboratoires Lafon—screened molecules in search of analgesics and discovered a new molecule, *adrafinil* [(diphenylmethyl)sulfinyl-2acetohydroxamic acid]. The pharmacological testing of adrafinil was performed later by pharmacologists from Laboratoires Lafon Duteil and Rambert who found out that mice treated with adrafinil exhibited hyperactivity (Duteil et al. 1979). Adrafinil was then tested by Michel Jouvet and coworkers on cats and later on by Milhaud and Klein on monkeys. An increase of electroencephalographic wakefulness was found by the first group and an increase of the nocturnal activity by the second one. In 1977–1978, Jouvet prescribed adrafinil to narcoleptic patients with inconsistent results (Billiard et al. 2007).

Meanwhile, in 1976, an active metabolite of adrafinil, modafinil [2-(diphenylmethyl)sulfinylacetamide] (Fig. 1), was discovered and this new molecule appeared more efficient than adrafinil. Modafinil went through the same steps of development leading to the demonstration of a dose-dependent increase in locomotor activity in mice (Duteil et al. 1990), an increase of electroencephalographic wakefulness in cats (Lin et al. 1992), an increase of electroencephalographic wakefulness (Lagarde and Milhaud 1990), and an increase in nocturnal activity and in behavioral arousal without stereotyped behavior (Hermant et al. 1991) in rhesus monkeys. As soon as early 1983 Jouvet prescribed modafinil to narcoleptic and idiopathic hypersomnia patients. The results surpassed expectations. In 1984 Laboratoire Lafon decided to start clinical trials in both healthy volunteers and narcoleptics. First studies of modafinil on night sleep and daytime sleepiness in healthy volunteers were conducted by Goldenberg et al. (1987) and Saletu et al. (1989). Goldenberg and her colleagues found decreased total sleep time, decreased NREM stages 3 and 4, no modification of REM sleep, and no rebound phenomenon after a single evening dose of modafinil 200 mg or placebo in parallel groups. In addition, sleep latency on the multiple sleep latency test (MSLT) increased in every single session following a single dose of modafinil, 200 mg at

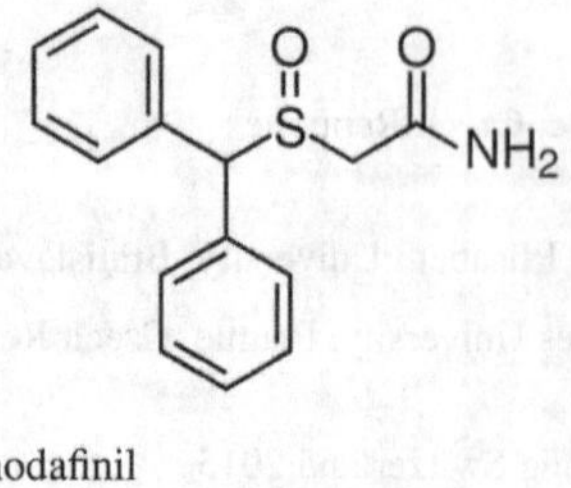

Fig. 1 Chemical structure of modafinil

10 a.m. Saletu with his coworkers compared the effect of modafinil on nocturnal sleep in 100 and 200 mg dosage with D-amphetamine 10 and 20 mg and placebo. Drop in sleep efficiency was much smaller in modafinil compared to amphetamines. Studies in healthy volunteers were accompanied by open-label trials in subjects with narcolepsy and idiopathic hypersomnia performed in different research centers—the first study and the first clinical publication on modafinil came from Jouvet's center in Lyon (Bastuji and Jouvet 1988). These studies and the first multicenter, randomized, placebo-controlled study (Billiard et al. 1994) led to the official registration of modafinil in France in June 1992 followed by its commercial availability there, from September 1994 (Billiard et al. 2007).

Further research including preclinical, phase I trials and multicenter, randomized, placebo-controlled studies were conducted by Cephalon who originally leased the rights from Lafon in 1993, but eventually purchased the company in 2001. Modafinil received orphan drug status in the USA in 1993 and became there commercially available in 1998 (Thorpy 2007). Other European countries made modafinil available in late 1990s. In 2007, Cephalon began to market the R-enantiomer of modafinil called armodafinil in the USA. After protracted patent litigation and negotiations, generic versions of modafinil became available in the USA in 2012. In 2011 Teva took over the Cephalon Company with all products including modafinil and armodafinil.

2 Modafinil Pharmacology and Mode of Action

2.1 Pharmacology

Classical central nervous stimulants (wake-promoting agents) are based on amphetamine. Amphetamine has a simple chemical structure resembling endogenous catecholamines and its major mode of action is an increase of catecholamine (dopamine and norepinephrine but also to a lesser extent serotonin) release and inhibition of catecholamine reuptake. This results in an increase in catecholamine concentration in the synaptic cleft and enhances postsynaptic stimulation. The presynaptic modulations by amphetamines are mediated by specific catecholamine transporters. These transporters—the dopamine transporter (DAT) and the norepinephrine transporter (NET)—move normally dopamine and norepinephrine from the outside to the inside of the cell and amphetamines can reverse the direction of this transport. The side effect of amphetamines is mediated mostly by release of norepinephrine, which stimulates indirectly alpha- and beta-adrenergic receptors. Alpha-adrenergic stimulation produces vasoconstriction, increasing thereby systolic and diastolic blood pressure. In large doses tachycardia and cardiac arrhythmia may occur. Other side effects include mild gastrointestinal disturbances, anorexia, dryness of the mouth, insomnia and restlessness, headache, palpitations, anxiety, and vasomotor disturbances. D-amphetamine isomers are more active than isomers of the L-type, and have more effects on dopaminergic synapses than on other monoaminergic synapses (Nishino and Mignot 2011; Mignot 2012).

2.2 Mechanism of Action

The precise mechanism of modafinil wakefulness promotion is unknown. Modafinil has weak to negligible interactions with receptors for norepinephrine, serotonin, dopamine, GABA, adenosine, histamine-3, melatonin, and benzodiazepines. Modafinil also does not inhibit activity of MAO-B or phosphodiesterases II–V. Modafinil is not a direct- or indirect-acting dopamine receptor agonist. However, in vitro, modafinil binds to the DAT and inhibits dopamine reuptake. This activity has been associated in vivo with increased extracellular dopamine levels in some brain regions of animals. In genetically engineered mice lacking the DAT, modafinil lacked wake-promoting activity, suggesting that this activity is DAT dependent. Modafinil appears to reduce GABA levels and increase glutamate and serotonin levels in the cortex indirectly, but these activities do not appear to be involved in the clinical effect of modafinil. Modafinil administration increases C-fos activity in the wake-promoting hypothalamic brain regions including hypocretin neurons, in tuberomammillary and suprachiasmatic nuclei, and at higher doses in the striatum and cingulate cortex. Optical enantiomers of modafinil have similar pharmacological actions in animals (FDA 2007; Monderer and Thorpy 2011).

2.3 Pharmacokinetics and Metabolism

Modafinil is a *racemic compound*, whose enantiomers have different pharmacokinetics.

Modafinil is rapidly absorbed (plasma concentrations peaking in 2–4 h) and slowly cleared. It bounds to plasma proteins at 60 % and its volume distribution is 0.8–0.9 l/kg, suggesting that the compound is readily able to penetrate into tissues. Its half-life ranges from 9 to 14 h. The *R enantiomer* of modafinil (armodafinil) has a longer half-life (10–15 h) than S enantiomer (3–4 h).

60–90 % of modafinil is converted in liver to inactive metabolites with subsequent renal elimination. Metabolism primarily utilizes cytochrome P-450 3A4/5. Induction of metabolizing enzymes, most importantly P-450 (CYP) 3A4, has been observed in vitro and in vivo after extended administration of modafinil at 400 mg/day.

Modafinil exhibit linear kinetics upon multiple dosing of 200–600 mg/day once daily in healthy volunteers. Apparent steady states of racemic modafinil are reached after 2–4 days of dosing (Nishino and Mignot 2011; FDA 2007).

3 Clinical Involvement of Modafinil and Armodafinil

EDS is a symptom and can be related to many conditions, but only some of these can be treated symptomatically by modafinil. The following diseases will be mentioned in detail below: narcolepsy type 1 and 2, idiopathic hypersomnia, residual hypersomnia in patients with adequately treated obstructive sleep apnea (OSA), hypersomnia associated with Parkinson's disease (PD), multiple sclerosis (MS), posttraumatic hypersomnia, and shift work disorder.

Armodafinil is targeting the same indications as the racemic form of the drug—modafinil.

3.1 Narcolepsy

Narcolepsy is a disabling lifelong sleep disorder with EDS as the main symptom. There are two forms (independent nosological entities with different pathophysiology) of narcolepsy. Narcolepsy type 1 (narcolepsy with cataplexy) and narcolepsy type 2 (narcolepsy without cataplexy) with the same clinical manifestation of EDS. Narcolepsy type 1 symptoms include not only sleepiness but also abnormal rapid-eye movement (REM) sleep manifestations, including cataplexy (short-lasting symmetrical loss of muscle strength caused by emotional trigger), sleep paralysis, and hypnagogic hallucinations. The disappearance of hypothalamic hypocretin-producing neurons at the disease onset is the main etiopathogenic factor of narcolepsy type 1. Narcolepsy type 2 is clinically characterized by sleepiness, and absence of cataplexies. Its etiopathogenesis is unknown. Narcolepsy type 1 and type 2 belong to the group of central disorders of hypersomnolence according to the International Classification of Sleep Disorders, third edition (American Academy of Sleep Medicine 2014).

EDS is the most inconveniencing symptom in narcolepsy. Both forms of narcolepsy share the same objective criteria of EDS in MSLT: mean sleep latency <8 min and the presence of REM sleep in two or more measurements within MSLT, REM sleep episode within 15 minutes of sleep onset on the preceding nocturnal polysomnogram may replace one of the SOREMPs on the MSLT (American Academy of Sleep Medicine 2014). MSLT consists of five 20 min long opportunities to sleep for 15 min. Clinical picture of EDS in both types includes episodes of sleepiness followed by sleep in conditions favorable or even unfavorable for sleeping, sleep attacks with and without preceding sleepiness, and automatic behavior. Naps are usually of short duration and the patient wakes up feeling refreshed. There is no causal treatment of sleepiness and of REM sleep manifestations in narcolepsy.

3.1.1 Modafinil in Narcolepsy

The first multicenter, randomized, placebo-controlled trial of modafinil was performed in patients suffering from narcolepsy type 1 in four research centers in France and Canada. Subjects were either drug naive or had discontinued psychostimulant medication for at least 14 days prior to the study. Modafinil was administered in a double-blind crossover design, at a dosage of 300 mg versus placebo. The duration of the study was 12 weeks with the 4-week treatment/placebo periods. Sleep logs did not show any modification of night sleep, but a reduction of both daytime sleepiness and overwhelming episodes of sleep. Cataplexy remained unchanged. There was a significant improvement of the sleep latencies measured by maintenance of wakefulness test (MWT) in modafinil compared to placebo (Billiard et al. 1994).

Subsequently in Canada, 75 patients with narcolepsy enrolled into a 6-week, three-period, randomized, crossover, placebo-controlled trial. Patients received placebo, modafinil 200 mg, or modafinil 400 mg in divided doses (morning and noon). Evaluations occurred at baseline and at the end of each 2-week period. Compared with placebo, modafinil 200 and 400 mg significantly increased the mean sleep latency on the MWT by 40 and 54 %, with no significant difference between the two doses. Modafinil, 200 and 400 mg, also reduced the combined number of daytime sleep episodes and periods of severe sleepiness noted in sleep logs. The likelihood of falling asleep as measured by the Epworth Sleepiness Scale (ESS) was equally reduced by both modafinil dose levels. There was no effect on nocturnal sleep. Neither dose interfered with the patients' ability to nap voluntarily during the day or with nocturnal sleep quantity or quality. Modafinil caused no changes in blood pressure or heart rate in either normotensive or hypertensive patients. The only significant adverse effects were observed at the 400 mg dose associated with more frequent nausea and nervousness than either placebo or the 200 mg dose (Broughton et al. 1997).

The U.S. Modafinil in Narcolepsy Multicenter Study Group conducted two double-blind, placebo-controlled studies in narcolepsy (U.S. Modafinil in Narcolepsy Multicenter Study Group 1998, 2000); they were similarly designed, with identical end points that measured both objective and subjective sleepiness, as well as overall clinical condition. Each trial (one conducted in 18 research centers and one in 21 centers) used as primary end points the MWT and the Clinical Global Impression of Change (CGI-C). MSLT and ESS were the secondary end points. The duration of each U.S. narcolepsy trial lasted for 9 weeks with visits scheduled every 3 weeks. A total of 558 patients were included in the efficacy analyses of these trials, and randomized to placebo, modafinil 200 mg, or modafinil 400 mg. Eighty-one percent of patients had also cataplexy, but only patients without anticataplectic treatment were included. Combined results of these two studies showed that mean sleep latency on the MWT increased by more than 2 min in each treatment group, compared with a decrease of 0.7 min in the placebo group. A significantly higher percentage of modafinil patients showed improvement in overall clinical condition in the CGI-C (61–66 % vs. 37 % for placebo). The 18-center study used only a 1-day titration period for the 400 mg group and a higher percentage of patients in the 400 mg group withdrew due to adverse events compared with the 200 mg and placebo groups (12 % vs. 1 % and 0 %, respectively). A more refined step-up protocol lasting 9 days was planned in the subsequent study and only 1 % of the 400 mg group withdrew due to adverse effects. Similar objective improvements were seen on the MSLT and on the ESS. No dose–response effect was seen for the 400 mg dose compared with the 200 mg dose in either study. There were no significant changes in nocturnal sleep parameters. The 21-center study included a 2-week treatment discontinuation phase to determine the effect of withdrawal from modafinil. During the discontinuation period, subjects experienced a loss of improvement in wakefulness that had been seen over the course of the study. No symptoms of tolerance or abrupt withdrawal were reported. The consistency of response on the same measures between trial and improvement across variety of measures within each trial were significant. Absolute changes from baseline on the MWT and MSLT may appear small; however, these tests are done in settings designed to maximize the likelihood of sleep onset. Under these conditions, small increases in sleep latency can represent clinically significant improvements in wakefulness (U.S. Modafinil in Narcolepsy Multicenter Study Group 1998, 2000; Monderer and Thorpy 2011).

The results of the U.S. modafinil trials corroborate those by a group of Canadian investigators in a randomized, double-blind, placebo-controlled, 6-week trial. In this trial, consisting of three 2-week crossover phases, significant improvements were seen on the MWT and ESS at 200 and 400 mg doses, given twice daily in the morning and at noon. This trial also failed to show dose-dependent therapeutic effect (Moldofsky et al. 2000).

Further studies were performed to determinate the optimal dosing protocols for modafinil in narcolepsy (Monderer and Thorpy 2011). While the original US placebo-controlled studies showed no dose–response effect for 400 mg compared to 200 mg, a later study, using a modified version of the MWT that included an evening test session, demonstrated significantly improved evening wakefulness with the 400 mg dose (whether in a single dose or divided dose) compared to the 200 mg dose. The greatest improvement in evening wakefulness was seen with the 400 mg split-dose regimen, as compared to the 200 and 400 mg once-daily regimen (Schwartz et al. 2003a). Next study looking at dosing effects of 600 mg split-dose regimen (400 mg in the morning and 200 mg in the early afternoon) was found to achieve more sustained wakefulness throughout the day compared to 400 mg once daily (Schwartz et al. 2004). A third study showed that split-dose regimens of either 400 mg in the morning and 200 mg at noon or 200 mg both in the morning and at noon were superior to a single dose of 200 mg in the morning (Schwartz et al. 2005).

A small double-blind crossover study with 300 mg modafinil showed a decrease of EDS measured by MWT and also improvement of psychomotor performances involving attention (Besset et al. 1993). This result is in line with new finding of improvement of driving abilities in patients suffering from narcolepsy and idiopathic hypersomnia (Philip et al. 2014).

3.1.2 Armodafinil in Narcolepsy

Patients receiving armodafinil 150 and 250 mg/day in a large randomized, placebo-controlled study of 196 patients with narcolepsy experienced increased MWT mean sleep latency compared to placebo. These changes were seen at all time points for the 150 mg dose, but statistical significance was not reached at weeks 8 and 12 for the 250 mg group. The CGI-S showed improvement at the final visit in both armodafinil groups as well as the ESS scores. Memory, attention, and fatigue showed statistically significant improvement with both doses of armodafinil (Harsh et al. 2006). Armodafinil administered for 12 months or more was generally well tolerated and improved wakefulness and the overall clinical condition in patients with narcolepsy (Schwartz et al. 2010).

Long-term efficacy of modafinil and armodafinil in narcolepsy was well documented by open-label studies (Besset et al. 1996; Black et al. 2010).

Modafinil and armodafinil are recommended as EDS treatment in narcolepsy by the American Academy of Sleep Medicine (AASM) (Morgenthaler et al. 2007). Modafinil is recommended as EDS treatment in narcolepsy by European Federation of Neurological Societies (Billiard et al. 2006).

3.2 Idiopathic Hypersomnia

Idiopathic hypersomnia (IH) is also a primary disorder of hypersomnolence differing from narcolepsy by the lack of REM sleep disturbances, but the MSLT criterion of mean sleep latency <8 min remains the same (American Academy of Sleep Medicine 2014). IH is a rare lifelong sleep disorder. Persisting EDS cannot be compensated by sleep. IH patients suffer from the impaired quality of life and psychosocial consequences as narcolepsy patients (Ozaki et al. 2012). However, some studies in IH patients used criteria that did not include MSLT tests. The first trial with modafinil in IH was performed as an open-label study in Lyon, France (Bastuji and Jouvet 1988). 18 patients participated in the study and modafinil was administered in the morning and at noon. The dose varied from 200 to 500 mg/day according to the patient's weight and the severity of the symptoms. The number of drowsiness and sleep episodes during daytime was significantly reduced in 15 patients. It took 25 years to have next prospective trials in IH.

Recently, a multicenter study has been performed in 31 adult patients with idiopathic hypersomnia without long sleep time, 14 on modafinil and 17 on placebo. Modafinil 200 mg given in the morning improved ESS and CGI-C compared to placebo and led to a nonsignificant increase in the mean sleep latency on the MWT (Mayer et al. 2013).

There are two large observational studies available (Anderson et al. 2007; Ali et al. 2009). Both studies show the good response rate between 60 and 70 %. Recent French study compared benefits and risks of modafinil in consecutive patients suffering from IH and narcolepsy with cataplexy. The improvement of ESS was similar and sudden loss of efficacy and habituation were rare in both. Patients with IH reported similar but more frequent adverse effects with modafinil than narcolepsy patients: nervousness (14 %), palpitations (13 %), and headache (11 %) (Lavault et al. 2011).

Modafinil is mentioned in AASM practice parameters for the treatment of narcolepsy and hypersomnias of central origin as an optional therapy for treatment of daytime sleepiness due to IH (Morgenthaler et al. 2007).

A randomized, crossover, double-blind placebo-controlled trial has been conducted among 13 patients with narcolepsy and 14 patients with IH. Patients were randomly assigned to receive modafinil (400 mg) or placebo for 5 days prior to the driving test. Each treatment period was separated by at least 3 weeks of washout. Modafinil improved driving performance judged on the number of Inappropriate Line Crossings and Standard Deviation of Lateral Position of the vehicle as well as the mean sleep latency on the MWT. MWT mean sleep latency correlated with the mean number of Inappropriate Line Crossings (Philip et al. 2014). This finding is very important because patients with EDS are at high risk for driving accidents, and physicians are concerned by the effect of alerting drugs on driving skills of sleepy patients.

3.3 Residual Hypersomnia in Patients with Adequately Treated Obstructive Sleep Apnea (OSA)

Some patients with OSA report persistent sleepiness despite optimal treatment of their sleep apnea (usually continuous positive airway pressure—CPAP) (Sforza and Krieger 1992). Modafinil and armodafinil are the only drugs successfully studied in alleviating EDS in OSA.

A 4-week randomized, double-blind, placebo-controlled, parallel group study evaluated the efficacy of modafinil (the first week 200 mg/day, the second to fourth week 400 mg/day) in patients with OSA (respiratory disturbances index—RDI $\geq$15) and residual EDS while compliant with CPAP therapy (CPAP usage $\geq$4 h per night on 70 % of nights). Mean changes from baseline in ESS at weeks 1 and 4 displayed a greater improvement in EDS in patients on modafinil compared to those on placebo. At week 4, 51 % of subjects on modafinil had an ESS within normal range ($<$10) as compared to 27 % in the placebo group. Furthermore, at week 4, the mean sleep latency on MSLT improved from a baseline of 7.4–8.6 min in modafinil group as compared to a decrease from 7.5 to 7.2 min in placebo group. CGI-C ratings were significantly more improved in the modafinil group as compared with the placebo group. There was a small increase in the arousal index in patients receiving modafinil compared to controls, but no change in either group in the number of hours of CPAP use (Pack et al. 2001). During the follow-up 12-week open-label study the significant improvement seen on the ESS during the 4-week study was maintained and more than 93 % of patients clinically improved on the CGI-C from weeks 2–12 of the open-label study. However, unlike the previous study, there was a small but significant drop in the mean nightly CPAP use (from 6.3 to 5.9 h) (Schwartz et al. 2003c).

A smaller and shorter randomized, double-blind, placebo-controlled crossover study in subjects suffering from OSA compliant with CPAP with 400 mg of modafinil and placebo showed only improvement on MWT and no effect on either the MSLT or ESS and on quality of life (SF-36) or Functional Outcomes of Sleep Questionnaire (FOSQ) or cognitive performance. Patients on modafinil again mildly reduced the CPAP use (Kingshott et al. 2001).

A more recent, 12-week double-blind, placebo-controlled study was conducted with 309 OSA patients compliant with CPAP who were randomized to receive either 400 or 200 mg of modafinil or placebo. Wakefulness was significantly improved on the MSLT with both the 400 and 200 mg of modafinil as compared to control on weeks 4, 8, and 12. The ESS scores decreased by 4.5 in both modafinil groups in contrast to the placebo group, which had a 1.8-point decrease. CGI-C improved in 61 and 68 % of patients on modafinil, respectively, as compared with 37 % on placebo. Vigilance, general productivity, and activity level subscale scores on FOSQ improved with modafinil. The therapy did not change hours of nightly CPAP use (Black and Hirshkowitz 2005).

Two large 12-week randomized, double-blind, placebo-controlled studies evaluated armodafinil in the treatment of residual EDS in patients with OSA on CPAP. Patients had ESS score $\geq$10 and apnea/hypopnea index $\leq$10 in polysomnogram with CPAP and sufficient regular CPAP use (Roth et al. 2006; Hirshkowitz et al. 2007). The results of these two studies (651 patients) were analyzed together in the subsequent paper (Roth et al. 2008). Armodafinil improved wakefulness as measured by mean sleep latency on MWT at week 4, 8, and 12. Armodafinil also improved late-day (afternoon) wakefulness on MWT at the final visit. ESS was improved at all visits in the armodafinil group and at week 12, 49 % of patients in armodafinil group had an ESS $\leq$ 10 compared to 26 % in the placebo group. The quality (not the speed) of long-term memory was improved compared to placebo; however, the power of attention and continuity of attention between treatment groups were not significantly different. Global fatigue scores were improved in both armodafinil groups. However a significant reduction of hours of nightly CPAP use in the armodafinil group was observed. The night sleep variables were in both groups the same (Roth et al. 2008).

A long-term open-label study with armodafinil (50–250 mg/day) in OSA patients showed permanent efficacy and adverse effect of mild-to-moderate intensity (Black et al. 2010).

A recent randomized placebo-controlled crossover trial showed in patients with untreated mild-to-moderate OSA that modafinil significantly improved subjective sleepiness. The size of this effect is clinically relevant at 3–4 ESS points of improvement. Driving simulator performance and reaction time also improved on modafinil (Chapman et al. 2014).

One must take into account that modafinil does not treat the primary disease and that the improvement of residual sleepiness may diminish compliance to the positive airway pressure therapy and patients need usual regular supervision. Modafinil is recommended as the treatment of residual EDS in OSA by AASM (Morgenthaler et al. 2006); in the USA modafinil is registered for this indication, but not in European Community (EU).

3.4 Hypersomnia Secondary to Parkinson's Disease (PD)

Significant hypersomnolence documented by MSLT has been reported in some cases of PD.

Modafinil was found to be effective in the treatment of EDS in PD (ESS and CGI-C) by two double-blind studies with doses of 100–200 mg/day (Hogl et al. 2002; Adler et al. 2003), but a larger double-blind placebo-controlled study failed to find significant improvement of ESS or MSLT with modafinil 200–400 mg/day (Ondo et al. 2005).

Later a small randomized, open-label 8-week study of 19 subjects with PD demonstrated that although modafinil may be effective in reducing physical fatigability, it did not improve fatigue symptoms and did not change ESS (Lou et al. 2009).

Modafinil 100 mg twice a day was safe and modestly effective for the treatment of EDS in the elderly according to naturalistic open-label 3 weeks study in 10 PD patients (Lokk 2010). This result is not congruent with other open-label studies showing the lack of the effect of modafinil on EDS in PD (Nieves and Lang 2002).

Due to inconsistent results modafinil cannot be recommended generally for the treatment of EDS and fatigue in PD (Sheng et al. 2013). Nevertheless, modafinil has mild side effects, does not modify PD course, and thus can be taken in consideration as off-label therapy in PD subjects handicapped by EDS.

3.5 Posttraumatic Hypersomnia

Hypersomnolence appears to be common consequence of traumatic brain injury (TBI), with one meta-analysis suggesting a frequency of 28 % of TBI patients. In some cases this may be caused by injury to the hypocretin/orexin neurons or other wake-promoting neural systems.

A double-blind, placebo-controlled crossover trial, where 53 participants with TBI were randomly assigned to receive up to 400 mg of modafinil or placebo, showed sporadic statistically significant differences, but there was no clear beneficial pattern from modafinil for any of the 12 outcomes (Jha et al. 2008). Another prospective, double-blind, randomized, placebo-controlled study with 100 and 200 mg of modafinil in 20 patients with TBI who had fatigue or EDS or both included the ESS, the Fatigue Severity Scale, actigraphy, polysomnography, MWT, and Psychomotor Vigilance Test (PVT). The results indicate that modafinil is effective and well tolerated in the treatment of posttraumatic EDS but not the fatigue (Kaiser et al. 2010).

More studies are needed to elucidate the effectiveness of modafinil in traumatic brain injury.

3.6 Multiple Sclerosis (MS)

Fatigue is a common and disabling feature of multiple sclerosis and anecdotally patients with MS suffer from EDS.

A single-blind study with 72 patients found improvement on fatigue measures with 200 mg of modafinil compared to placebo, but not with the 400 mg dose. ESS was reduced in both 200 and 400 mg arms, but ESS was already in normal range (Rammohan et al. 2002). This result is supported by another small study (Brioschi et al. 2009) but not by another one (Ledinek et al. 2013). A recent retrospective study displayed that modafinil was most effective for patients with fatigue and also EDS (Littleton et al. 2010). Nevertheless, randomized placebo-controlled studies failed to demonstrate the effect of modafinil on fatigue in MS (Moller et al. 2011; Stankoff et al. 2005). Niepel and coworkers speculate that the anti-fatigue effect of modafinil may reflect the activation of the noradrenergic locus coeruleus, since this wakefulness-promoting nucleus is damaged in MS (Niepel et al. 2013).

The evidence of effectiveness of modafinil on fatigue and EDS in MS is not yet sufficient for modafinil to be recommended (Sheng et al. 2013), but in selected patients suffering from EDS in MS modafinil may be taken in consideration as an off-label therapy.

Two double-blind, placebo-controlled studies showed the effect of either 8 weeks of modafinil or 1 week of armodafinil on cognitive performance in patients with MS (Bruce et al. 2012; Lange et al. 2009). The effect of modafinil and armodafinil on cognitive functions in MS merits future research.

3.7 Shift Work Disorder

Shift work disorder is characterized by complaints of insomnia or excessive sleepiness during work hours that take place, at least in part, during the usual sleep time.

Three randomized, double-blind, placebo controlled trials were conducted to evaluate the effect of 200–300 mg of modafinil given prior to their night shift. The first study was short-lasting and simulated the real life in laboratory conditions and two remaining were 12-week studies in real conditions. These studies showed improvement in vigilance measured by modified MWT, by PVT and CGI-C, and improvement in both the quality of life and FOSQ (Walsh et al. 2004; Czeisler et al. 2005; Erman et al. 2007).

Armodafinil studied in similar 12-week randomized controlled study confirmed similarly beneficial effect and EDS reduction on the way home after night shift. The study also showed improvement of long-term memory and attention (Czeisler et al. 2009).

The improvements from baseline in efficacy assessments started at month 1 and were maintained throughout the open-label study lasting 12 months and more (Black et al. 2010).

Modafinil is registered as the treatment of shift work disorder in the USA but not in the EU. It should be pointed out that the best treatment of the shift work disorder is to quit such job.

3.8 Depression

Sleep disturbances including EDS and fatigue are common symptoms of depression which may persist despite treatment with antidepressant medication.

Modafinil is an effective augmentation strategy for acute depressive episodes, including for symptoms of fatigue, in both unipolar and bipolar affective disorders. The meta-analysis of 6 randomized controlled studies, with a total of 910 patients with major depressive disorder or bipolar depression, revealed effects of modafinil on improvements in overall depression scores and remission rates. The treatment effects were evident in both major depressive disorder and bipolar depression, with no difference between both disorders. Modafinil showed a significant positive effect on fatigue symptoms. The adverse events were no different from placebo (Goss et al. 2013).

3.9 Other Indications

Myotonic dystrophy is frequently associated with severe EDS and two studies noted some improvement, but recently published randomized, double-blind, placebo-controlled study reported no significant effect measured by MWT (Orlikowski et al. 2009).

Although there have been some clinical trials on the effect of modafinil on fatigue and EDS associated with post-polio syndrome, attention deficit hyperactivity disorder, schizophrenia, and cocaine addiction, they are beyond the scope of this chapter.

Modafinil had no effect on fatigue related to cancer and primary brain tumors and should not be prescribed outside a clinical trial setting (Spathis et al. 2014; Boele et al. 2013).

4 Safety

Modafinil and armodafinil are generally well tolerated, with side effects mostly mild-to-moderate intensity. The most frequent undesirable effects are headache, nausea, insomnia and loss of appetite, and nervousness. Some studies in OSA subjects reported mild elevation of blood pressure and more patients taking modafinil have required antihypertensive drugs. Likewise small but consistent increase in blood pressure has been seen in armodafinil. Most studies have not reported any ECG changes with modafinil and armodafinil. It is recommended that both drugs should not be used in patients with a history of left ventricular hypertrophy or in patients with mitral valve prolapse who have experienced the mitral valve prolapse syndrome when previously receiving CNS stimulants. Overall, the cardiovascular profile of modafinil and armodafinil is more favorable than with other stimulants available.

Psychiatric symptoms, including mania, delusions, hallucinations, and suicidal ideation, have been experienced in association with modafinil and armodafinil use.

In clinical trials no serious cases of skin rash were reported. Nevertheless rare cases of serious or life-threatening rash such as Stevens-Johnson syndrome, toxic epidermal necrolysis, and drug rash with eosinophilia and systemic symptoms have been reported in children and adults taking modafinil. Since it is not possible to predict whether a rash will become serious, ar/modafinil must be discontinued with the first sign of rash (Monderer and Thorpy 2011; FDA 2007).

It is considered that modafinil and armodafinil have low potential of abuse even in high dosage and long-term experience of administration of modafinil is encouraging (Monderer and Thorpy 2011).

There are no important interactions with other drugs. Patients on other psychostimulants can be safely diverted to modafinil (Schwartz et al. 2003b).

Modafinil and armodafinil are category C drugs for pregnancy. It is recommended that pregnant women avoid taking this medication. The amount of modafinil/armodafinil excreted in mother's milk is unknown (Monderer and Thorpy 2011; FDA 2007).

Safety and effectiveness in pediatric patients, below age 16, have not been established. In a controlled 6-week study, 165 pediatric patients (aged 5–17 years) with narcolepsy were treated with modafinil ($n = 123$) or placebo ($n = 42$). There were no statistically significant differences favoring modafinil over placebo in prolonging sleep latency as measured by MSLT or in perceptions of sleepiness as determined by the CGI-C (FDA 2007). Modafinil and armodafinil are not registered for pediatric population in the USA nor in Europe. The age limitation has been criticized by expert group based on their own experience (Lecendreux et al. 2012). This criticism is, among other reasons, caused by the fact that drugs against EDS allowed for children are only amphetamines and their derivatives with well-known elevated risk of abuse.

In terms of overdose, modafinil seems to be a safe drug. In clinical trials, a total of 151 protocol-specified doses ranging from 1,000 to 1,600 mg/day have been administered to 32 subjects, including 13 subjects who received doses of 1,000 or 1,200 mg/day for 7–21 consecutive days. In addition, several intentional acute overdoses occurred; the two largest being 4,500 and 4,000 mg taken by two subjects participating in foreign depression studies. None of these study subjects experienced any unexpected or life-threatening effects. Adverse events that were reported at these doses included excitation or agitation, insomnia, and slight or moderate elevations in hemodynamic parameters. Other observed high-dose effects in clinical studies have included anxiety, irritability, aggressiveness, confusion, nervousness, tremor, palpitations, sleep disturbances, nausea, diarrhea, and decreased prothrombin time. From post-marketing experience, there have been no reports of fatal overdoses involving modafinil alone (doses up to 12 g).

5 Nonmedical Use of Modafinil

Modafinil is used by special forces (e.g., army), by solitary sailors in the race, etc., for improving and prolonging the ability to stay awake.

Conclusion

Modafinil and its R-enantiomer armodafinil are wakefulness and alertness enhancing agents frequently used in modern sleep medicine and seems to be the best medication of its kind available for treatment of EDS due to narcolepsy. Both have good safety profile, but they are approved only for the use in adults and in women using birth control means. Modafinil is approved in the USA for the treatment of EDS associated with narcolepsy and with treated OSA and shift work disease but in Europe for narcolepsy only. The therapeutic potential of modafinil and armodafinil seems to be larger than indicated by the label.

Acknowledgments The chapter was prepared with the support of Czech Ministry of Health grant NT13238-4 and Charles University grants PRVOUK-P26 and PRVOUK-P34

References

Adler CH, Caviness JN, Hentz JG, Lind M, Tiede J (2003) Randomized trial of modafinil for treating subjective daytime sleepiness in patients with Parkinson's disease. Mov Disord 18:287–293

Ali M, Auger RR, Slocumb NL, Morgenthaler TI (2009) Idiopathic hypersomnia: clinical features and response to treatment. J Clin Sleep Med 5:562–568

American Academy of Sleep Medicine (2014) International classification of sleep disorders, 3rd edn. American Academy of Sleep Medicine, Darien

Anderson KN, Pilsworth S, Sharples LD, Smith IE, Shneerson JM (2007) Idiopathic hypersomnia: a study of 77 cases. Sleep 30:1274–1281

Bastuji H, Jouvet M (1988) Successful treatment of idiopathic hypersomnia and narcolepsy with modafinil. Prog Neuropsychopharmacol Biol Psychiatry 12:695–700

Besset A, Tafti M, Villemin E, Billiard M (1993) The effects of modafinil (300 mg) on sleep, sleepiness and arousal in narcoleptic patients. Neurophysiol Clin 23:47–60

Besset A, Chetrit M, Carlander B, Billiard M (1996) Use of modafinil in the treatment of narcolepsy: a long term follow-up study. Neurophysiol Clin 26:60–66

Billiard M, Besset A, Montplaisir J, Laffont F, Goldenberg F, Weill JS, Lubin S (1994) Modafinil: a double-blind multicentric study. Sleep 17:S107–S112

Billiard M, Bassetti C, Dauvilliers Y, Dolenc-Groselj L, Lammers GJ, Mayer G, Pollmacher T, Reading P, Sonka K (2006) EFNS guidelines on management of narcolepsy. Eur J Neurol 13:1035–1048

Billiard M, Nicolet A, Dauvilliers Y, Carlander B (2007) Modafinil: the European experience. In: Bassetti CL, Biliard M, Mignot E (eds) Narcolepsy and hypersomnia. Informa Healthcare, New York

Black JE, Hirshkowitz M (2005) Modafinil for treatment of residual excessive sleepiness in nasal continuous positive airway pressure-treated obstructive sleep apnea/hypopnea syndrome. Sleep 28:464–471

Black JE, Hull SG, Tiller J, Yang R, Harsh JR (2010) The long-term tolerability and efficacy of armodafinil in patients with excessive sleepiness associated with treated obstructive sleep apnea, shift work disorder, or narcolepsy: an open-label extension study. J Clin Sleep Med 6:458–466

Boele FW, Douw L, de Groot M, van Thuijl HF, Cleijne W, Heimans JJ, Taphoorn MJ, Reijneveld JC, Klein M (2013) The effect of modafinil on fatigue, cognitive functioning, and mood in primary brain tumor patients: a multicenter randomized controlled trial. Neuro Oncol 15:1420–1428

Brioschi A, Gramigna S, Werth E, Staub F, Ruffieux C, Bassetti C, Schluep M, Annoni JM (2009) Effect of modafinil on subjective fatigue in multiple sclerosis and stroke patients. Eur Neurol 62:243–249

Broughton RJ, Fleming JA, George CF, Hill JD, Kryger MH, Moldofsky H, Montplaisir JY, Morehouse RL, Moscovitch A, Murphy WF (1997) Randomized, double-blind, placebo-controlled crossover trial of modafinil in the treatment of excessive daytime sleepiness in narcolepsy. Neurology 49:444–451

Bruce J, Hancock L, Roberg B, Brown A, Henkelman E, Lynch S (2012) Impact of armodafinil on cognition in multiple sclerosis: a randomized, double-blind crossover pilot study. Cogn Behav Neurol 25:107–114

Chapman JL, Kempler L, Chang CL, Williams SC, Sivam S, Wong KK, Yee BJ, Grunstein RR, Marshall NS (2014) Modafinil improves daytime sleepiness in patients with mild to moderate obstructive sleep apnoea not using standard treatments: a randomised placebo-controlled crossover trial. Thorax 69:274–279

Czeisler CA, Walsh JK, Roth T, Hughes RJ, Wright KP, Kingsbury L, Arora S, Schwartz JR, Niebler GE, Dinges DF (2005) Modafinil for excessive sleepiness associated with shift-work sleep disorder. N Engl J Med 353:476–486

Czeisler CA, Walsh JK, Wesnes KA, Arora S, Roth T (2009) Armodafinil for treatment of excessive sleepiness associated with shift work disorder: a randomized controlled study. Mayo Clin Proc 84:958–972

Duteil J, Rambert FA, Pessonnier J, Gombert R, Assous E (1979) A possibe alpha-adrenergic mechanism for drug (CRL 40028)-induced hyperactivity. Eur J Pharmacol 59:121–123

Duteil J, Rambert FA, Pessonnier J, Hermant JF, Gombert R, Assous E (1990) Central alpha 1-adrenergic stimulation in relation to the behaviour stimulating effect of modafinil; studies with experimental animals. Eur J Pharmacol 180:49–58

Erman MK, Rosenberg R, Modafinil Shift Work Sleep Disorder Study Group (2007) Modafinil for excessive sleepiness associated with chronic shift work sleep disorder: effects on patient functioning and health-related quality of life. Prim Care Companion J Clin Psychiatry 9:188–194

FDA (2007) NDA 20–717 PROVIGIL® (modafinil) tablets FDA approved labeling. http://www.accessdata.fda.gov/drugsatfda_docs/label/2007/020717s020s013s018lbl.pdf. Accessed 21 Apr 2014

Goldenberg F, Weil JS, Vonfrenckeel V (1987) Effects of modafinil on diurnal variation of objective sleepiness in normal subjects. Sleep Res 16:91

Goss AJ, Kaser M, Costafreda SG, Sahakian BJ, Fu CH (2013) Modafinil augmentation therapy in unipolar and bipolar depression: a systematic review and meta-analysis of randomized controlled trials. J Clin Psychiatry 74:1101–1107

Harsh JR, Hayduk R, Rosenberg R, Wesnes KA, Walsh JK, Arora S, Niebler GE, Roth T (2006) The efficacy and safety of armodafinil as treatment for adults with excessive sleepiness associated with narcolepsy. Curr Med Res Opin 22:761–774

Hermant JF, Rambert FA, Duteil J (1991) Awakening properties of modafinil: effect on nocturnal activity in monkeys (*Macaca mulatta*) after acute and repeated administration. Psychopharmacology (Berl) 103:28–32

Hirshkowitz M, Black JE, Wesnes K, Niebler G, Arora S, Roth T (2007) Adjunct armodafinil improves wakefulness and memory in obstructive sleep apnea/hypopnea syndrome. Respir Med 101:616–627

Hogl B, Saletu M, Brandauer E, Glatzl S, Frauscher B, Seppi K, Ulmer H, Wenning G, Poewe W (2002) Modafinil for the treatment of daytime sleepiness in Parkinson's disease: a double-blind, randomized, crossover, placebo-controlled polygraphic trial. Sleep 25:905–909

Jha A, Weintraub A, Allshouse A, Morey C, Cusick C, Kittelson J, Harrison-Felix C, Whiteneck G, Gerber D (2008) A randomized trial of modafinil for the treatment of fatigue and excessive daytime sleepiness in individuals with chronic traumatic brain injury. J Head Trauma Rehabil 23:52–63

Kaiser PR, Valko PO, Werth E, Thomann J, Meier J, Stocker R, Bassetti CL, Baumann CR (2010) Modafinil ameliorates excessive daytime sleepiness after traumatic brain injury. Neurology 75:1780–1785

Kingshott RN, Vennelle M, Coleman EL, Engleman HM, Mackay TW, Douglas NJ (2001) Randomized, double-blind, placebo-controlled crossover trial of modafinil in the treatment of residual excessive daytime sleepiness in the sleep apnea/hypopnea syndrome. Am J Respir Crit Care Med 163:918–923

Lagarde D, Milhaud C (1990) Electroencephalographic effects of modafinil, an alpha-1-adrenergic psychostimulant, on the sleep of rhesus monkeys. Sleep 13:441–448

Lange R, Volkmer M, Heesen C, Liepert J (2009) Modafinil effects in multiple sclerosis patients with fatigue. J Neurol 256:645–650

Lavault S, Dauvilliers Y, Drouot X, Leu-Semenescu S, Golmard JL, Lecendreux M, Franco P, Arnulf I (2011) Benefit and risk of modafinil in idiopathic hypersomnia vs. narcolepsy with cataplexy. Sleep Med 12:550–556

Lecendreux M, Bruni O, Franco P, Gringras P, Konofal E, Nevsimalova S, Paiva T, Partinen M, Peeters E, Peraita-Adrados R, Plazzi G, Poli F (2012) Clinical experience suggests that modafinil is an effective and safe treatment for paediatric narcolepsy. J Sleep Res 21:481–483

Ledinek AH, Sajko MC, Rot U (2013) Evaluating the effects of amantadin, modafinil and acetyl-L-carnitine on fatigue in multiple sclerosis–result of a pilot randomized, blind study. Clin Neurol Neurosurg 115(Suppl 1):S86–S89

Lin JS, Roussel B, Akaoka H, Fort P, Debilly G, Jouvet M (1992) Role of catecholamines in the modafinil and amphetamine induced wakefulness, a comparative pharmacological study in the cat. Brain Res 591:319–326

Littleton ET, Hobart JC, Palace J (2010) Modafinil for multiple sclerosis fatigue: does it work? Clin Neurol Neurosurg 112:29–31

Lokk J (2010) Daytime sleepiness in elderly Parkinson's disease patients and treatment with the psychostimulant modafinil: a preliminary study. Neuropsychiatr Dis Treat 6:93–97

Lou JS, Dimitrova DM, Park BS, Johnson SC, Eaton R, Arnold G, Nutt JG (2009) Using modafinil to treat fatigue in Parkinson disease: a double-blind, placebo-controlled pilot study. Clin Neuropharmacol 32:305–310

Mayer G, Benes H, Bitterlich M, Young P, Rodenbeck A (2013) Modafinil for the treatment of idiopathic hypersomnia. Results of a randomized, double blind, placebo controlled study. Sleep 36(Suppl):A255

Mignot EJ (2012) A practical guide to the therapy of narcolepsy and hypersomnia syndromes. Neurotherapeutics 9:739–752

Moldofsky H, Broughton RJ, HILL JD (2000) A randomized trial of the long-term, continued efficacy and safety of modafinil in narcolepsy. Sleep Med 1:109–116

Moller F, Poettgen J, Broemel F, Neuhaus A, Daumer M, Heesen C (2011) HAGIL (Hamburg Vigil Study): a randomized placebo-controlled double-blind study with modafinil for treatment of fatigue in patients with multiple sclerosis. Mult Scler 17:1002–1009

Monderer R, Thorpy MJ (2011) Modafinil/armodafinil in the treatment of excessive daytime sleepiness. In: Thorpy MJ, Billiard M (eds) Sleepiness: causes, consequences, and treatment. Cambridge University Press, Cambridge

Morgenthaler TI, Kapen S, Lee-Chiong T, Alessi C, Boehlecke B, Brown T, Coleman J, Friedman L, Kapur V, Owens J, Pancer J, Swick T (2006) Practice parameters for the medical therapy of obstructive sleep apnea. Sleep 29:1031–1035

Morgenthaler TI, Kapur VK, Brown T, Swick TJ, Alessi C, Aurora RN, Boehlecke B, Chesson AL Jr, Friedman L, Maganti R, Owens J, Pancer J, Zak R (2007) Practice parameters for the treatment of narcolepsy and other hypersomnias of central origin. Sleep 30:1705–1711

Niepel G, Bibani RH, Vilisaar J, Langley RW, Bradshaw CM, Szabadi E, Constantinescu CS (2013) Association of a deficit of arousal with fatigue in multiple sclerosis: effect of modafinil. Neuropharmacology 64:380–388

Nieves AV, Lang AE (2002) Treatment of excessive daytime sleepiness in patients with Parkinson's disease with modafinil. Clin Neuropharmacol 25:111–114

Nishino S, Mignot E (2011) Wake-promoting medications: basic mechanisms and pharmacology. In: Kryger MH, Roth T, Dement WC (eds) Principles and practice of sleep medicine. Elsevier, St. Louis

Ondo WG, Fayle R, Atassi F, Jankovic J (2005) Modafinil for daytime somnolence in Parkinson's disease: double blind, placebo controlled parallel trial. J Neurol Neurosurg Psychiatry 76:1636–1639

Orlikowski D, Chevret S, Quera-Salva MA, Laforet P, Lofaso F, Verschueren A, Pouget J, Eymard B, Annane D (2009) Modafinil for the treatment of hypersomnia associated with myotonic muscular dystrophy in adults: a multicenter, prospective, randomized, double-blind, placebo-controlled, 4-week trial. Clin Ther 31:1765–1773

Ozaki A, Inoue Y, Hayashida K, Nakajima T, Honda M, Usui A, Komada Y, Kobayashi M, Takahashi K (2012) Quality of life in patients with narcolepsy with cataplexy, narcolepsy without cataplexy, and idiopathic hypersomnia without long sleep time: comparison between patients on psychostimulants, drug-naive patients and the general Japanese population. Sleep Med 13:200–206

Pack AI, Black JE, Schwartz JR, Matheson JK (2001) Modafinil as adjunct therapy for daytime sleepiness in obstructive sleep apnea. Am J Respir Crit Care Med 164:1675–1681

Philip P, Chaufton C, Taillard J, Capelli A, Coste O, Leger D, Moore N, Sagaspe P (2014) Modafinil improves real driving performance in patients with hypersomnia: a randomized double-blind placebo-controlled crossover clinical trial. Sleep 37:483–487

Rammohan KW, Rosenberg JH, Lynn DJ, Blumenfeld AM, Pollak CP, Nagaraja HN (2002) Efficacy and safety of modafinil (Provigil) for the treatment of fatigue in multiple sclerosis: a two centre phase 2 study. J Neurol Neurosurg Psychiatry 72:179–183

Roth T, White D, Schmidt-Nowara W, Wesnes KA, Niebler G, Arora S, Black J (2006) Effects of armodafinil in the treatment of residual excessive sleepiness associated with obstructive sleep apnea/hypopnea syndrome: a 12-week, multicenter, double-blind, randomized, placebo-controlled study in nCPAP-adherent adults. Clin Ther 28:689–706

Roth T, Rippon GA, Arora S (2008) Armodafinil improves wakefulness and long-term episodic memory in nCPAP-adherent patients with excessive sleepiness associated with obstructive sleep apnea. Sleep Breath 12:53–62

Saletu B, Frey R, Krupka M, Anderer P, Grunberger J, Barbanoj MJ (1989) Differential effects of a new central adrenergic agonist–modafinil–and D-amphetamine on sleep and early morning behaviour in young healthy volunteers. Int J Clin Pharmacol Res 9:183–195

Schwartz JR, Feldman NT, Bogan RK, Nelson MT, Hughes RJ (2003a) Dosing regimen effects of modafinil for improving daytime wakefulness in patients with narcolepsy. Clin Neuropharmacol 26:252–257

Schwartz JR, Feldman NT, Fry JM, Harsh J (2003b) Efficacy and safety of modafinil for improving daytime wakefulness in patients treated previously with psychostimulants. Sleep Med 4:43–49

Schwartz JR, Hirshkowitz M, Erman MK, Schmidt-Nowara W (2003c) Modafinil as adjunct therapy for daytime sleepiness in obstructive sleep apnea: a 12-week, open-label study. Chest 124:2192–2199

Schwartz JR, Nelson MT, Schwartz ER, Hughes RJ (2004) Effects of modafinil on wakefulness and executive function in patients with narcolepsy experiencing late-day sleepiness. Clin Neuropharmacol 27:74–79

Schwartz JR, Feldman NT, Bogan RK (2005) Dose effects of modafinil in sustaining wakefulness in narcolepsy patients with residual evening sleepiness. J Neuropsychiatry Clin Neurosci 17:405–412

Schwartz JR, Khan A, McCall WV, Weintraub J, Tiller J (2010) Tolerability and efficacy of armodafinil in naive patients with excessive sleepiness associated with obstructive sleep apnea, shift work disorder, or narcolepsy: a 12-month, open-label, flexible-dose study with an extension period. J Clin Sleep Med 6:450–457

Sforza E, Krieger J (1992) Daytime sleepiness after long-term continuous positive airway pressure (CPAP) treatment in obstructive sleep apnea syndrome. J Neurol Sci 110:21–26

Sheng P, Hou L, Wang X, Huang C, Yu M, Han X, Dong Y (2013) Efficacy of modafinil on fatigue and excessive daytime sleepiness associated with neurological disorders: a systematic review and meta-analysis. PLoS ONE 8:e81802

Spathis A, Fife K, Blackhall F, Dutton S, Bahadori R, Wharton R, O'Brien M, Stone P, Benepal T, Bates N, Wee B (2014) Modafinil for the treatment of fatigue in lung cancer: results of a placebo-controlled, double-blind, randomized trial. J Clin Oncol 32:1882–1888

Stankoff B, Waubant E, Confavreux C, Edan G, Debouverie M, Rumbach L, Moreau T, Pelletier J, Lubetzki C, Clanet M (2005) Modafinil for fatigue in MS: a randomized placebo-controlled double-blind study. Neurology 64:1139–1143

Thorpy M (2007) Modafinil: the U.S. experience. In: Bassetti CL, Billiard M, Mignot E (eds) Narcolepsy and hypersomnia. Informa Healthcare, New York

U.S. Modafinil in Narcolepsy Multicenter Study Group (1998) Randomized trial of modafinil for the treatment of pathological somnolence in narcolepsy. US Modafinil in Narcolepsy Multicenter Study Group. Ann Neurol 43:88–97

U.S. Modafinil in Narcolepsy Multicenter Study Group (2000) Randomized trial of modafinil as a treatment for the excessive daytime somnolence of narcolepsy: US Modafinil in Narcolepsy Multicenter Study Group. Neurology 54:1166–1175

Walsh JK, Randazzo AC, Stone KL, Schweitzer PK (2004) Modafinil improves alertness, vigilance, and executive function during simulated night shifts. Sleep 27:434–439

Sodium Oxybate (Xyrem®): A New and Effective Treatment for Narcolepsy with Cataplexy

Ahmed S. BaHammam, David Nelson Neubauer, and Seithikurippu R. Pandi-Perumal

Abstract Sodium oxybate (SXB; Xyrem®) is the sodium salt of gamma hydroxybutyric acid (GHB). SXB has been approved in the United States for the treatment of cataplexy and excessive daytime sleepiness (EDS) in patients with narcolepsy and in the European Union for the treatment of narcolepsy with cataplexy. Randomized clinical trials have shown that SXB is highly effective in reducing the frequency of cataplexy and in reducing EDS in patients with narcolepsy. In addition, SXB improved sleep architecture by increasing delta power and the duration of slow-wave sleep and by reducing nocturnal awakenings in patients with narcolepsy with cataplexy. Global function and health-related quality of life were also improved by SXB. The majority of the above effects are dose-related. SXB is administered orally after mixing with water. It is rapidly absorbed and eliminated with a mean elimination half-life of 30–60 min, and its duration of action is 2–4 h. Therefore, SXB must be administered twice during the night to consolidate 6–8 h of nocturnal sleep. In general, SXB is well tolerated, and the most commonly reported adverse events are nausea, vomiting, dizziness, and urinary incontinence. SXB is a drug with potential for misuse and abuse. In addition, SXB has a narrow safety margin with a risk of toxicity. Therefore, strict risk-management strategies and rigorous adherence to the titration schedule are essential.

A.S. BaHammam (✉)
College of Medicine, National Plan for Science and Technology, University Sleep Disorders Center, King Saud University, Box 225503 Riyadh 11324, Saudi Arabia
e-mail: ashammam2@gmail.com; ashammam@ksu.edu.sa

D.N. Neubauer
Department of Psychiatry and Behavioral Science, Johns Hopkins University School of Medicine, Baltimore, Maryland 21287-7413, USA

S.R. Pandi-Perumal
Department of Population Health, Center for Healthful Behavior Change (CHBC), Division of Health and Behavior, Clinical & Translational Research Institute, New York University Medical Center, 227 East 30th Street (between 2nd and 3rd Ave), Floor # 6 – 632E, New York, NY 10016, USA

A. Guglietta (ed.), *Drug Treatment of Sleep Disorders*, Milestones in Drug Therapy,
DOI 10.1007/978-3-319-11514-6_11

1 Introduction

Narcolepsy is a sleep disorder characterized by excessive daytime sleeping (EDS) and uncontrollable attacks of sleep associated with abnormal rapid eye movement (REM) sleep manifestations, such as sudden loss of muscle tone (cataplexy), hypnagogic/hypnopompic hallucinations (vivid hallucinations upon falling asleep or waking), sleep paralysis (unpleasant generalized paralysis just before or while falling asleep or upon waking), and sleep onset REM (SOREM) (Dauvilliers et al. 2007). Despite the daytime sleepiness, disrupted nocturnal sleep is a common manifestation of narcolepsy. Cataplexy is specific to narcolepsy and is the most accurate diagnostic marker of the disease. It is characterized by a sudden, typically bilateral, partial, or complete loss of muscle tone that is provoked by emotional stimuli, such as laughter, excitement, or fright (Dauvilliers et al. 2007). Cataplexy episodes may be limited to specific muscles, such as facial muscles or limbs, and may manifest as slurred speech, jaw tremor, head or jaw dropping, dropping of objects or unlocking of the knees, or may be generalized attacks that manifest as a collapse while conscious (Dauvilliers et al. 2007). Each episode can last from less than one second to several minutes and may occur at frequencies of less than one per year to several episodes per day. In the majority of cases, narcolepsy symptoms begin during teenage or young adult years. However, the disorder may first be evident in very young children or in middle-aged adults (Wise 1998). Studies have shown that 65–75 % of patients with narcolepsy have cataplexy (BaHammam and Alenezi 2006; Silber et al. 2002). The prevalence of narcolepsy with cataplexy is approximately 25–50 per 100,000 people, and the incidence rate is approximately 0.74 per 100,000 person-years (BaHammam and Alenezi 2006; Silber et al. 2002). Narcolepsy often interferes with every aspect of life including work and social settings (Dauvilliers et al. 2003, 2007; Longstreth et al. 2007).

The disorder is highly associated with the human leukocyte antigen (HLA) DQB1*06:02 and is believed to be an autoimmune disease. Great advances have been made in understanding this, as yet, mysterious disorder. Reduced levels of the recently described neuropeptide hypocretin (orexin) have been reported in the nervous system (Merino-Andreu and Martinez-Bermejo 2009). Hypocretin/orexin is a neurotransmitter that regulates arousal, wakefulness, and appetite. Sleep and wakefulness are controlled by multiple brain structures and neurotransmitters (Fig. 1; Sakurai 2007). Pathology within those structures or in their neurotransmitter systems manifests as sleep disorders. Loss of hypocretin/orexin-producing neurons in the lateral and posterior hypothalamus is associated with an inability to maintain prolonged periods of wakefulness or sleep and the intrusion of rapid eye movement (REM) sleep, or REM sleep signs, during wakefulness.

Although there is no cure for narcolepsy, the last decade has witnessed valuable improvements in the treatment of narcolepsy symptoms. Pharmacological management of EDS primarily uses non-sympathomimetic stimulants including modafinil or armodafinil and sympathomimetic stimulants, such as amphetamine, methamphetamine, dexamphetamine, and methylphenidate, which increase wakefulness.

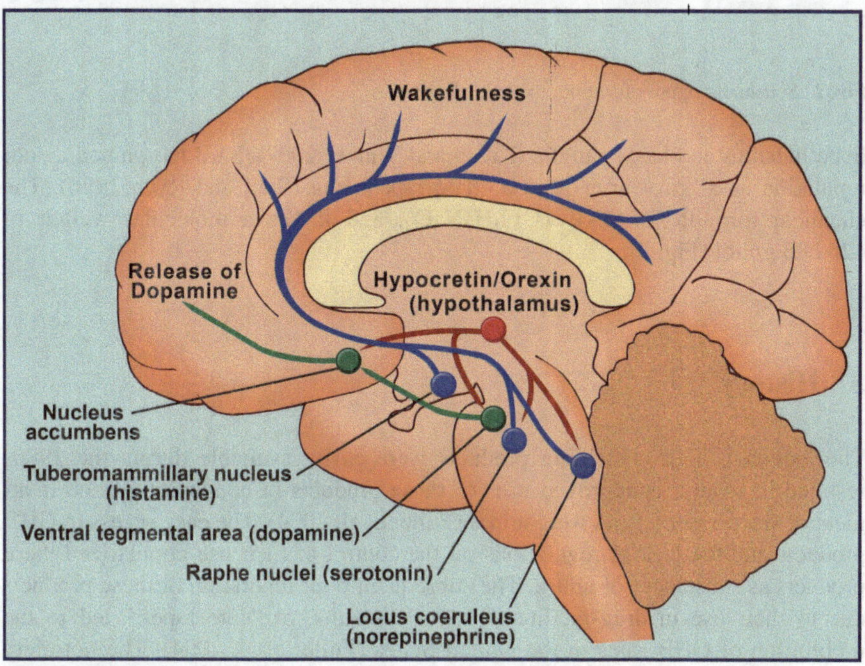

Fig. 1 A diagram showing the main projections of hypocretin/orexin neurons. Hypocretin/orexin neurons located in the lateral and posterior hypothalamus control sleep and wakefulness by sending excitatory projections to the monoaminergic and cholinergic nuclei in the brain stem and hypothalamic regions including the locus ceruleus (expressing noradrenaline), the tuberomammillary nucleus (expressing histamine), and the raphe nuclei (expressing serotonin). Orexin neurons are also associated with the reward system via the ventral tegmental area (nucleus accumbens, containing dopamine)

Narcolepsy with cataplexy may be treated with medications including tricyclic antidepressants, serotonin reuptake inhibitors, and psychostimulants; however, none of these medications are approved by the Food and Drug Administration (FDA) for the treatment of cataplexy. Sodium oxybate (SXB; Xyrem®) is the only FDA-approved treatment for cataplexy.

In this chapter, we review the indications, pharmacology, efficacy and safety of, and practical experience with SXB for the treatment of narcolepsy with cataplexy.

2 Pharmacology

SXB (Xyrem®) is an orphan drug that was approved in 2002 for the treatment of narcolepsy with cataplexy. The active drug in SXB is sodium oxybate, the sodium salt of gamma hydroxybutyric acid (GHB). GHB is an endogenous cerebral inhibitory neuromodulator/neurotransmitter with highest concentrations found in the

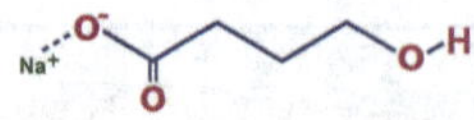

Fig. 2 Schematic figure of sodium oxybate (Xyrem)

hypothalamus and basal ganglia (Keam and Walker 2007). SXB has gained a poor reputation as a "rave drug" and a "date rape" drug (Stell and Ryan 1996). The chemical formula for SXB is $C_4H_7NaO_3$, and it has a molecular weight of 126.086 g/mol (Fig. 2).

3 History

The fact that GHB-containing products were easily available during the 1980s resulted in several unapproved uses of those products in bodybuilding, the treatment of sleeplessness, and weight loss (Fuller et al. 2004). The easy access to GHB products and the lack of regulations on the control of their use popularized these products as substances of abuse. The subsequent poor reputation of these products due to their use in drug-facilitated sexual assaults, or "date rapes," led to the prohibition of GHB sales in the United States (Fuller et al. 2004). The notoriety of GHB and the legal efforts to control the distribution of the product prevented the clinical development of SXB as a treatment for narcolepsy in the United States for several years. Therefore, as a condition of its approval and to promote safe use of SXB and alleviate concerns over possible diversions and abuse following product approval, the FDA required the manufacturer of Xyrem® (sodium oxybate; Jazz Pharmaceuticals, Inc., Palo Alto, CA) to establish and maintain a risk management program (Fuller et al. 2004). Components of the program include the use of a single central pharmacy for direct dispensing of the drug to patients and the gathering of adverse events data, including deaths, that are reported to the FDA, a specially trained pharmacy staff, a method for tracking prescription shipments, and an initial post-marketing surveillance program (Fuller et al. 2004). In addition, this system requires physician registration and verification of reading of educational materials by patients (Fuller et al. 2004). In July 2002, Xyrem® was approved by the FDA as an orphan drug for the treatment of cataplexy in patients with narcolepsy as Xyrem® (sodium oxybate) oral solution. GBH is covered by Schedule I penalties for illicit use; however, for medical purposes, it is under Schedule III via the Xyrem® Success Program, a restricted drug-distribution system (Fuller et al. 2004).

4 Indications

SXB was recently approved to treat patients aged 16 years and above who are diagnosed with narcolepsy and symptoms of cataplexy or excessive daytime sleepiness (EDS). SXB is distinguished from other drugs used in the treatment of narcolepsy by being effective in controlling both EDS and cataplexy. The FDA approved SXB for the treatment of adult narcoleptic patients with cataplexy in 2002. SXB gained approval by the European Medicines Agency (EMEA) in Europe in 2005 for the treatment of narcolepsy with cataplexy. In 2005, it was approved in the United States for the treatment of EDS in patients with narcolepsy also (Zaharna et al. 2010). SXB is not approved for the treatment of other narcolepsy symptoms, such as nocturnal sleep disturbances, sleep paralysis, or hypnagogic hallucinations. The utilization of SXB for cataplexy–narcolepsy is low; only 26,000 patients across 15 countries received SXB for this indication between 2002 and March 2008. This low uptake may be due in part to the low number of trials that examined SXB for the treatment of narcolepsy with cataplexy in small groups of patients or due to the many debates regarding the safety of SXB (Wang et al. 2009; Zvosec et al. 2009). There are other unapproved (off-label) indications for SXB, such as fibromyalgia and insomnia. However, in this chapter, we will discuss the utility of SXB in the approved indication (i.e., in patients with narcolepsy with cataplexy). It is worth mentioning here that SXB has been used on an off-label basis for treating children with narcolepsy with cataplexy; however, there is limited information on the long-term efficacy and safety of SXB in this population (Mansukhani and Kotagal 2012; Murali and Kotagal 2006).

5 Mechanism of Action

SXB was initially developed as an anesthetic agent. However, it was noted that the effects of SXB on sleep were different to other anesthetic agents in that it increased slow wave sleep (stage N3) and REM sleep and decreased in stage N1 sleep, wake after sleep onset, and the number of awakenings suggesting a distinctive mode of action (Pardi and Black 2006). Based on the fact that nocturnal sleep is typically disturbed in patients with narcolepsy with cataplexy, SXB was used initially with the assumption that the SXB-mediated increase in slow wave and REM sleep would reduce the likelihood to sleep during the day and daytime sleepiness (Broughton and Mamelak 1979). The precise mechanism by which SXB produces an effect on cataplexy is unknown. However, it increases turnover of 5-hydroxytryptamine, interacts with opioid systems, and may act as a gamma-aminobutyric acid receptor type B ($GABA_B$ receptor) agonist. The latter action on $GABA_B$ receptor was thought to be the mechanism underlying the sedative effect of SXB (Koek and France 2008). However, an effect of SXB on $GABA_B$ receptors is not accepted by all investigators as the primary reason for the improvement in patients with narcolepsy and cataplexy (Mignot 2012). Huang and Guilleminault conducted a study to evaluate the actions of baclofen and SXB, two medications with $GABA_B$ receptor agonist properties, on symptoms of narcolepsy in drug naïve teenagers with narcolepsy and cataplexy (Huang and Guilleminault 2009). Both drugs increased total sleep time and slow wave sleep duration; however, only SXB showed an effect on daytime sleepiness and cataplexy at 3 months. The authors concluded that the improvement in total nocturnal sleep time did not affect daytime sleepiness and cataplexy and that the mechanism by which SXB improved cataplexy and sleepiness was secondary to properties beyond a direct $GABA_B$ receptor agonist effect (Huang and Guilleminault 2009).

Due to its metabolic pathway, it shows minimal pharmacokinetic interactions with other drugs. Nevertheless, SXB is expected to potentiate the central nervous system and respiratory depressant effects of drugs including alcohol, narcotic analgesics, benzodiazepines, and other drugs with sedative effects (Li et al. 1998).

6 Dosing, Administration, and Adverse Events

SXB is administered orally after mixing with water. It is rapidly absorbed and eliminated with a mean elimination half-life of 30–60 min, and its duration of action is 2–4 h (Owen 2008). Therefore, SXB must be administered twice during the night to consolidate 6–8 h of nocturnal sleep. For use, SXB is provided with a measuring syringe and two 90-ml dosing cups with child-resistant caps. Each measured dose of SXB must be dispensed into the dosing cup and diluted with 60 ml of water prior to ingestion. Because food significantly reduces the bioavailability of SXB (its absorption is slowed by food), patients should not eat for at least 2–3 h before taking the first dose of SXB. It is recommended that both doses of SXB should be prepared at the same time upon retiring to bed.

In adults, we typically initially administer a dose of 4.5 g divided into two equivalent doses: the first at bedtime and the second approximately 3–4 h later. The dose is increased gradually on a weekly basis by 1.5 g per night based on each individual's response until symptoms are controlled, side effects appear, or the maximum nightly dose of 9.0 g is achieved. The typical dose for adults is 6–9 g per night. Doses higher than 9 g per night have not been studied and should not normally be administered.

6.1 *Precaution to Patients*

The sedative effect of SXB is very strong initially; therefore, it is important to warn the patient of the importance of administrating the first dose while in bed and to ask another individual to watch the patient during the first few nights in particular for snoring or gasping.

6.2 Adverse Events

SXB is generally well tolerated with mild-to-moderate side effects that are dose related. There is concern over the narrow margin between efficacious and toxic doses of SXB (Robinson and Keating 2007). In general, the incidence of adverse events increases with dose, and most adverse events subside when the dose is reduced. In a pooled analysis of 717 patients with narcolepsy, the most frequently reported adverse events associated with administration of SXB were headache (22 %), nausea (21 %), dizziness (17 %), nasopharyngitis (8 %), somnolence (8 %), vomiting (8 %), and urinary incontinence ((7 %); Xyrem® (sodium oxybate) oral solution, Jazz Pharmaceuticals Inc. [online]).

In a recently published meta-analysis of studies for all harms including studies using SXB at 9 g/night versus placebo (Alshaikh et al. 2012), patients receiving SXB showed statistically more adverse events compared with placebo including nausea (3 studies, risk ratio (RR): 7.74, 95 % CI: 3.2, 19.2), vomiting (2 studies, RR: 11.8, 95 % CI: 1.6, 89.4), and dizziness (3 studies, RR: 4.3, 95 % CI: 1.1, 16.4;). Incidences of enuresis were not significantly different from placebo groups (2 studies, RR: 2.6, 95 % CI: 0.8, 9.8); nevertheless, there was a tendency towards less incidences in the placebo groups. In the US Xyrem study, 10 patients (7.4 %) withdrew due to adverse events (The U.S. Xyremâ Multicenter Study Group 2002). Side effects were significantly more frequent in the SXB recipients compared with placebo-treated individuals and included nausea, vomiting, dizziness, and urinary incontinence (The U.S. Xyremâ Multicenter Study Group 2002). In the international Xyrem trial, 21 (9.2 %) patients withdrew due to adverse events; however, in that trial, there was no difference in the incidence of urinary incontinence between the SXB and placebo recipients (Xyrem International Study Group 2005a). In the combination therapy trial, adverse events were reported in 70 % of placebo recipients, 54 % of patients receiving modafinil monotherapy, 60 % of patients receiving SXB monotherapy, and in 79 % of patients receiving SXB + modafinil combination therapy (Black and Houghton 2006). In this combination therapy trial, one, two, four, and six patients withdrew from the placebo, modafinil monotherapy, SXB monotherapy, and SXB + modafinil combination therapy groups, respectively, due to adverse events (Black and Houghton 2006). Serious side effects were uncommon. In the combination therapy trial, one patient developed a serious psychotic disorder related to narcissistic personality disorder (Black and Houghton 2006). In the US trial, acute confusion was reported in one patient at a dose of SXB 6 g/night (The U.S. Xyremâ Multicenter Study Group 2002). Other rarely reported dose-related adverse events include parasomnias, such as sleep walking, nocturnal eating, and catathrenia (nocturnal groaning) (Wallace et al. 2011; Poli et al. 2012).

Long-term follow-up data on adverse events are limited. A 12-month extension study reported adverse events in 93 % of patients including dizziness, headache, nausea, urinary incontinence, viral infection, somnolence, and pain (U.S. Xyrem Multicenter Study Group 2004). Dizziness was the only adverse event that was statistically more common in the SXB group (U.S. Xyrem Multicenter Study Group 2004).

7 Efficacy

SXB is one of the most effective treatments for narcolepsy with cataplexy. Several studies have assessed the efficacy of SXB in treating narcolepsy symptoms. The first study that utilized of SXB in the treatment of narcolepsy dates back to 1979 (Broughton and Mamelak 1979). Two recent meta-analyses reviewed the efficacy of SXB in patients with narcolepsy with cataplexy, and both demonstrated that SXB resulted in significant reductions in cataplexy and daytime sleepiness (Alshaikh et al. 2012; Boscolo-Berto et al. 2012). Table 1 presents a summary of these randomized trials.

7.1 Cataplexy

Four randomized placebo-controlled trials have shown that SXB reduced the frequency of cataplexy attacks in a dose-related manner when compared with placebo (U.S. Xyrem Multicenter Study Group 2004; The U.S. Xyremâ Multicenter Study Group 2002; Scrima et al. 1989; Xyrem International Study Group 2005b). The US Xyrem® Multicenter Study Group conducted a 4-week study of 136 patients recruited from various American centers who were experiencing a median of 21 cataplexy attacks per week (The U.S. Xyremâ Multicenter Study Group 2002). Prior to randomization, existing anti-cataplexy medications were withdrawn gradually. The patients were then randomized to treatment with placebo or SXB at 3, 6, or 9 g per night. After 4 weeks of treatment, the number of weekly cataplexy attacks decreased by 4 (28 %), 7 (49 %), 10 (49 %), and 16 (69 %), respectively, compared with baseline. The difference between placebo and the 9 g per night group was statistically significant. A 12-month extension of the US Xyrem® Multicenter Study reported data on 118 patients who received SXB at doses of 3 g (14 %), 4.5 g (36 %), 7.5 g (11 %), and 9 g (30 %) per night. The number of cataplexy attacks was compared with the baseline recorded from the 4-week double-blind trial. Cataplexy attacks were reduced by a mean of 23.65 per week in the first month ($p < 0.001$) and by a mean of 35.48 per week at the end of the study period (U.S. Xyrem Multicenter Study Group 2003). In a subsequent 2-week double-blind study of patients who were on SXB for an average of 22 months, the US Xyrem® Multicenter Study Group randomized 55 patients with cataplexy to receive the same dose of SXB or placebo to establish the long-term efficacy of SXB for the treatment of cataplexy (U.S. Xyrem Multicenter Study Group 2004). The abrupt cessation of SXB in the placebo patients resulted in a significant increase in the number of cataplexy attacks (median = 21; P, 0:001) compared with patients who remained on

Table 1 Characteristics of the published randomized clinical trials that have studied the effects of SXB in patients with narcolepsy

Study	Type of trial	*n*	Setting	Duration of trial in weeks (longest duration of FU)	Trial arms (dose in grams per night)
Xyrem Int. Group 2005 (Xyrem International Study Group 2005a)	RCT	228	42 sleep clinics in the USA, Canada, and Europe	8 (8)	1. SXB (4.5) 2. SXB (6) 3. SXB (9) 4. Placebo
U.S. Xyrem Multicenter Study Group (U.S. Xyrem Multicenter Study Group 2004)	RCT	55	14 clinical sites (location NR)	2 (2)	1. SXB (3) 2. SXB (4.5) 3. SXB (6) 4. SXB (7.5) 5. SXB (9) 6. Placebo
The U.S. Xyremâ Multicenter Study Group (2002)	RCT	136	18 clinical sites (location NR)	4 (4)	1. SXB (3) 2. SXB (6) 3. SXB (9) 4. Placebo
Black and Houghton (2006)	RCT	278 (222 ITT)	44 clinical sites in USA, Canada, and Europe	4 (8)	1. SXB (6 titrated to 9)/modafinil 2. Modafinil/placebo 3. SXB (6 titrated to 9)/placebo 4. Placebo/placebo
Lammers et al. (1993)	Crossover RCT*	24	Leiden University Hospital, Netherlands	4 (4)	1. SXB (60 mg/kg/night) 2. Placebo
Scrima et al. (1989)	Crossover RCT*	20	Sleep Disorders Ceber, University of Arkansas for Medical Sciences	4 (12)	1. SXB (50 mg/kg/night) 2. Placebo

SXB (median = 0). However, unlike antidepressant withdrawal, there was no evidence of rebound cataplexy upon abrupt discontinuation of treatment, and the return of cataplexy attacks was gradual in the placebo patients, who reported a median of 4.2 and 11.7 cataplexy attacks during the first and second weeks, respectively (U.S. Xyrem Multi-Center Study Group 2003).

Two studies have demonstrated beneficial effects on cataplexy attacks with even smaller doses of SXB (4.5 g/night; Xyrem International Study Group 2005b; Scrima et al. 1989). In the International Xyrem trial, which was an 8-week study of nightly SXB, the median cataplexy frequency reduced by 57–85 % at all doses. When compared with placebo, a significant reduction in cataplexy frequency occurred after 4 weeks of SXB treatment at different doses (44, 52, and 62 % at 4.5, 6, and 9 g per night, respectively; Xyrem International Study Group 2005b).

Based on the available evidence, the European Federation of Neurological Societies (EFNS) guidelines suggest that SXB is now the first line of treatment for patients with narcolepsy with cataplexy (Billiard et al. 2006).

7.2 EDS

Based on the two published meta-analyses, SXB causes dose-related reduction in subjective daytime sleepiness (as measured by the Epworth Sleepiness Scale [ESS]) (Alshaikh et al. 2012; Boscolo-Berto et al. 2012). Randomized trials have shown that EDS decreased and sleep latency increased in the SXB arm compared with placebo when evaluated objectively using the maintenance of wakefulness test (MWT). MWT is a validated objective measure of the ability to stay awake over a defined time that measures the mean latency to fall asleep during four to five sessions of trying to stay awake (Littner et al. 2005). Two studies reported a benefit of SXB on EDS measured using the MWT ($n = 101$ and 91 subjects; Black and Houghton 2006; Xyrem International Study Group 2005a). The previous two studies used different MWT protocols. Although the Xyrem International Study group used the 40-min version of the protocol (Xyrem International Study Group 2005a), Black and Houghton used the 20-min version of the protocol (Black and Houghton 2006), which may influence the sleep latencies obtained, particularly in patients who do not fall asleep quickly. The improvement was dose-related; however, the benefits documented in the MWT were statistically significant only at the highest dose (9 g/night). The improvement in EDS appeared after 8 weeks of treatment (Xyrem International Study Group 2005a; Black and Houghton 2006).

Current evidence suggests that SXB is more effective than modafinil alone in alleviating daytime sleepiness. In a double-blind, placebo-controlled, multicenter study, 270 adult patients with narcolepsy (who did not necessarily have cataplexy with narcolepsy) taking 200–600 mg of modafinil daily for the treatment of excessive daytime sleepiness were assigned randomly to 1 of 4 treatment groups: SXB placebo plus modafinil placebo, SXB plus modafinil placebo, modafinil plus SXB placebo, or SXB plus modafinil. SXB was administered at 6 g per night for 4 weeks and was then increased to 9 g per night for an additional 4 weeks. Daytime sleepiness was assessed subjectively using the ESS score and objectively using the MWT. In the placebo group, the mean sleep latency was 7 min compared with 10 min in the modafinil monotherapy group, 12 min in the SXB monotherapy group, and 13 min in the combined (SXB + modafinil) therapy group. ESS score showed a similar trend with no change in the placebo group, an increase of one point in the modafinil monotherapy group, and a reduction of three and four points in the SXB monotherapy and the combined (SXB + modafinil) therapy groups, respectively. However, the combined therapy was associated with a greater incidence of adverse events.

7.3 *Effects on Sleep Architecture*

As discussed earlier, sleep is frequently disturbed in patients with narcolepsy. Sleep studies have documented several changes in patients with narcolepsy including sleep interruption, prolonged awakening after sleep onset, increased stage N1, increased stage shifts, and reduced stage N3 (Baker et al. 1986). Several studies have shown that nocturnal administration of SXB improves subjective and objective measures of nocturnal sleep and sleep architecture with significant increases in stage N3 and delta power (Lammers et al. 1993; Scrima et al. 1989, 1990; Black et al. 2009, 2010; Alshaikh et al. 2011; Broughton and Mamelak 1979, 1980; Mamelak et al. 2004; Scharf et al. 1985). Two recent randomized trials have shown that SXB resulted in significant dose-related decreases in stage N1 and wake after sleep onset and nighttime awakenings and increases in slow wave sleep and total sleep time (Black et al. 2009, 2010).

7.4 *Other Effects*

The Clinical Global Impression of Change (CGI) scores, which are commonly used measures of symptom severity, treatment response, and the efficacy of treatments, were dichotomized to responders as "very much improved" or "much improved." "Much improved" and "very much improved" were statistically significant for all doses (4.5, 6 and 9 g/night) compared with placebo (Xyrem International Study Group 2005a). Interestingly, in the combination-therapy trial, "very much improved" was only observed in the SXB monotherapy and combination therapy (SXB + modafinil) arms (Black and Houghton 2006).

Only one study has examined quality of life indicators (Weaver and Cuellar 2006). Improved quality of life was observed for SXB at 9 g/night versus placebo in all subscales of the Functional Outcomes of Sleep Questionnaire except for intimacy and sexual relationships. A statistically significant and clinically relevant result was noted for SXB at 9 g/night versus placebo.

For the treatment of hypnagogic hallucinations and sleep paralysis, the evidence remains unclear. Changes in the incidence of hypnagogic hallucinations have not been observed in published trials, and a reduction in sleep paralysis attacks was observed only in the group receiving 6 g of SXB per night (Xyrem International Study Group 2005a, b). However, importantly, in both the international Xyrem trial

(Xyrem International Study Group 2005a, b) and the US trial (The U.S. Xyremâ Multicenter Study Group 2002), the baseline frequencies of hypnagogic hallucinations and sleep paralysis attacks were low with a median of 1–3 (Xyrem International Study Group 2005a, b) or a mean of 3–8 (The U.S. Xyremâ Multicenter Study Group 2002) attacks per week.

8 Practical Concerns and Precautions

8.1 High Sodium Load

In patients receiving SXB 4.5–9 g/day, sodium intake will be increased by 0.75–1.6 g/day. Therefore, the treating physicians should be cautious in patients with heart failure, hypertension, or impaired renal function where sodium intake reductions should be considered (Robinson and Keating 2007).

8.2 Weight Loss

Hypocretin/orexin stimulates arousal and feeding. Narcolepsy is typically associated with weight gain particularly in children even when food intake is reduced (Sellayah et al. 2011). Hypocretin/orexin plays an important role in the regulation of body weight via effects on brown adipose tissue differentiation and function. It has been shown in a hypocretin/orexin-null mouse model that the depletion of hypocretin/orexin is associated with brown fat hypoactivity, which leads to reduced energy expenditure (Sellayah et al. 2011).

It has been reported that the administration of SXB is associated with weight loss in some patients (Husain et al. 2009; Alshaikh et al. 2011). However, the exact mechanism underlying this weight loss remains unknown. The improvement in sleep architecture and the increase in slow-wave sleep after SXB use have been suggested as possible mechanisms that induce circadian rhythms in hormonal secretion and improved energy metabolism and weight loss (Husain et al. 2009). Weight loss can be very useful in narcoleptics who are overweight or obese. However, weight loss has been reported in some narcoleptics who are not overweight (Alshaikh et al. 2011).

8.3 Respiratory Depression and Sleep-Disordered Breathing

The co-occurrence of obstructive sleep apnea (OSA) in narcoleptics has been reported to be relatively high (9–25 %); therefore, caution is advised when treating narcoleptics with concurrent OSA with SXB (Sansa et al. 2010; BaHammam and Alenezi 2006). SXB has the potential to induce respiratory depression and apnea. There have been reports of respiratory suppression with GHB abuse, in particular when combined with alcohol (Mason and Kerns 2002). There is a theoretical risk of increasing sleep-disordered breathing (SDB) or inducing sleep hypoventilation in narcoleptic patients on SXB. Previous trials that assessed the efficacy and safety of SXB excluded patients with SDB. Available data on the effect of SXB on the apnea-hypopnea index (AHI) are limited and conflicting (Seeck-Hirschner et al. 2009; George et al. 2010). Additionally, no reports have addressed the safety of SXB in patients with severe OSA (AHI >30/h). A short-term study with a 12-week follow-up in a group of elderly patients (mean age 61.5 years) with Parkinson's disease and without SDB demonstrated good tolerability to SXB (Ondo et al. 2008). In another study, George et al. randomized 48 patients with mild-to-moderate OSA (mean AHI 23.5/h) to receive a low dose of SXB (4.5 g/night) or placebo who were examined using polysomnography (PSG) at baseline and after 14 days (George et al. 2011). SXB did not increase AHI, obstructive apnea events, or central apneas and did not lower SaO2 (George et al. 2011). However, other reports have raised the concern that the effect of SXB on SDB may be unpredictable and variable even in patients with mild OSA (Seeck-Hirschner et al. 2009). Therefore, caution is advised when treating narcoleptics with concurrent SDB, particularly those with severe OSA, and physicians should confirm that patients with concurrent OSA are compliant with positive airway pressure (PAP) therapy before initiating SXB. It is important to stress that the physician should confirm 100 % compliance with CPAP therapy in narcoleptics with concurrent OSA before prescribing SXB. In addition, the patients should understand that SXB may worsen AHI and oxygen saturation if PAP is not used. We recommend that patients at a high risk for OSA or sleep hypoventilation should be monitored with PSG and CO_2 monitoring prior to initiating SXB (Alshaikh et al. 2011). Those with significant SDB should receive PAP therapy before initiating SXB. In our practice, we perform PSG on all patients on SXB once the optimal dose has been reached (Alshaikh et al. 2011).

8.4 SXB and Death

There was concern regarding whether the administration of SXB may increase mortality (Wang et al. 2009; Zvosec et al. 2009). The results of a post-marketing safety study of the period between 2002 and 2008 reported that 30 fatalities had occurred in patients prescribed SXB; however, the majority of deaths were determined to be unrelated to the medication (Wang et al. 2009). Worldwide, there have been a total of 227 deaths in patients who received SXB via prescription from the period of market introduction in 2002 through to May 31, 2011 (Wang et al. 2011). However, deaths were not proven to be related to the use of SXB (Feldman 2009, 2011). A subsequent correction (Wang et al. 2011) of the original data reported

(Wang et al. 2009) was published where the investigators reanalyzed and updated the data and corrected the previously reported numbers. The reanalysis did not show significantly increased mortality in patients taking SXB (Wang et al. 2011). The FDA stated that it was impossible to determine whether the cause of death was attributable to SXB (Xyrem) because many of the death reports were poorly documented or incomplete (FDA Drug Safety Communication 2012, Xyrem®). A number of deaths occurred in patients who were reported to be concurrently taking medications that could depress the central nervous system and respiratory system, such as neuroleptics, benzodiazepines, opioids, and others, and in other patients, alcohol was ingested while taking SXB (FDA Drug Safety Communication 2012, Xyrem®). Moreover, many deaths occurred in patients who were prescribed SXB at doses that exceeded the recommended maximum dose, those who underwent a rapid dose titration that did not adhere to the recommended protocol in the product label, or in patients who were prescribed SXB for unapproved uses, such as fibromyalgia, insomnia, or migraine (FDA Drug Safety Communication 2012, Xyrem®). Nevertheless, the above data should serve as a reminder to physicians to use SXB with great caution.

8.5 *Other Concerns*

SXB is expected to potentiate the central nervous system and respiratory depressant effects of drugs including alcohol, narcotic analgesics, benzodiazepines, and other drugs with sedative effects (Li et al. 1998). Therefore, extreme caution is required in patients receiving these medications.

On rare occasions, psychiatric complications have been reported in patients taking SXB including anxiety and depression (Russell et al. 2011). Therefore, caution is required prior to initiating SXB in patients with psychiatric comorbidities. In these cases, appropriate psychiatric assessment is recommended.

9 Key Points

- SXB (Xyrem®) is an orphan drug that was approved by the FDA for the treatment of narcolepsy with cataplexy in 2002 and for the treatment of excessive daytime sleepiness in narcolepsy in 2005. In Europe, the EMEA approved SXB for the treatment of narcolepsy with cataplexy in 2005. The active drug in SXB is sodium oxybate, the sodium salt of gamma hydroxybutyric acid (GHB).
- SXB has demonstrated clear clinical benefits in patients with narcolepsy with cataplexy in placebo-controlled trials and in reports of case series including some long-term follow-up trials with clear clinical benefits for both cataplexy and daytime sleepiness.

- SXB is generally well tolerated and the most common drug-related adverse events are nausea, vomiting, dizziness, and urinary incontinence.
- SXB is a drug with a potential for misuse and abuse. In addition, SXB has a narrow safety margin with a risk of toxicity. Therefore, strict risk-management strategies and strict adherence to the titration schedule is essential.

Acknowledgments This work was supported by the National Plan for Science and Technology program by the King Saud University Project in Saudi Arabia.

Competing Interests/Disclosure Statement S.R. Pandi-Perumal is a stockholder and the President and Chief Executive Officer of Somnogen Canada Inc., a Canadian Corporation. He declares that he has no competing interests that could be perceived to influence the content of this article. AS BaHammam declares that he has no proprietary, financial, professional, or any other personal interest of any nature in any product or services and/or company that could be construed or considered to be a potential conflict of interest and that may have influenced the views expressed in this chapter.

References

Alshaikh MK, Gacuan D, George S, Sharif M, Bahammam AS (2011) Long-term follow-up of patients with narcolepsy-cataplexy treated with sodium oxybate (xyrem). Clin Neuropharmacol 34(1):1–4. doi:10.1097/WNF.0b013e318203d415

Alshaikh MK, Tricco AC, Tashkandi M, Mamdani M, Straus SE, BaHammam AS (2012) Sodium oxybate for narcolepsy with cataplexy: systematic review and meta-analysis. J Clin Sleep Med 8(4):451–458. doi:10.5664/jcsm.2048

BaHammam AS, Alenezi AM (2006) Narcolepsy in Saudi Arabia. Demographic and clinical perspective of an under-recognized disorder. Saudi Med J 27(9):1352–1357, 20060245' [pii]

Baker TL, Guilleminault C, Nino-Murcia G, Dement WC (1986) Comparative polysomnographic study of narcolepsy and idiopathic central nervous system hypersomnia. Sleep 9 (1 Pt 2):232–242

Billiard M, Bassetti C, Dauvilliers Y, Dolenc-Groselj L, Lammers GJ, Mayer G, Pollmacher T, Reading P, Sonka K (2006) EFNS guidelines on management of narcolepsy. Eur J Neurol 13 (10):1035–1048. doi:10.1111/j.1468-1331.2006.01473.x, ENE1473 [pii]

Black J, Houghton WC (2006) Sodium oxybate improves excessive daytime sleepiness in narcolepsy. Sleep 29(7):939–946

Black J, Pardi D, Hornfeldt CS, Inhaber N (2009) The nightly administration of sodium oxybate results in significant reduction in the nocturnal sleep disruption of patients with narcolepsy. Sleep Med 10(8):829–835. doi:10.1016/j.sleep.2009.05.004, S1389-9457(09)00223-8 [pii]

Black J, Pardi D, Hornfeldt CS, Inhaber N (2010) The nightly use of sodium oxybate is associated with a reduction in nocturnal sleep disruption: a double-blind, placebo-controlled study in patients with narcolepsy. J Clin Sleep Med 6(6):596–602

Boscolo-Berto R, Viel G, Montagnese S, Raduazzo DI, Ferrara SD, Dauvilliers Y (2012) Narcolepsy and effectiveness of gamma-hydroxybutyrate (GHB): a systematic review and meta-analysis of randomized controlled trials. Sleep Med Rev 16(5):431–443. doi:10.1016/j.smrv.2011.09.001

Broughton R, Mamelak M (1979) The treatment of narcolepsy-cataplexy with nocturnal gamma-hydroxybutyrate. Can J Neurol Sci 6(1):1–6

Broughton R, Mamelak M (1980) Effects of nocturnal gamma-hydroxybutyrate on sleep/waking patterns in narcolepsy-cataplexy. Can J Neurol Sci 7(1):23–31

Dauvilliers Y, Billiard M, Montplaisir J (2003) Clinical aspects and pathophysiology of narcolepsy. Clin Neurophysiol 114(11):2000–2017, S1388245703002037 [pii]

Dauvilliers Y, Arnulf I, Mignot E (2007) Narcolepsy with cataplexy. Lancet 369(9560):499–511. doi:10.1016/S0140-6736(07)60237-2, S0140-6736(07)60237-2 [pii]

FDA Drug Safety Communication (2012) Warning against use of Xyrem (sodium oxybate) with alcohol or drugs causing respiratory depression. Available from http://www.fda.gov/Drugs/DrugSafety/ucm332029.htm. Accessed 2 Apr 2014

Feldman NT (2009) Xyrem safety: the debate continues. Sleep Med 10(4):405–406. doi:10.1016/j.sleep.2009.02.002

Feldman NT (2011) Sodium oxybate post-marketing surveillance. J Clin Sleep Med 7(4):417. doi:10.5664/JCSM.1216

Fuller DE, Hornfeldt CS, Kelloway JS, Stahl PJ, Anderson TF (2004) The Xyrem risk management program. Drug Saf 27(5):293–306

George CF, Feldman N, Inhaber N, Steininger TL, Grzeschik SM, Lai C, Zheng Y (2010) A safety trial of sodium oxybate in patients with obstructive sleep apnea: acute effects on sleep-disordered breathing. Sleep Med 11(1):38–42. doi:10.1016/j.sleep.2009.06.006, S1389-9457(09)00386-4 [pii]

George CF, Feldman N, Zheng Y, Steininger TL, Grzeschik SM, Lai C, Inhaber N (2011) A 2-week, polysomnographic, safety study of sodium oxybate in obstructive sleep apnea syndrome. Sleep Breath 15(1):13–20. doi:10.1007/s11325-009-0320-0

Huang YS, Guilleminault C (2009) Narcolepsy: action of two gamma-aminobutyric acid type B agonists, baclofen and sodium oxybate. Pediatr Neurol 41(1):9–16. doi:10.1016/j.pediatrneurol.2009.02.008, S0887-8994(09)00082-4 [pii]

Husain AM, Ristanovic RK, Bogan RK (2009) Weight loss in narcolepsy patients treated with sodium oxybate. Sleep Med 10(6):661–663. doi:10.1016/j.sleep.2008.05.012, S1389-9457(08)00207-4 [pii]

Keam S, Walker MC (2007) Therapies for narcolepsy with or without cataplexy: evidence-based review. Curr Opin Neurol 20(6):699–703. doi:10.1097/WCO.0b013e3282f22ad9, 00019052-200712000-00017 [pii]

Koek W, France CP (2008) Cataleptic effects of gamma-hydroxybutyrate (GHB) and baclofen in mice: mediation by GABA(B) receptors, but differential enhancement by N-methyl-d-aspartate (NMDA) receptor antagonists. Psychopharmacology (Berl) 199(2):191–198. doi:10.1007/s00213-008-1160-5

Lammers GJ, Arends J, Declerck AC, Ferrari MD, Schouwink G, Troost J (1993) Gammahydroxybutyrate and narcolepsy: a double-blind placebo-controlled study. Sleep 16(3):216–220

Li J, Stokes SA, Woeckener A (1998) A tale of novel intoxication: seven cases of gamma-hydroxybutyric acid overdose. Ann Emerg Med 31(6):723–728

Littner MR, Kushida C, Wise M, Davila DG, Morgenthaler T, Lee-Chiong T, Hirshkowitz M, Daniel LL, Bailey D, Berry RB, Kapen S, Kramer M (2005) Practice parameters for clinical use of the multiple sleep latency test and the maintenance of wakefulness test. Sleep 28 (1):113–121

Longstreth WT Jr, Koepsell TD, Ton TG, Hendrickson AF, van Belle G (2007) The epidemiology of narcolepsy. Sleep 30(1):13–26

Mamelak M, Black J, Montplaisir J, Ristanovic R (2004) A pilot study on the effects of sodium oxybate on sleep architecture and daytime alertness in narcolepsy. Sleep 27(7):1327–1334

Mansukhani MP, Kotagal S (2012) Sodium oxybate in the treatment of childhood narcolepsy-cataplexy: a retrospective study. Sleep Med 13(6):606–610. doi:10.1016/j.sleep.2011.10.032

Mason PE, Kerns WP 2nd (2002) Gamma hydroxybutyric acid (GHB) intoxication. Acad Emerg Med 9(7):730–739

Merino-Andreu M, Martinez-Bermejo A (2009) Narcolepsy with and without cataplexy: an uncommon disabling and unrecognized disease. An Pediatr (Barc) 71(6):524–534. doi:10.1016/j.anpedi.2009.07.028, S1695-4033(09)00467-6 [pii]

Mignot EJ (2012) A practical guide to the therapy of narcolepsy and hypersomnia syndromes. Neurotherapeutics 9(4):739–752. doi:10.1007/s13311-012-0150-9

Murali H, Kotagal S (2006) Off-label treatment of severe childhood narcolepsy-cataplexy with sodium oxybate. Sleep 29(8):1025–1029

Ondo WG, Perkins T, Swick T, Hull KL Jr, Jimenez JE, Garris TS, Pardi D (2008) Sodium oxybate for excessive daytime sleepiness in Parkinson disease: an open-label polysomnographic study. Arch Neurol 65(10):1337–1340. doi:10.1001/archneur.65.10.1337, 65/10/1337 [pii]

Owen RT (2008) Sodium oxybate: efficacy, safety and tolerability in the treatment of narcolepsy with or without cataplexy. Drugs Today (Barc) 44(3):197–204. doi:10.1358/dot.2008.44.3.1162240, 1162240 [pii]

Pardi D, Black J (2006) Gamma-Hydroxybutyrate/sodium oxybate: neurobiology, and impact on sleep and wakefulness. CNS Drugs 20(12):993–1018

Poli F, Ricotta L, Vandi S, Franceschini C, Pizza F, Palaia V, Moghadam KK, Banal D, Vetrugno R, Thorpy MJ, Plazzi G (2012) Catathrenia under sodium oxybate in narcolepsy with cataplexy. Sleep Breath 16(2):427–434. doi:10.1007/s11325-011-0520-2

Robinson DM, Keating GM (2007) Sodium oxybate: a review of its use in the management of narcolepsy. CNS Drugs 21(4):337–354, 2147 [pii]

Russell IJ, Holman AJ, Swick TJ, Alvarez-Horine S, Wang YG, Guinta D (2011) Sodium oxybate reduces pain, fatigue, and sleep disturbance and improves functionality in fibromyalgia: results from a 14-week, randomized, double-blind, placebo-controlled study. Pain 152(5):1007–1017. doi:10.1016/j.pain.2010.12.022

Sakurai T (2007) The neural circuit of orexin (hypocretin): maintaining sleep and wakefulness. Nat Rev Neurosci 8(3):171–181. doi:10.1038/nrn2092, nrn2092 [pii]

Sansa G, Iranzo A, Santamaria J (2010) Obstructive sleep apnea in narcolepsy. Sleep Med 11 (1):93–95. doi:10.1016/j.sleep.2009.02.009, S1389-9457(09)00283-4 [pii]

Scharf MB, Brown D, Woods M, Brown L, Hirschowitz J (1985) The effects and effectiveness of gamma-hydroxybutyrate in patients with narcolepsy. J Clin Psychiatry 46(6):222–225

Scrima L, Hartman PG, Johnson FH Jr, Hiller FC (1989) Efficacy of gamma-hydroxybutyrate versus placebo in treating narcolepsy-cataplexy: double-blind subjective measures. Biol Psychiatry 26(4):331–343

Scrima L, Hartman PG, Johnson FH Jr, Thomas EE, Hiller FC (1990) The effects of gamma-hydroxybutyrate on the sleep of narcolepsy patients: a double-blind study. Sleep 13 (6):479–490

Seeck-Hirschner M, Baier PC, von Freier A, Aldenhoff J, Goder R (2009) Increase in sleep-related breathing disturbances after treatment with sodium oxybate in patients with narcolepsy and mild obstructive sleep apnea syndrome: two case reports. Sleep Med 10(1):154–155. doi:10.1016/j.sleep.2007.11.018, S1389-9457(07)00430-3 [pii]

Sellayah D, Bharaj P, Sikder D (2011) Orexin is required for brown adipose tissue development, differentiation, and function. Cell Metab 14(4):478–490. doi:10.1016/j.cmet.2011.08.010

Silber MH, Krahn LE, Olson EJ, Pankratz VS (2002) The epidemiology of narcolepsy in Olmsted County, Minnesota: a population-based study. Sleep 25(2):197–202

Stell JM, Ryan JM (1996) Ecstasy and neurodegeneration. gamma-Hydroxybutyrate is a new recreational drug that may lead to loss of consciousness. BMJ 313(7054):424

The U.S. Xyremâ Multicenter Study Group (2002) A randomized, double blind, placebo-controlled multicenter trial comparing the effects of three doses of orally administered sodium oxybate with placebo for the treatment of narcolepsy. Sleep 25(1):42–49

U.S. Xyrem Multicenter Study Group (2003) A 12-month, open-label, multicenter extension trial of orally administered sodium oxybate for the treatment of narcolepsy. Sleep 26(1):31–35

U.S. Xyrem Multicenter Study Group (2004) Sodium oxybate demonstrates long-term efficacy for the treatment of cataplexy in patients with narcolepsy. Sleep Med 5(2):119–123. doi:10.1016/j.sleep.2003.11.002, S1389945703002491 [pii]

U.S. Xyrem Multi-Center Study Group (2003) The abrupt cessation of therapeutically administered sodium oxybate (GHB) does not cause withdrawal symptoms. J Toxicol Clin Toxicol 41 (2):131–135

Wallace DM, Maze T, Shafazand S (2011) Sodium oxybate-induced sleep driving and sleep-related eating disorder. J Clin Sleep Med 7(3):310–311. doi:10.5664/JCSM.1082

Wang YG, Swick TJ, Carter LP, Thorpy MJ, Benowitz NL (2009) Safety overview of postmarketing and clinical experience of sodium oxybate (Xyrem): abuse, misuse, dependence, and diversion. J Clin Sleep Med 5(4):365–371

Wang YG, Swick TJ, Carter LP, Thorpy MJ, Benowitz NL (2011) Sodium oxybate: updates and correction to previously published safety data. J Clin Sleep Med 7(4):415–416. doi:10.5664/JCSM.1214

Weaver TE, Cuellar N (2006) A randomized trial evaluating the effectiveness of sodium oxybate therapy on quality of life in narcolepsy. Sleep 29(9):1189–1194

Wise MS (1998) Childhood narcolepsy. Neurology 50(2 Suppl 1):S37–42

Xyrem International Study Group (2005a) A double-blind, placebo-controlled study demonstrates sodium oxybate is effective for the treatment of excessive daytime sleepiness in narcolepsy. J Clin Sleep Med 1(4):391–397

Xyrem International Study Group (2005b) Further evidence supporting the use of sodium oxybate for the treatment of cataplexy: a double-blind, placebo-controlled study in 228 patients. Sleep Med 6(5):415–421. doi:10.1016/j.sleep.2005.03.010, S1389-9457(05)00074-2 [pii]

Xyrem® (sodium oxybate) oral solution. Jazz Pharmaceuticals Inc [online]. Available from: http://www.xyrem.com/healthcare-professionals/safety-information. Accessed Jan 2014

Zaharna M, Dimitriu A, Guilleminault C (2010) Expert opinion on pharmacotherapy of narcolepsy. Expert Opin Pharmacother 11(10):1633–1645. doi:10.1517/14656566.2010.484021

Zvosec DL, Smith SW, Hall BJ (2009) Three deaths associated with use of Xyrem. Sleep Med 10 (4):490–493. doi:10.1016/j.sleep.2009.01.005, S1389-9457(09)00013-6 [pii]

Cell Therapy to Treat Narcolepsy

Oscar Arias-Carrión

Abstract The sleep disorder narcolepsy may now be considered a neurodegenerative disease, as there is a massive reduction in the number of neurons containing the neuropeptide, hypocretin/orexin (Hcrt). Hcrt neurons are solely located in the hypothalamus, particularly in its perifornical, dorsomedial, and lateral portions. Hcrt fibers widely project throughout the brain and generally have excitatory effects on their postsynaptic cells. Patients with narcolepsy have a severe reduction in the levels of Hcrt in the cerebrospinal fluid (CSF), a finding consistent with Hcrt neuronal loss. Experimental models have been generated in order to study the physiology of the Hcrt system and narcolepsy. In consequence, the Hcrt neuronal loss provokes depletion in CSF Hcrt levels and displays behavioral narcoleptic-like symptoms. Given that the Hcrt depletion is the hallmark of narcolepsy, these models represent the opportunity to explore the potential use of transplants as a therapeutical tool in order to treat narcolepsy.

1 Introduction

The hypocretins (Hcrt-1 and -2, also named as orexin-A and -B, respectively) are two neuropeptides derived from the same precursor whose expression is restricted to a few thousand neurons of the lateral hypothalamus (Trivedi et al. 1998; Hervieu et al. 2001; España et al. 2005). Hcrt-1 consists of 33 amino acids, whereas Hcrt-2 is a 28 amino acid molecule.

Hcrt neurons are located between the rat fornix and the mammillothalamic tracts in the lateral hypothalamus (LH) from where hypocretins fiber project throughout the brain and spinal cord, including several areas implicated in regulations of the sleep/wake cycle (Peyron et al. 1998, Sakurai et al. 1998; Trivedi et al. 1998; Chemelli et al. 1999; Hervieu et al. 2001). Hcrt fibers widely project throughout the brain and spinal cord and generally have excitatory effects on serotonergic (Hagan et al. 1999; Brown et al. 2002), noradrenergic (Bayer et al. 2002), histaminergic

O. Arias-Carrión (✉)
Unidad de Trastornos del Movimiento y Sueño (TMS), Hospital General Dr. Manuel Gea González/IFC-UNAM, Calzada de Tlalpan 4800, Delegación Tlalpan, 14080 México DF, Mexico
e-mail: arias@ciencias.unam.mx

A. Guglietta (ed.), *Drug Treatment of Sleep Disorders*, Milestones in Drug Therapy,
DOI 10.1007/978-3-319-11514-6_12

(Bayer et al. 2001), and cholinergic neurons in basal forebrain (Eggermann et al. 2001), laterodorsal tegmental nucleus (Takahashi et al. 2002), as well as thalamocortical neurons of thalamus (Bayer et al. 2002). Basically, Hcrt neurons regulate arousal and have been shown to be implicated in food reward and drug-seeking behavior (Sakurai 2007).

2 Narcolepsy and Hypocretin Neurons: The Link

Narcolepsy is a debilitating neurological disease characterized by excessive daytime sleepiness, premature transitions to rapid eye movement sleep (named "sleep-onset REM periods"), and cataplexy (sudden bilateral skeletal muscle weakness without impairment of conscience) (for review see Siegel and Boehmer 2006). Hcrt was linked to human narcolepsy since it was discovered in a mutation in the Hcrt receptor in mice (Chemelli et al. 1999) and dogs (Lin et al. 1999). Once it was speculated that a deficiency in the Hcrt system might be the explanation for human narcolepsy, diverse groups explored this possible link. Indeed, Hcrt system was linked to human narcolepsy.

In a study of postmortem brains of human narcoleptics, for instance, a massive reduction in the number of Hcrt-containing cells (85–98 %) was discovered compared with healthy controls (Peyron et al. 2000; Thannickal et al. 2000). On the other hand, narcoleptic patients have reduced cerebrospinal fluid (CSF) levels of Hcrt (Dalal et al. 2001; Kanbayashi et al. 2002; Mignot et al. 2002; Nishino et al. 2000, 2001; Ripley et al. 2001), a finding consistent with the loss of Hcrt-containing neurons as mentioned above. CSF measurements of Hcrt provide a valuable diagnostic tool for narcolepsy, separating narcolepsy from other sleep and neurological disorders (Mignot et al. 2002; Ripley et al. 2001).

3 Experimental Models of Narcolepsy

There are experimental models that mimic the sleep disorder narcolepsy; for instance, Hcrt/orexin genes knockout mice (Chemelli et al. 1999), canines with a mutation in the Hcrt-2 receptor (Lin et al. 1999), or mice with a targeted destruction of the Hcrt neurons (Hara et al. 2001) that exhibit symptoms of narcolepsy.

Recently, our group and others have generated an experimental model using a toxin which represents a very reliable procedure. This model consists of a ribosome-inactivating protein saporin (SAP) (Stirpe et al. 1992) that is conjugated to the hypocretin/orexin receptor that binds ligand hypocretin-2/orexin-B (Hcrt-2) to lesion Hcrt receptor-bearing neurons. It is known that the LH contains a high concentration of Hcrt receptor mRNA (Trivedi et al. 1998) as well as immunoreactivity (Hervieu et al. 2001). This fact indicates the presence of the Hcrt receptor on Hcrt neurons. When the Hcrt-2/SAP is injected into LH of rats, the toxin lesions Hcrt neurons, and produces behavioral symptoms that are characteristic of narcolepsy.

Hcrt-2/SAP binds to cells containing the Hcrt receptor but does not bind to cells that do not contain the Hcrt receptor. This indicates the specificity of the toxin. For instance, the diurnal rhythm of wakefulness (W) and slow-wave sleep (SWS) in a rat lesion is attenuated. This is because of an increase in sleep during the lights-off period. Hcrt-2/SAP increases SWS over the 24 h period. Total time spent in SWS and rapid eye movement sleep (REMS) was found to correlate with a decline in number of Hcrt neurons.

Using this model, Gerashchenko et al. (2001) showed that Hcrt-2/SAP induced more SWS and REM sleep at night and multiple periods of abnormal behavioral arrest during purposeful behavior. Indeed, this experimental model of narcolepsy provides a method of investigating the contribution of the Hcrt system to the regulation of the sleep–wake cycle and its relationship with narcolepsy. Gerashchenko et al. (2003a) also reported that two concentrations (90 ng or 490 ng/0.5 μl) of the Hcrt-2/SAP injected directly to the lateral hypothalamus caused a significant Hcrt cell loss. Narcoleptic-like sleep behavior was produced by both concentrations of this toxin (Gerashchenko et al. 2003a) (Fig. 1).

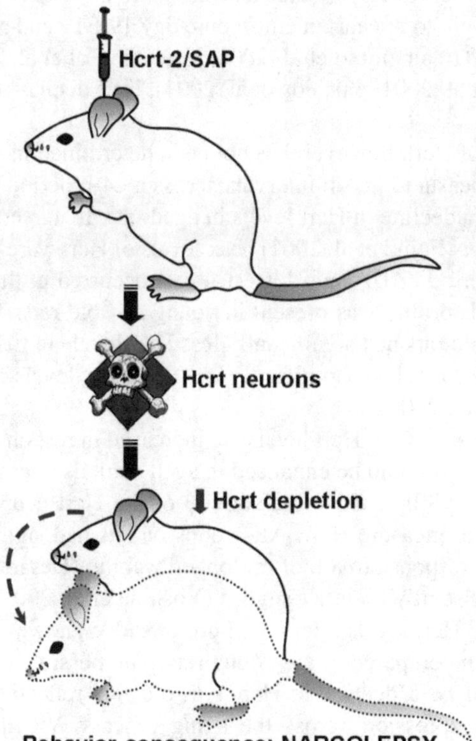

Fig. 1 Experimental procedure showing the injection of the Hcrt-2/SAP into lateral hypothalamus (LH). The Hcrt-2/SAP induces Hcrt neuronal loss leading to a diminution in levels of Hcrt in CSF. Narcolepsy is the behavioral consequence as a result of the Hcrt neuronal loss and diminution levels in CSF

As a conclusion, the use of Hcrt-2/SAP induced characteristic of narcolepsy, such as sleep fragmentation, sleep-onset REM sleep periods, increased NREM sleep, and REM sleep time during the normally active lights-off period (Gerashchenko et al. 2001, 2003a).

The relationship between Hcrt neuronal loss and changes in levels of the peptide was not known. Using the Hcrt-2/SAP to lesion neurons in the LH, a 50 % reduction in CSF Hcrt levels was found when 73 % of the number of Hcrt neurons was lost (Gerashchenko et al. 2003b). The sleep deprivation method is used to test the homeostatic mechanism of sleep regulation. It was also found that in lesioned rats, the Hcrt levels were not increased by 6 h prolonged W, indicating that surviving neurons were not able to increase the output of Hcrt into CSF to compensate for the Hcrt neuronal loss.

In the same study, the authors showed that Hcrt levels in CSF measured at different times of the day–night cycle showed that control rats had significantly higher Hcrt levels at ZT0 (+72.9 %) compared to a different time point (ZT8). The rats with a 72 % Hcrt neuronal loss did not show a significant difference between ZT0 and ZT8. Surprisingly, a significant correlation was found between Hcrt neurons and Hcrt levels at ZT0 (Gerashchenko et al. 2003b).

Hcrt neurons begin to appear on embryonic day 19 E19 and are fully developed by postnatal day 20 (Yamamoto et al. 2000; Van Den Pol et al. 2001). As reported by others (Fujiki et al. 2001; Yoshida et al. 2001), Hcrt display a diurnal rhythm in young animals.

Such a rhythm of Hcrt, however, has not been determined in aged animals. CSF Hcrt levels were measured at 4-h intervals across a 24-h period to test the hypothesis that there was a decline in Hcrt levels in aged rats. In agreement with previous studies in young rats (Fujiki et al. 2001) peak levels of Hcrt were found at the end of the wake-active period (ZT0), and lowest levels occurred at the end of the sleep period (ZT12). This profile was present in young and old rats. The authors found, however, that the old rats had significantly less Hcrt levels in CSF compared to the young rats. The aged rats had significantly less CSF Hcrt levels compared to young rats (Desarnaud et al. 2004).

As mentioned above, CSF Hcrt levels are increased in response to prolonged W; we tested if Hcrt levels could be enhanced in aged animals after prolonged W. Rats were kept awake for 8 h to drive the activity of the Hcrt neurons, and CSF was collected in order to measure Hcrt. All groups of rats had significantly increased CSF Hcrt levels in response to 8 h of prolonged waking (Desarnaud et al. 2004), a finding that is consistent with other studies (Yoshida et al. 2001).

The overall CSF Hcrt levels after 8 h of prolonged W, however, were still lower in the old rats when compared to the young rats. The possible explanation for this phenomenon might be a decline in Hcrt-1 in old rats reflecting a diminution in prepro-Hcrt gene expression across the aging process. We therefore measured prepro-Hcrt mRNA levels in the posterior hypothalamus of young and old rats by Northern blot analysis. No difference in mRNA between age group was found (Desarnaud et al. 2004).

4 Therapies for Narcolepsy

What are the approaches that currently are used to treat narcolepsy? The goal of all therapeutic approaches in narcolepsy is to control the narcoleptic symptoms and to allow the patient to continue full participation in familial and professional daily activities. Pharmacological treatments for narcolepsy include the use of amphetamines, modafinil, whereas cataplexy is treated with tricyclic antidepressants such as clomipramine (Mignot et al. 2002; Scammell 2003; Mignot and Nishino 2005).

Another possible experimental approach involves Hcrt cell replacement therapy. We have shown recently the use of cell Hcrt transplantation might represent a new approach to treat this disease (Arias-Carrión et al. 2004, 2006; Arias-Carrión and Murillo-Rodríguez 2014).

5 Transplants as a New Therapy for Narcolepsy

Neural transplantation is one of the most promising approaches for the treatment of Parkinson's disease (PD), a major neurodegenerative disorder with a prevalence as frequent as that of narcolepsy (Arias-Carrión et al. 2004). Neural transplantation involves implantation of living neuronal tissue into a host system. Several studies using animal models have demonstrated that grafted tissue survives, integrates within the host brain, and provides functional recovery following brain inventions (Drucker-Colín and Verdugo-Díaz 2004; Lindvall et al. 2004). This type of study for PD began in the latter half of the 1970s. Initially, DA neurons from animal fetuses were used as donors, and then paraneurons such as chromaffin cells were used (Drucker-Colín and Verdugo-Díaz 2004). Based on a large number of experimental animal studies, neural transplantation has been applied clinically (Drucker-Colín and Verdugo-Díaz 2004; Lindvall et al. 2004). Studies in patients with PD after intrastriatal transplantation of human fetal mesencephalic tissue, rich in postmitotic dopaminergic neurons, have provided proof of principle that neuronal replacement can work in the human brain (Lindvall et al. 2004). The grafted neurons survive and reinnervate the striatum for as long as 10 years despite an ongoing disease process, which destroys the patient's own dopaminergic neurons (Drucker-Colín and Verdugo-Díaz 2004). The grafts are able to normalize striatal dopamine release (Drucker-Colín and Verdugo-Díaz 2004) and to reverse the impairment of cortical activation underlying akinesia. Thus, grafted dopaminergic neurons can become functionally integrated into neuronal circuitries in the brain (Drucker-Colín and Verdugo-Díaz 2004; Lindvall et al. 2004). Several open-label trials have reported clinical benefit (for revision, see Lindvall et al. 2004). Some patients have been able to withdraw from L-dopa treatment for several years and resume an independent life (Drucker-Colín and Verdugo-Díaz 2004; Lindvall et al. 2004). Beneficial effects have been demonstrated, and autopsy cases have shown that many transplanted cells were able to survive in the human brain for long periods. These findings contributed a great deal to the research in regeneration of the central nervous system.

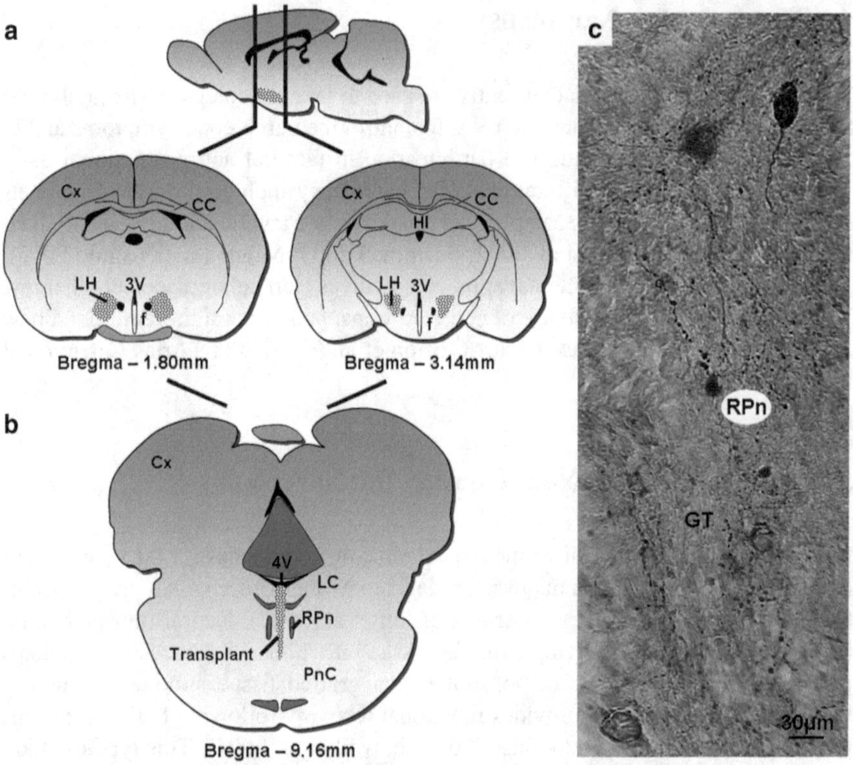

Fig. 2 (**a**) Representation of the brain sections taken from donor containing the hypothalamus for tissue preparation for transplant. (**b**) Targeted area within the brainstem for Hcrt-grafted neurons. (**c**) Microphotography of Hcrt survival cells in target area at 36 days after grafting. 3V: 3rd ventricle; 4V: 4th ventricle; Cx: cortex; CC: corpus callosum; f; fornix; LC; locus coeruleus; LH: lateral hypothalamus; PnC: pontine reticular nucleus; RPn: raphe pontis nucleus

Thus it would be interesting to know whether a graft of hypocretin neurons into a host brain could survive (Arias-Carrión et al. 2004, 2006; Arias-Carrión and Murillo-Rodríguez 2014). We demonstrated that Hcrt neurons suspension cells derived from posterior hypothalamus of 8–10-day old rat pups can survive when transplanted into the pons (a region of the brain that is innervated by hypocretin axons, but where the hypocretin somata are not present) in adults rats (Arias-Carrión et al. 2004, 2006; Arias-Carrión and Murillo-Rodríguez 2014). In our preliminary study, we found that well-defined hypocretin-immunoreactive somata with processes and varicosities were present in the graft zone 36 days after implantation of the cell suspension, suggesting that hypocretin neurons obtained from rat pups can be grafted into an adult host brain (Arias-Carrión et al. 2004). These somata were similar in size and appearance to adult rat hypocretin-immunoreactive neurons (Fig. 2).

Next, we investigated the time course of survival of grafted Hcrt neurons into the pons of adult rats (Arias-Carrión et al. 2006). Control rats received a transplant that consisted of cells from the cerebellum where no Hcrt neurons are present. All adult host rats were sacrificed 1, 3, 6, 9, 12, 24, or 36 days after grafting. Inmunohistochemistry was used to identify and count the presence of the Hcrt-grafted neurons in the target area. The tally of Hcrt neurons present in the graft zone 1 day post-grafting was considered to be the baseline. From day 3 to 36 post-transplant there was a steady decline in the number of Hcrt neurons. We also noted that on day 36, the Hcrt neurons that survived in the pons had morphological features that were similar to mature Hcrt neurons in the adult lateral hypothalamus, suggesting that these neurons might be functionally active. Control rats that received grafts of cerebellar tissue did not show Hcrt neurons in the target area. These results demonstrate that there is a progressive decline in the number of transplanted neurons, but a significant percentage of Hcrt neurons do survive until day 36. Some evidence exists that hypocretin neuron transplantation in rats somewhat diminished narcolepsy-like sleep behavior (Arias-Carrión and Murillo-Rodríguez 2014), but no studies in people have been planned. These studies, however, highlights the potential use of transplants as a therapeutical tool in order to treat narcolepsy.

Conclusions

Narcolepsy is a sleep disorder characterized by sleep attacks. Experimental models have been generated in order to understand the physiology of this disease. Recent evidence has concluded narcolepsy in humans and animal models the result from the failure of cellular signaling mediated by Hcrt.

Recently, we have described that Hcrt cells are able to be transplanted with the aim to generate an alternative therapeutical tool to treat narcolepsy. We have showed here that grafted cells express the machinery for Hcrt release, and possess the morphological feature of an Hcrt neuron. Recently, we reported for the first time that transplantation of Hcrt neurons into the LH of Hcrt2/SAP-lesioned rats diminishes narcoleptic-like sleep behavior. The following points, however, have to be demonstrated in the future: (a) electrophysiological properties of fully mature Hcrt neuron; (b) grafted Hcrt neurons should reestablish a dense, functional Hcrt releasing terminal network; and (c) grafts have to become functionally integrated into host circuitries.

Acknowledgments The author would like to thank the people working at the Unidad de Trastornos del Movimiento y Sueño (TMS), Hospital General Dr. Manuel Gea González.

References

Arias-Carrión O, Drucker-Colín R, Murillo-Rodríguez E (2006) Survival rates through time of hypocretin grafted neurons within their projection site. Neurosci Lett 404:93–97

Arias-Carrión O, Murillo-Rodríguez E (2014) Effects of hypocretin/orexin cell transplantation on narcoleptic-like sleep behavior in rats. PLoS One 9(4):e95342

Arias-Carrión O, Murillo-Rodriguez E, Xu M, Blanco-Centurion C, Drucker-Colín R, Shiromani PJ (2004) Transplant of hypocretin neurons into the pontine reticular formation: preliminary results. Sleep 27:1465–1470

Bayer L, Eggermann E, Saint-Mleux B, Machard D, Jones BE, Muhlethaler M, Serafin M (2002) Selective action of orexin (hypocretin) on nonspecific thalamocortical projection neurons. J Neurosci 22:7835–7839

Bayer L, Eggermann E, Serafin M, Saint-Mleux B, Machard D, Jones B, Muhlethaler M (2001) Orexins (hypocretins) directly excite tuberomammillary neurons. Eur J Neurosci 14:1571–1575

Brown RE, Sergeeva OA, Eriksson KS, Haas HL (2002) Convergent excitation of dorsal raphe serotonin neurons by multiple arousal systems (orexin/hypocretin, histamine and noradrenaline). J Neurosci 22:8850–8859

Chemelli RM, Willie JT, Sinton CM, Elmquist JK, Scammell T, Lee C, Richardson JA, Williams SC, Xiong Y, Kisanuki Y, Fitch TE, Nakazato M, Hammer RE, Saper CB, Yanagisawa M (1999) Narcolepsy in orexin knockout mice: molecular genetics of sleep regulation. Cell 98:437–451

Dalal MA, Schuld A, Haack M, Uhr M, Geisler P, Eisensehr I, Noachtar S, Pollmacher T (2001) Normal plasma levels of orexin A (hypocretin-1) in narcoleptic patients. Neurology 56:1749–1751

Desarnaud F, Murillo-Rodriguez E, Lin L, Xu M, Gerashchenko D, Shiromani SN, Nishino S, Mignot E, Shiromani PJ (2004) The diurnal rhythm of hypocretin in young and old F344 rats. Sleep 27:851–856

Drucker-Colín R, Verdugo-Díaz L (2004) Cell transplantation for Parkinson's disease: present status. Cell Mol Neurobiol 24:301–316

Eggermann E, Serafin M, Bayer L, Machard D, Saint-Mleux B, Jones BE, Muhlethaler M (2001) Orexins/hypocretins excite basal forebrain cholinergic neurones. Neuroscience 108:177–181

España RA, Reis KM, Valentino RJ, Berridge CW (2005) Organization of hypocretin/orexin efferents to locus coeruleus and basal forebrain arousal-related structures. J Comp Neurol 481:160–178

Fujiki N, Yoshida Y, Ripley B, Honda K, Mignot E, Nishino S (2001) Changes in CSF hypocretin-1 (orexin a) levels in rats across 24 hours and in response to food deprivation. Neuroreport 12:993–997

Gerashchenko D, Blanco-Centurion C, Greco MA, Shiromani PJ (2003a) Effects of lateral hypothalamic lesion with the neurotoxin hypocretin2-saporin on sleep in Long Evans rats. Neuroscience 116:223–235

Gerashchenko D, Kohls MD, Greco M, Waleh NS, Salin-Pascual R, Kilduff TS, Lappi DA, Shiromani PJ (2001) Hypocretin-2-saporin lesions of the lateral hypothalamus produce narcoleptic-like sleep behavior in the rat. J Neurosci 21:7273–83

Gerashchenko D, Murillo-Rodriguez E, Lin L, Xu M, Hallett L, Nishino S, Mignot E, Shiromani PJ (2003b) Relationship between CSF hypocretin levels and hypocretin neuronal loss. Exp Neurol 184:1010–1016

Hagan JJ, Leslie RA, Patel S, Evans ML, Wattam TA, Holmes S, Benham CD, Taylor SG, Routledge C, Hemmati P, Munton RP, Ashmeade TE, Shah AS, Hatcher JP, Hatcher PD, Jones DN, Smith MI, Piper DC, Hunter AJ, Porter RA, Upton N (1999) Orexin A activates locus coeruleus cell firing and increases arousal in the rat. Proc Natl Acad Sci USA 96:10911–10916

Hara J, Beuckmann CT, Nambu T, Willie JT, Chemelli RM, Sinton CM, Sugiyama F, Yagami K, Goto K, Yanagisawa M, Sakurai T (2001) Genetic ablation of orexin neurons in mice results in narcolepsy, hypophagia, and obesity. Neuron 30:345–354

Hervieu GJ, Cluderay JE, Harrison DC, Roberts JC, Leslie RA (2001) Gene expression and protein distribution of the orexin-1 receptor in the rat brain and spinal cord. Neuroscience 103:777–797

Kanbayashi T, Inoue Y, Chiba S, Aizawa R, Saito Y, Tsukamoto H, Fujii Y, Nishino S, Shimizu T (2002) CSF hypocretin-1 (orexin-A) concentrations in narcolepsy with and without cataplexy and idiopathic hypersomnia. J Sleep Res 11:91–93

Lin L, Faraco J, Li R, Kadotani H, Rogers W, Lin X, Qiu X, de Jong PJ, Nishino S, Mignot E (1999) The sleep disorder canine narcolepsy is caused by a mutation in the hypocretin (orexin) receptor 2 gene. Cell 98:365–376

Lindvall O, Lokaia Z, Martinez-Serrano A (2004) Stem cell therapy for human neurodegenerative disorders-how to make it work. Nat Med 10:S42–S50

Mignot E, Lammers GJ, Ripley B, Okun M, Nevsimalova S, Overeem S, Vankova J, Black J, Harsch J, Bassetti C, Schrader H, Nishino S (2002) The role of cerebrospinal fluid hypocretin measurement in the diagnosis of narcolepsy and other hypersomnias. Arch Neurol 59:1553–1562

Mignot E, Nishino S (2005) Emerging therapies in narcolepsy-cataplexy. Sleep 28:754–763

Nishino S, Ripley B, Overeem S, Lammers GJ, Mignot E (2000) Hypocretin (orexin) deficiency in human narcolepsy. Lancet 355:39–40

Nishino S, Ripley B, Overeem S, Nevsimalova S, Lammers GJ, Vankova J, Okun M, Rogers W, Brooks S, Mignot E (2001) Low cerebrospinal fluid hypocretin (Orexin) and altered energy homeostasis in human narcolepsy. Ann Neurol 50:381–388

Peyron C, Faraco J, Rogers W, Ripley B, Overeem S, Charnay Y, Nevsimalova S, Aldrich M, Reynolds D, Albin R, Li R, Hungs M, Pedrazzoli M, Padigaru M, Kucherlapati M, Fan J, Maki R, Lammers GJ, Bouras C, Kucherlapati R, Nishino S, Mignot E (2000) A mutation in a case of early onset narcolepsy and a generalized absence of hypocretin peptides in human narcoleptic brains. Nat Med 6:991–997

Peyron C, Tighe DK, Van den Pol AN, De Lecea L, Heller HC, Sutcliffe JG, Kilduff TS (1998) Neurons containing hypocretin (orexin) project to multiple neuronal systems. J Neurosci 18:9996–10015

Ripley B, Overeem S, Fujiki N, Nevsimalova S, Uchino M, Yesavage J, Di Monte D, Dohi K, Melberg A, Lammers GJ, Nishida Y, Roelandse FW, Hungs M, Mignot E, Nishino S (2001) CSF hypocretin/orexin levels in narcolepsy and other neurological conditions. Neurology 57:2253–2258

Sakurai T, Amemiya A, Ishii M, Matsuzaki I, Chemelli RM, Tanaka H, Williams SC, Richarson JA, Kozlowski GP, Wilson S, Arch JR, Buckingham RE, Haynes AC, Carr SA, Annan RS, McNulty DE, Liu WS, Terrett JA, Elshourbagy NA, Bergsma DJ, Yanagisawa M (1998) Orexins and orexin receptors: a family of hypothalamic neuropeptides and G protein-coupled receptors that regulate feeding behavior. Cell 92:573–585

Sakurai T (2007) The neural circuit of orexin (hypocretin): maintaining sleep and wakefulness. Nat Rev Neurosci 8:171–81

Scammell TE (2003) The neurobiology, diagnosis, and treatment of narcolepsy. Ann Neurol 53:154–166

Siegel JM, Boehmer LN (2006) Narcolepsy and the hypocretin system—where motion meets emotion. Nat Clin Pract Neurol 2:548–556

Stirpe F, Barbieri L, Battelli MG, Soria M, Lappi DA (1992) Ribosome-inactivating proteins from plants: present status and future prospects. Biotechnology (N Y) 10:405–12

Takahashi K, Koyama Y, Kayama Y, Yamamoto M (2002) Effects of orexin on the laterodorsal tegmental neurones. Psychiatry Clin Neurosci 56:335–336

Thannickal TC, Moore RY, Nienhuis R, Ramanathan L, Gulyani S, Aldrich M, Cornford M, Siegel JM (2000) Reduced number of hypocretin neurons in human narcolepsy. Neuron 27:469–474

Trivedi P, Yu H, MacNeil DJ, Van der Ploeg LH, Guan XM (1998) Distribution of orexin receptor mRNA in the rat brain. FEBS Lett 438:71–75

Van Den Pol AN, Patrylo PR, Ghosh PK, Gao XB (2001) Lateral hypothalamus: early developmental expression and response to hypocretin (orexin). J Comp Neurol 433:349–363

Yamamoto Y, Ueta Y, Hara Y, Serino R, Nomura M, Shibuya I, Shirahata A, Yamashita H (2000) Postnatal development of orexin/hypocretin in rats. Mol Brain Res 78:108–119

Yoshida Y, Fujiki N, Nakajima T, Ripley B, Matsumura H, Yoneda H, Mignot E, Nishino S (2001) Fluctuation of extracellular hypocretin-1 (orexin a) levels in the rat in relation to the light-dark cycle and sleep-wake activities. Eur J Neurosci 14:1075–81

Part IV
Drugs to Treat Circadian Sleep Disorders

Tasimelteon

David N. Neubauer, Ahmed S. BaHammam, and Seithikurppu R. Pandi-Perumal

Abstract The role of the melatonin in relation with circadian rhythms and the sleep/wake cycle is well established. It is thought to play a role in controlling the sleep–wake cycle through physiological processes regulated in the suprachiasmatic nucleus of the hypothalamus. In normal subjects, the endogenous circadian period is a little over 24 h, but is entrained to the 24-h day through exposure to light. In the absence of light or light perception (totally blind), the synchronization is lost and the circadian rhythm follows the intrinsic non-24 h clock, resulting in a non-24 h sleep–wake disorder. Non-24 h sleep–wake disorder is characterized by a misalignment of the 24-h light/dark (LD) cycle and a non-entrained sleep/wake cycle propensity resulting in asymptomatic periods alternating with episodes of insomnia, excessive daytime sleepiness (EDS), or a combination of both. Tasimelteon (VEC-162) is an orally bioavailable melatonin receptor agonist of the melatonin MT1 and MT2 receptors recently approved by the U.S. Food and Drug Administration (FDA) for the treatment of non-24 h sleep–wake rhythm disorder.

D.N. Neubauer (✉)
Johns Hopkins Bayview Medical Center, 4940 Eastern Avenue, Box 151, Baltimore, MD 21224, USA
e-mail: neubauer@jhmi.edu

A.S. BaHammam
Department of Medicine, College of Medicine, The University Sleep Disorders Center, King Saud University, Riyadh, Saudi Arabia

National Plan for Science and Technology, King Saud University, Riyadh, Saudi Arabia

S.R. Pandi-Perumal
Division of Health and Behavior, Department of Population Health, Clinical and Translational Research Institute, Center for Healthful Behavior Change, New York University Langone Medical Center, New York, NY, USA

Somnogen Canada Inc., Toronto, ON, Canada

A. Guglietta (ed.), *Drug Treatment of Sleep Disorders*, Milestones in Drug Therapy,
DOI 10.1007/978-3-319-11514-6_13

1 Introduction

Tasimelteon (VEC-162) is a melatonin receptor agonist recently approved by the U.S. Food and Drug Administration (FDA) for the treatment of non-24 h sleep–wake rhythm disorder (AASM 2014). It is marketed by Vanda Pharmaceuticals with the brand name Hetlioz. The compound initially was developed by Bristol-Myers Squibb (BMS-214778) and in 2004 was licensed to Vanda. Previous investigations with doses ranging from 10 to 100 mg included the potential treatment of transient insomnia associated with shifted times, primary insomnia, and major depression (Dhillon and Clarke 2014). Presently tasimelteon is available as a capsule with a single 20 mg dose. Initially the medication is provided only through a single pharmacy, although it is not a controlled substance and is not considered to have abuse potential (VandaPharmaceuticals 2014).

The role of the melatonin in relation with circadian rhythms and the sleep/wake cycle is well established (Lewy 2003; Cardinali et al. 2010; Hardeland et al. 2006; Pandi-Perumal et al. 2005, 2006, 2007). In normal circumstances endogenous melatonin is produced and secreted by the pineal gland under control of the anterior hypothalamic suprachiasmatic nuclei (SCN). Neurons within the SCN function as the master timekeepers of the circadian system with entrainment to the photoperiod through melanopsin photochemical effects within intrinsically photosensitive retinal ganglion cells (ipRGCs) and transmission through the retinal-hypothalamic tract (RHT). Normally in humans, melatonin levels are low throughout the daytime and into the early evening hours but then gradually rise as a person's natural bedtime approaches. Melatonin plateaus at a higher level throughout the sleep period and subsequently declines by the morning daylight hours. The melatonin onset and offset hours may depend on an individual's circadian predisposition (advanced/lark vs. delayed/night owl), sleep/wake habits, and the pattern of exposure to artificial light and darkness.

In humans, the circadian system promotes maximum arousal early in the evening. Sleep onset at one's habitual bedtime is facilitated by the rising melatonin, which decreases the SCN arousal signal. The SCN has high concentrations of the melatonin MT1 and MT2 receptor subtypes. Agonist activity at the MT1 subtype appears to decrease the circadian-promoted arousal while stimulation at the MT2 subtype reinforces circadian rhythmicity. The effects of exogenous melatonin on sleep and the circadian rhythm phase have been extensively investigated.

While meta-analysis studies of melatonin taken at bedtime have not demonstrated robust effects on sleep onset or maintenance, considerable evidence supports the phase-shifting effects of strategically timed melatonin and the therapeutic potential in treating clinical problems associated with abnormal intrinsic circadian phase patterns or challenges resulting from extrinsic schedule manipulations (Buscemi et al. 2005; Lewy 2003).

2 Non-24 Hour Sleep–Wake Rhythm Disorder

The International Classification of Sleep Disorders, 3rd Edition, includes the most recently updated nosology of circadian rhythm sleep–wake disorders. This category now includes delayed sleep–wake phase disorder, advanced sleep–wake phase disorder, irregular sleep–wake phase disorder, shift work disorder, jet lag disorder, and non-24 h sleep–wake rhythm disorder, which sometimes is labeled free-running disorder, nonentrained disorder, hypernychthemeral syndrome, or simply NON-24. The essential diagnostic criteria include the history consistent with a misalignment of the 24-h light/dark (LD) cycle and a nonentrained sleep/wake cycle propensity resulting in asymptomatic periods alternating with episodes of insomnia, excessive daytime sleepiness (EDS), or a combination of both. The symptoms should persist for at least 3 months, not be attributable to another disorder or the use of a medication or substance, and be supported by daily sleep logs and actigraphy for at least 2 weeks showing a progressive sleep–wake cycle delay with a circadian pattern greater than 24 h (AASM 2014).

The periodicity of the alternating pattern in NON-24 depends upon an individual's intrinsic circadian timing which may range from slightly greater than 24 h to approximately 25 h and in rare circumstances may be slightly less than 24 h (Sack and Lewy 2001). The repeating cycle may be as short as a few weeks or as long as a few months. A precise assessment of the pattern may be obtained through a dim light melatonin onset (DLMO) test or sequential measurements of urinary 6-sulfatoxymelatonin (aMT6s) (Keijzer et al. 2013; Pandi-Perumal et al. 2007; Smits et al. 2010). The population predominantly associated with NON-24 is people who are totally blind experiencing no light perception due to their inability to become entrained to the photoperiod. It is estimated that the majority of totally blind individuals meet criteria for NON-24 (Dressman et al. 2012). Rarely, sighted individuals also display this type of free-running progressive phase delay. Several studies have demonstrated the benefits of exogenous melatonin administration to free-running blind individuals to help stabilize their circadian systems allowing more consistent nighttime sleep and daytime alertness (Lockley et al. 2000; Lewy et al. 2003).

3 Pharmacology

3.1 *Chemistry*

Tasimelteon is a crystalline powder with a molecular weight of 245.32 and a molecular formula of $C_{15}H_{19}NO_2$. The chemical designation is ((1R-trans)-*N*-[[2-[2,3-dihydro-4-benzofuranyl]cyclopropoyl]methyl]propanamide) (Fig. 1).

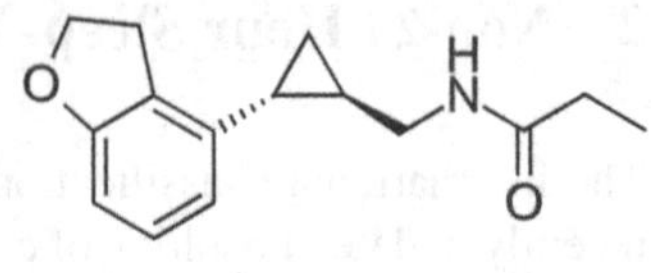

Fig. 1 Schematic figure of tasimelteon

3.2 *Pharmacodynamics*

Tasimelteon is a melatonin receptor agonist with greater affinity for the MT2 compared with the MT1 subtype. Presumably tasimelteon helps to synchronize the circadian system and consequently improve nighttime sleep and daytime wakefulness through action at these melatonin receptor subtypes.

3.3 *Pharmacokinetics*

Tasimelteon displays linear pharmacokinetics over a wide dosage range. It is rapidly absorbed reaching a peak concentration about 0.5–3 h after oral administration in the fasting state. Absorption is delayed with a decreased peak concentration following a high fat meal and it is recommended that it be taken without food. The apparent volume of distribution in young healthy subjects is approximately 56–126 L and at therapeutic concentrations it is approximately 90 % protein bound. The compound undergoes extensive metabolism primarily through the CYP1A2 and CYP3A4 pathways. Oxidation and oxidative dealkylation open the dihydrofuran ring ultimately forming carboxylic acid. Phenolic glucuronidation provides additional phase II metabolism. The major metabolites are at least 13-fold less active at melatonin receptors than the parent compound. Elimination is predominantly through the kidneys. Tasimelteon has an observed mean elimination half-life of 1.3 ± 0.4 h and a range of 1.3 ± 0.5 to 3.7 ± 2.2 h for the mean terminal elimination half-lives for the main metabolites. Repeated dosing does not result in significant accumulation (VandaPharmaceuticals 2014).

Studies in special populations show that the tasimelteon exposure is approximately double in elderly subjects and is about 20–30 % greater among females compared with males. The exposure is about 40 % less in smokers due to the induction of CYP1A2. No dosage adjustment is necessary for people with renal impairment. Similarly, no dosage adjust is necessary for individuals with mild to moderate hepatic impairment, although the exposure is approximately double with moderate hepatic impairment. It has not been evaluated and is not recommended for patients with severe hepatic impairment (VandaPharmaceuticals 2014).

Drug interaction studies show the potential for significant effects with the concomitant use of tasimelteon and drugs inhibiting or inducing the CYP1A2 or CYP3A4 metabolic pathways. Strong CYP1A2 inhibitors (e.g., fluvoxamine) should be avoided due to a marked increase in tasimelteon exposure. Strong CYP3A4 inhibitors (e.g., ketoconazole) may increase the tasimelteon to a modest extent. A strong CYP3A4 and moderate CYP2C19 inducer (e.g., rifampin) will decrease tasimelteon by about 90 % so should be avoided. The tasimelteon exposure may be reduced in smokers due to CYP1A2 induction. The combination of tasimelteon 20 mg with a single dose of alcohol demonstrated a trend for additive effects on psychomotor tests. Tasimelteon has not been shown to produce significant pharmacokinetic changes in other medications. The efficacy of tasimelteon may be reduced in patients concomitantly taking beta adrenergic receptor antagonists (VandaPharmaceuticals 2014).

4 Clinical Studies

Vanda has reported the results of two key trials of tasimelteon in the treatment of non-24 h sleep–wake rhythm disorder in totally blind individuals. The first study was called Safety and Efficacy of Tasimelteon (SET) and the second study, described as Randomized-withdrawal study of the Efficacy and Safety of Tasimelteon (RESET), examined discontinuation effects.

The SET study was a multicenter, double-masked, placebo-controlled trial of 84 totally blind patients ages 21–84 years (Lockley et al. 2013a). The subjects were given tasimelteon 20 mg at a fixed clock time about 1 h prior to their preferred bedtime for up to 6 months. Outcome measures included circadian rhythm assessments of urinary excretion of aMT6s and cortisol. Subjective estimates of nighttime sleep timing and daytime naps were recorded and were used to create a Non-24 Clinical Response Scale (N24CRS) score, for which a clinical response of ≥3 was considered evidence of entrainment. Due to varying severity of nighttime sleep difficulty and daytime sleepiness according to circadian alignment, efficacy was calculated by comparing the active drug and placebo subjects for the sleep time on the 25 % most symptomatic nights and 25 % most symptomatic days (napping). Compared with placebo, tasimelteon subjects experienced significantly greater increases in nighttime sleep and reductions in daytime nap time. A responder analysis based on increases of at least 45 min of nighttime sleep and decreases of at least 45 min of daytime nap time found that 29 % of tasimelteon subjects and 12 % in the placebo group achieved these criteria. Improvement also was assessed with the Clinical Global Impression of Change and a calculation of the Mid-point of Sleep Timing (MoST). During the study 79 subjects were assessed for entrainment. Compared with the placebo subjects, those taking tasimelteon were significantly more likely to be entrained as evidenced by the urinary aMT6s and cortisol patterns. The tasimelteon group also demonstrated significant improvement in N24CRS scores, MoST, and measures of nighttime and daytime sleep duration.

The RESET study was a multicenter, double-masked, placebo-controlled trial (Lockley et al. 2013b). The potential subjects were totally blind individuals given open-label tasimelteon 20 mg 1 hour prior to their bedtime during a 3 month period. Individuals demonstrating evidence of entrainment (confirmed with urinary aMT6s and cortisol) were randomized to continue the tasimelteon dose or were given placebo for 2 months. Assessments included the urinary measures and subjective estimates of nighttime and daytime nap sleep. Twenty entrained subjects (ages 28–70 years) participated in RESET. The tasimelteon subjects maintained entrainment to a significantly greater extent than the placebo group. Tasimelteon subjects had total nighttime sleep of 67.2 min more than placebo subjects on their worst quartile of nights. The total daytime sleep duration was 59.4 min longer in placebo subjects. The investigators concluded that tasimelteon discontinuation among these tasimelteon-entrained subjects resulted in a loss of entrainment and subsequent decreases in nighttime sleep and increases in daytime sleep.

Prior to the non-24 h sleep–wake disorder studies in totally blind subjects, Vanda performed clinical trials with tasimelteon (VEC-162) measuring circadian parameters in healthy individuals, in healthy subjects with induced transient insomnia, in the treatment of primary insomnia, and in the treatment of major depression (Rajaratnam et al. 2009; Lankford 2011). Positive results were reported for phase II ($n = 39$) and III ($n = 411$) sleep-time shift induced transient insomnia with evidence for improved sleep initiation and maintenance for tasimelteon (Rajaratnam et al. 2009). A phase III study ($n = 322$) of primary insomnia subjects measuring both subjective and objective (polysomnographic) outcomes showed significant improvements in latency to persistent sleep with tasimelteon 20 and 50 mg compared with placebo. Further development of tasimelteon for major depression was discontinued in 2013.

5 Safety

Vanda reports that in the clinical trials tasimelteon was safe and well tolerated. There were no significant next-day residual effects evident in the clinical trials. The adverse reactions reported in the clinical trials that occurred with an incidence greater than 5 % and were at least twice as frequent compared with placebo were headache (17 %), alanine aminotransferase increase (10 %), nightmares or unusual dreams, and upper respiratory (7 %) or urinary tract infections (7 %) (VandaPharmaceuticals 2014).

6 Regulatory Issues

Tasimelteon was granted orphan drug status for the treatment of non-24 h sleep wake disorder in blind individuals with no light perception in the USA in 2010 and by the European Commission in 2011. Vanda filed a New Drug Application with the US FDA in 2013 and the FDA approved the tasimelteon 20 mg dose for the treatment of non-24 h sleep–wake disorder in January, 2014. At the time of this writing tasimelteon has not yet been approved in other countries.

7 Prescribing Guidelines

The FDA approved prescribing information recommends that tasimelteon 20 mg be taken at the same time every night without food. The dose can be skipped if it cannot be taken at the usual time. The benefits of the medication may not be evident for several weeks or months of daily use due to individual differences in circadian rhythm patterns. Since tasimelteon can cause somnolence with potential performance impairment, individuals should limit their activities to preparing for bed after taking the medication. The safety and effectiveness of the medication have not been evaluated in pediatric populations. Tasimelteon has not been studied in pregnant women; however, animal study outcomes have led to a Pregnancy C categorization. It is unknown whether the compound is excreted in human milk. There are no contraindications to the use of the drug. There is no evidence of abuse or dependence with the use of tasimelteon (VandaPharmaceuticals 2014).

8 Discussion

Tasimelteon is a welcome addition to the US pharmacopoeia. As a melatonin receptor agonist it joins melatonin, available unregulated in the USA and by prescription in the EU, and ramelteon, approved in the USA and some additional countries. While it is approved for treating non-24-h sleep–wake rhythm disorder and having been developed as an orphan drug for totally blind individuals, both pharmacodynamic and pharmacokinetic properties suggest potential benefits for sleep-onset insomnia and additional circadian rhythm sleep–wake disorders. It would not be surprising to see further clinical trials of this compound for additional indications.

Are there advantages to tasimelteon in comparison with freely available melatonin? There is no direct evidence with regard to efficacy, but some might argue that this FDA approved pharmaceutical would have better manufacturing standards relative to unregulated melatonin, for which there have been past questions of purity and concentration (Williamson et al. 1998). Tasimelteon also has a longer elimination half-life of about 1–3 h compared to approximately 30 min for melatonin. While a moderately prolonged duration of action may be beneficial, an excessively long half-life for a compound or any potent active metabolites could lead to a spillover effect that could compromise strategic phase shifting effects of a melatonin agonist. There also may be variations in clinical effects based on pharmacodynamic actions. At present it is unclear whether there will be significant clinical differences in the comparison of tasimelteon and ramelteon, especially recognizing varying patterns of MT1 and MT2 affinity between these two compounds. Lastly, it will be interesting to see how tasimelteon compares clinically with the prolonged-release Neurim Pharmaceutical melatonin formulation available in numerous countries, but not in the USA.

Conclusion
The U.S. FDA has approved the melatonin receptor agonist tasimelteon for the treatment of non-24-h sleep–wake disorder following evidence derived from studies with totally blind individuals diagnosed with the disorder. While this compound should be beneficial for related conditions, such as insomnia disorder with sleep onset difficulty and selected circadian rhythm sleep–wake disorders, it remains to be seen whether tasimelteon will be granted additional formal indications. For the present, the release of this medication highlights the episodic insomnia and excessive sleepiness experienced by many blind individuals. Greater recognition and treatment of this clinical problem will result in improving nighttime sleep and daytime alertness, as well as enhanced quality of life.

References

AASM (2014) International classification of sleep disorders, 3rd edn

American Academy of Sleep Medicine (2014) International classification of sleep disorders, 3rd edn. American Academy of Sleep Medicine, Darien, IL

Buscemi N, Vandermeer B, Hooton N, Pandya R, Tjosvold L, Hartling L, Baker G, Klassen TP, Vohra S (2005) The efficacy and safety of exogenous melatonin for primary sleep disorders. A meta-analysis. J Gen Intern Med 20(12):1151–1158. doi:10.1111/j.1525-1497.2005.0243.x

Cardinali DP, Pandi-Perumal SR, Niles LP, Brown GM (2010) Melatonin and sleep: possible involvement of GABAergic mechanisms. In: Monti JM, Pandi-Perumal SR, Mohler H (eds) GABA and sleep. Springer, Switzerland

Dhillon S, Clarke M (2014) Tasimelteon: first global approval. Drugs 74(4):505–511. doi:10.1007/s40265-014-0200-1

Dressman MA, Licamele L, Feeney J, Polymeropoulos MH (2012) Seventy percent of totally blind people with sleep complaints are not entrained to the 24 hour clock. Vanda Pharmaceuticals Inc., Washington, DC

Hardeland R, Pandi-Perumal SR, Cardinali DP (2006) Melatonin. Int J Biochem Cell Biol 38 (3):313–316. doi:10.1016/j.biocel.2005.08.020

Keijzer H, Smits MG, Duffy JF, Curfs LM (2013) Why the dim light melatonin onset (DLMO) should be measured before treatment of patients with circadian rhythm sleep disorders. Sleep Med Rev 18(4):333–339. doi:10.1016/j.smrv.2013.12.001

Lankford DA (2011) Tasimelteon for insomnia. Expert Opin Investig Drugs 20(7):987–993. doi:10.1517/13543784.2011.583235

Lewy AJ (2003) Clinical applications of melatonin in circadian disorders. Dialogues Clin Neurosci 5(4):399–413

Lewy AJ, Emens J, Sack RL, Hasler BP, Bernert RA (2003) Zeitgeber hierarchy in humans: resetting the circadian phase positions of blind people using melatonin. Chronobiol Int 20 (5):837–852

Lockley SW, Skene DJ, James K, Thapan K, Wright J, Arendt J (2000) Melatonin administration can entrain the free-running circadian system of blind subjects. J Endocrinol 164(1):R1–R6

Lockley SW, Dressman MA, Xiao C, Fisher DM, Torres R, Lavedan C, Licamele L, Polymeropoulos MH (2013a) Tasimelteon treatment entrains the circadian clock and demonstrates a clinically meaningful benefit in totally blind individuals with non-24-hour circadian rhythms. In: 95th Annual meeting of the endocrine society, Abstract SUN-134

Lockley SW, Dressman MA, Xiao C, Licamele L, Polymeropoulos MH (2013b) RESET study demonstrates that tasimelteon maintains entrainment of melatonin and cortisol in totally blind individuals with non-24-hour circadian rhythms. In: 95th Annual meeting of the endocrine society, Abstract SUN-137

Pandi-Perumal SR, Zisapel N, Srinivasan V, Cardinali DP (2005) Melatonin and sleep in aging population. Exp Gerontol 40(12):911–925. doi:10.1016/j.exger.2005.08.009

Pandi-Perumal SR, Srinivasan V, Maestroni GJ, Cardinali DP, Poeggeler B, Hardeland R (2006) Melatonin: nature's most versatile biological signal? FEBS J 273(13):2813–2838. doi:10.1111/j.1742-4658.2006.05322.x

Pandi-Perumal SR, Srinivasan V, Spence DW, Cardinali DP (2007) Role of the melatonin system in the control of sleep: therapeutic implications. CNS Drugs 21(12):995–1018

Rajaratnam SM, Polymeropoulos MH, Fisher DM, Roth T, Scott C, Birznieks G, Klerman EB (2009) Melatonin agonist tasimelteon (VEC-162) for transient insomnia after sleep-time shift: two randomised controlled multicentre trials. Lancet 373(9662):482–491

Sack RL, Lewy AJ (2001) Circadian rhythm sleep disorders: lessons from the blind. Sleep Med Rev 5(3):189–206. doi:10.1053/smrv.2000.0147

Smits MG, Pandi-Perumal SR, Warren Spence D, Brown GM (2010) Dim light melatonin onset (DLMO) in psychiatric disorders. In: Pandi-Perumal SR, Kramer MCUP (eds) Sleep and mental illness. Cambridge University Press, Cambridge

VandaPharmaceuticals (2014) Tasimelteon Prescribing Information

Williamson BL, Tomlinson AJ, Mishra PK, Gleich GJ, Naylor S (1998) Structural characterization of contaminants found in commercial preparations of melatonin: similarities to case-related compounds from L-tryptophan associated with eosinophilia-myalgia syndrome. Chem Res Toxicol 11(3):234–240. doi:10.1021/tx970202h

Part V
Conclusive Remarks

The Past, Present, and Future of Drug Development and Treatment in Sleep Disorders

Antonio Guglietta

Abstract Sleep disorders are a group of medical conditions whose treatment has dramatically changed throughout history and recent years. In the past, drug treatment of these conditions was empirical and largely based on natural products such as herbal preparations to induce sleep whose real efficacy and safety, however, was mostly unknown or unproven. The first modern and scientifically based drugs such as chloral hydrate, bromide, and barbiturates to treat sleep disorders were developed around late nineteenth and early twentieth century. Subsequently, as more progresses were made in the understanding of this group of diseases, more specific, safer, and better drugs such as benzodiazepines were developed. However, the most important advancements in the area of drug treatment of sleep disorders were made in the recent years during which several and innovative drugs and drug products were developed. These drugs such as zolpidem, zolpidem sublingual formulations, doxepin, eszopiclone, ramelteon, melatonin slow release formulations, modafinil, and sodium oxybate have significantly improved the treatment of sleep disturbances. Nevertheless, there is still room for additional improvements in this area to address yet unmet medical needs and to further improve existing therapies. As scientific advancements in this area continue to be made, it is likely that in the future we will witness the development of new and better drugs and drug products to treat sleep disturbances. New molecules that are currently in the late phase of clinical development such as Lorediplon or orexin antagonists[1] are likely to reach the market while other innovative molecules and perhaps biologic products may also be developed along with innovative formulations and drug products made of combination of different active ingredients. The goal, strategy, and the way these new therapeutic products will be developed will probably also change in the future. Re-profiling and extension of indications of existing drugs will perhaps become

[1] After this book went to press, on Aug 13th, 2014, the US Food and Drug Administration approved Suvorexant tablets (Belsomra) to treat difficulty in falling and staying asleep (insomnia). Suvorexant is the first approved drug of the orexin antagonists class.

A. Guglietta (✉)
Independent Pharmaceutical Consultant, Magdalena Corcoll 16, Molins de Rei, Barcelona 08750, Spain
e-mail: consultant@guglietta.es

A. Guglietta (ed.), *Drug Treatment of Sleep Disorders*, Milestones in Drug Therapy,
DOI 10.1007/978-3-319-11514-6_14

more common and more drugs will be developed for treatment of insomnia and especially for the other conditions that make up the group of sleep disorders. At the same time, new treatments developed for subset of patients suffering from a given sleep disturbance are also likely to become more common and finally, the design and parameters to be evaluated in regulatory clinical studies of future sleep disorder drugs will probably also change to reflect the scientific advances that will be made in this area.

1 Introduction

The importance of sleep was already known among ancient civilizations. Egyptians built "*sleep temples*" where they would treat a variety of medical conditions by inducing a sort of hypnotic status while Greeks and Romans had their own Gods of sleep (*Hypnos* for the Greeks and *Somnos* for the Romans) (Thorpy 2010; Scott Littleton 2005; Horne 2007). For long time the only sleep disorder clearly recognized was insomnia while some of the other disturbances of sleep were only recently identified and characterized (see chapter "Classification of Sleep Disorders," Thorpy 2012; Schenck et al. 1986). Currently, several different sleep disorders are recognized which, according to the American Academy of Sleep Medicine, can be grouped into seven major categories: Insomnia Disorders, Sleep Related Breathing Disorders, Central Disorders of Hypersomnolence, Circadian Rhythm Sleep–Wake Disorders, Parasomnias, Sleep Related Movement Disorders, and Other Sleep Disorders (see chapter "Classification of Sleep Disorders," American Academy of Sleep Medicine 2014). Some of these disorders may be associated with chronic diseases, mental disorders, health-risk-behaviors, limitation of daily functioning, injury, and mortality (Ram et al. 2010; Center for Disease Control 2011; Strine and Chapman 2005).

Throughout history, several remedies, both pharmacological and non-pharmacological, were used to treat these conditions (Thorpy 2010). The pharmacologic treatment of sleep disorders, in the past, was based mainly on empirical remedies coming mostly from natural products such as herbal preparations with the first specific chemically designed molecules to treat insomnia developed only in nineteenth and twentieth centuries (Thorpy 2010; López-Muñoz et al. 2005). The present era of drug treatment of sleep disorders is based on pharmaceutical products developed on the basis of scientific demonstration of efficacy and safety and some of the drugs that have been recently approved for treatment of sleep disorders have in many aspects revolutionized the drug management of this area and have been reviewed in depth in this book.

Despite all these advances however, there is still room for improvement in this area since we are far from having efficacious treatments for all sleep disorders or from having addressed all the clinical manifestations and medical needs that characterize this group of conditions. Therefore, as scientific progresses in this area are continuously being made, we can predict that in the future more drug therapeutic options will be available to treat this group of diseases.

The goal of this chapter is to briefly review how the drug development and drug treatment of sleep disorders have changed throughout time and to predict in which directions these two areas will move in the future.

2 The Past of Drug Development and Treatment of Sleep Disorders

2.1 Insomnia

2.1.1 Herbal and Natural Products

One of the oldest pharmacological remedy to induce sleep has been the use of herbs, plant extracts, and other natural products used either alone or in combinations. The list of herbs and natural products that claim to be beneficial in insomnia is very long and include chamomile, hops, valerian, passion flower, kava-kava, poppy seeds, and lavender among others (Dey and Dey 2013). Generally speaking these natural products may induce a relaxing status which can favor sleep. In selected cases and to a limited extent, these claims are supported by experimental data published in scientific journals (Salter and Brownie 2010). Although some discrepancy on their activity is reported in the literature (Taibi et al. 2009), valerian and hops are perhaps among the natural products for which more scientific evidence of efficacy in insomnia has been published in medical journals (Salter and Brownie 2010). In the majority of cases, however, this scientific evidence of efficacy is lacking and therefore the use of herbal and natural products as hypnotics is mostly empirical (Dey and Dey 2013; Meolie et al. 2005). Most importantly, in the majority of cases, the safety of most of herbal and natural products has not been sufficiently evaluated either. “Natural” definitely is not synonymous of safe and therefore it should not be assumed that these natural products are necessarily safe and without risks for health (Ernst 2006; Sanchez-Ortuño et al. 2009). There is a risk that some of these natural preparations may cause some serious toxicologic effect such as in the case of kava-kava, a well-known herbal remedy used to induce sleep, whose use has been associated with liver toxicity (Dunnick and Nyska 2013; Wheathley 2005). Therefore, with the exception of few selected preparations which, when used within the recommended dosage range, are generally considered safe and for which at least some scientific evidence of efficacy exists (Salter and Brownie 2010), the use of natural and herbal preparations for the treatment of insomnia should be discouraged particularly in those cases in which lack of demonstration of efficacy is associated with some concern about their safety. In any case, herbal and natural products to treat insomnia have been used for long time and are still commonly used today in phytotherapy or in folk medicine such as Chinese traditional medicine (Dey and Dey 2013; Sanchez-Ortuño et al. 2009; Wheathley 2005; Yeung et al. 2012).

2.1.2 Alkaloids and Chemically Synthesized Molecules

In the nineteenth century, several alkaloids isolated from a variety of plants were used as sedative and hypnotics alone or as part of drug cocktail preparations (Shorter 1997; López-Muñoz et al. 2005). These would include, among others, morphine isolated from opium and other compounds isolated from different species of the plants belonging to the *Solanaceae* family such as hyoscyamus, hyoscyamine, and hyoscine (López-Muñoz et al. 2005; Shorter 1997).

The first synthetically produced hypnotic was chloral hydrate which was synthesized in 1832 and introduced in medicine as hypnotic in 1869 when its efficacy to induce sleep was demonstrated (López-Muñoz et al. 2005). Due mainly to its practical advantages, chloral hydrate rapidly replaced alkaloids as hypnotic (Shorter 1997). After the introduction of chloral hydrate as hypnotic/sedative, several other synthetic compounds such as paraldehyde, sulphonal, and bromides were introduced on the market and used in medicine to sedate patients and induce sleep, and bromides became the most widely used sedatives in the second half of the nineteenth century despite their high toxicity (López-Muñoz et al. 2005).

At the beginning of the twentieth century, the barbiturates were introduced on the market and between the 1920s and the mid-1950 they became the only drugs used as sedatives and hypnotics (López-Muñoz et al. 2005). The first barbiturate to reach the market in 1904 was diethyl-barbituric acid which was then followed by several other molecules of the same family such as phenobarbital, pentobarbital, and thiopental (López-Muñoz et al. 2005). Although effective, the use of barbiturates in the treatment of insomnia was associated with serious side effects and often these drugs were also used for suicidal purposes (Johns 1977; Olesen et al. 2010; López-Muñoz et al. 2005).

A major step forward in the treatment of insomnia was the introduction on the market in the second half of the twentieth century of the benzodiazepines which act by binding to specific sites of the $GABA_A$ receptor. Several benzodiazepines were developed and commercialized throughout the years and became the drug of choice for the treatment of insomnia replacing therefore the barbiturates (López-Muñoz et al. 2005; Wick 2013). Benzodiazepines were effective drugs with a similar side effect profile as the barbiturates although less severe (Griffin et al. 2013).

2.2 *Narcolepsy*

Narcolepsy, a disorder mainly characterized by excessive daytime sleepiness and catalepsy, was first described in the second half of the nineteenth century (Fisher 1878; Gélineau 1880, 1881; Westphal 1877). In the past, this disorder was treated mainly with CNS stimulants such as amphetamines and amphetamine-like stimulants (Hirai and Nishino 2011; Nishino and Okuro 2010). Amphetamines were first used in the treatment of narcolepsy in 1935 (Prinzmetal and Bloomberg 1935) and some drugs of this group such as Dextro-amphetamine and Methylphenidate are approved by the FDA for treatment of this condition (FDA 2013). Although amphetamine-like stimulants are to some extent still used today to treat narcolepsy, they are being replaced by newer and more specific drugs (Hirai and Nishino 2011). The catalepsy manifestation of narcolepsy on the other hand, traditionally, has been treated mainly with tricyclic antidepressant but also in this case these drugs are being replaced by adrenergic/serotonergic selective reuptake inhibitors and other compounds such as sodium oxybate (Nishino and Okuro 2010).

2.3 *Other Sleep Disorders*

Apart from insomnia and narcolepsy, much less attention was reserved in the past to the pharmacotherapy of other sleep disorders. There are multiple reasons for this. First of all, some of these disorders are not serious enough and/or not frequent enough to justify drug treatment or drug discovery and development. On the other hand, some of these disorders such as the REM sleep behavior disorders were identified and characterized only recently (Schenck et al. 1986) and very little is known about their pathophysiology to allow the design of specific drugs. Therefore, in most cases in the past, pharmacotherapy of these disorders was rarely supported by solid scientific evidence (Wilson et al. 2010) and was based on the use of no specific drugs without the support of well-controlled studies and regulatory approval.

3 The Present of Drug Development and Treatment of Sleep Disorders

3.1 *Current Status of Sleep Medicine*

The present of sleep research and pharmacotherapy is being characterized by an explosion of new scientific and epidemiologic data leading to an increasing recognition of the high prevalence of sleep disturbances (Bixler 2009; Institute of Medicine 2006; Ram et al. 2010) and stressing their importance and impact on the life of a person and the society (Bixler 2009; CDC 2011; Ram et al. 2010; Strine and Chapman 2005). Sleep centers specialized in the research, evaluation, and management of sleep disorders have developed across the globe and the number of sleep scientific societies, foundations, and training programs for physicians has increased significantly (Shepard et al. 2005). In 1990, following a first major classification of sleep disorders published in 1979 (Sleep Disorders Classification Committee 1979), the American Academy of Sleep Medicine published the first International Classification of Sleep Disorder (Thorpy 1990). A second edition of this publication was then published in 2005 (Thorpy 2005) and a 3rd edition came out in 2014 (Thorpy 2014). The number of articles with the keyword "sleep" published in scientific journals reported by the US National Library of Medicine increased from approx. 37,000 for the period 1985–1999 to more than 90,000 in the period 2000–2013. Furthermore, regulatory agencies have issued new guidelines to assist investigators to develop new drugs for the treatment of sleep disorders (European Medicine Agency 2011). Overall, all these activities indicate an increasing interest in the area of sleep disorders which is today recognized as an independent and dynamic specialty of medicine by basic scientists, clinicians, media, government, and patients.

3.2 Classes of Pharmaceutical Products to Treat Sleep Disorders

The pharmaceutical industry has also been part of this dynamism. Taking advantages of the new findings in the area of sleep physiology and pharmacology, such as the discovery of the orexinergic system in the brain (Sakurai et al. 1998; de Lecea et al. 1998; Equihua et al. 2013), the characterization of the functional role of the $GABA_A$ receptor subunits (Nutt 2006; Nutt and Sathl 2010; Siegel and Steinmann 2012), and the advances of new pharmaceutical technologies, new drugs for the treatment of sleep disorders were recently developed such as zolpidem, zolpidem sublingual formulations, doxepin, eszopiclone, ramelteon, melatonin slow release formulations, modafinil, and sodium oxybate. These drugs have significantly improved the treatment of these conditions and some of them were reviewed in depth in other chapters of this book.

From a pharmaceutical perspective, the new therapeutic agents that were developed in the recent years for treatment of sleep disorders or those that are in advanced phase of development can be grouped into different classes.

New Chemical Entities of Small Molecules These are new small molecules that act on "old" or "new" targets known to play an important role in sleep disorders. Some of these molecules have already reached the market such as in the case of Ramelteon while others are in advanced phase of development.

New Formulations of Already Approved Drugs In many cases the therapeutic efficacy of a drug can be improved by modifying its formulation. An appropriate formulation can change certain characteristics of a molecule and may lead to an improved therapeutic profile of the active molecule. Examples of new therapeutic products that have been developed recently for sleep disorders which are reformulations of molecules that were already on the market include, among others, zolpidem sublingual formulations intended to accelerate the absorption of the active molecule and melatonin slow release formulations which allow a more sustained blood level of melatonin. In both cases, these new formulations overcome a weak point of the active molecule and provide a therapeutic advantage compared to the original molecule.

Enantiomers of Previously Approved Racemic Compounds Many times, from a chemistry standpoint, a pharmacological active molecule is actually a racemic compound. That is a mixture of two optical active compounds, termed enantiomers, which are mirror images of the same molecule and which rotate plane-polarized light by equal amount but in opposite directions (Davies and Teng 2003). Pharmacologically, enantiomers may have similar or different activity and often only one enantiomer is responsible for the therapeutic activity while the other may be totally inactive or may even have unwanted effects which may contribute to the toxic effect or the racemic form (Davies and Teng 2003). Therefore, chemical separation of the two enantiomers and subsequent development of the therapeutic active form may be advantageous since it may improve the therapeutic profile of the original racemic mixture. The development of an enantiomer as a therapeutic agent presents some peculiar features and challenges and specific regulatory guidelines exist to guide investigators through this process (EMA 1993, 2010; FDA 1992). Eszopiclone and Armodafinil, the active enantiomers of zopiclone and modafinil, respectively, represent two examples of enantiomers developed and approved recently for treatment of sleep disorders available today.

Generic Drugs Several drugs developed years ago for the treatment of sleep disorders have run out of their patent life and this has opened the door for the development of their generic form. Indeed, in the recent years, several generic versions of sleep medications such as zolpidem have been developed and are available on the market (FDA 2007). These generic versions of sleep disorder drugs have the advantage of making good therapeutic options for sleep disorders available to patients and clinicians at a lower and more affordable price than their branded counterparts.

4 The Future of Drug Development and Treatment of Sleep Disorders

As for any therapeutic area, it is very difficult to predict how drug development and drug treatment of sleep disorders will evolve in the future since this is a function of many and unpredictable variables including new discoveries, political, social, and regulatory environment, technical and business aspects of drug development, etc.

Nevertheless, it is likely that in the coming years sleep research will continue to play an important role in medicine and science. As the awareness on the prevalence and importance of sleep disorders increases (Institute of Medicine 2006; Ram et al. 2010; CDC 2011; Bixler 2009; Strine and Chapman 2005), it is likely that this area will capture more and more attention of the general public, scientists, pharmaceutical companies, clinicians, and patients which will stimulate research in this area and create a demand for new and better drugs.

An increased awareness on the prevalence and impact of sleep disorders on the life of an individual and the society is helpful but, by itself, not sufficient to stimulate the search and development of new drugs in this area. In order for new drugs to be developed, it is also necessary that sleep disorder conditions are perceived as a business opportunity by the pharmaceutical industry and that there is enough knowledge on the biochemical mechanisms underlying the sleep disturbances to make drug discovery and development profitable and feasible.

Providing that these conditions are met, in the future we may witness new and important advances in drug development and treatment of sleep disorders. Based on what we have already started to see today, we can predict that in the future a greater multiplicity of pharmaceutical products will reach the market which will be developed for the treatment of a wider variety of sleep conditions and/or subset of patients suffering from a given disorder. The design and parameters of the clinical studies needed to demonstrate a favorable benefit/risk ratio of future drugs will probably also change to reflect the advances made in the understanding and clinical manifestations of this group of diseases.

4.1 Multiplicity of Pharmaceutical Products

4.1.1 New Chemical Entities of Small Molecules

As our understanding of the biochemical mechanisms underlying individual sleep disorders expands, new molecular targets for drug discovery will be identified and new ideas will be generated on how to treat specific disorders. For example, the observation that patients with narcolepsy have a reduced number of orexinergic neurons in the hypothalamus and low level of orexin in the cerebrospinal fluid (Mayer 2014) opens up some new and exciting diagnostic and therapeutic possibilities for this condition since molecules that bind to orexin receptors and that act as agonists could be useful in its treatment. Low levels of histamine have also been reported in the cerebrospinal fluid of narcoleptic patients with or without low level of orexin-1 (Nishino et al. 2009) suggesting an implication of histamine in the pathogenesis of narcolepsy. Therefore, histamine H_3 inverse agonists such as pitolisant (Schwartz 2011) could in the future also represent a new drug option for the treatment of narcolepsy. To further support this possibility some encouraging clinical data of activity in narcolepsy and also in obstructive sleep apnea have been reported with pitolisant (Dauvilliers et al. 2013; Schwartz 2011)

4.1.2 Biological Products

Biological products such as peptides perhaps could also be developed in the future for treatment of sleep disorders such as for instance orexin or orexin-like peptides with agonistic activity for treatment of narcolepsy (Asahi et al. 2003). The development of peptides as drugs however can be very challenging since they are degraded quickly when taken orally, may have a short half-life, and do not easily cross the blood–brain barrier (Banga 2006; Craik et al. 2013). Nevertheless continuous advances are made in this area and new technologies such as synthesis of stable and more orally bioavailable analogs of peptides (Craik et al. 2013; Patel et al. 2014), use of shuttle molecules to cross the blood brain barrier (Niewoehner et al. 2014; Malakoutikhan et al. 2014), and development of nasal formulations of peptides (Patel et al. 2014; Campell et al. 2012) are being developed or perfected to overcome these problems.

4.1.3 Combination Products

Combining two or more active molecules in a single formulation has become very common in the treatment of certain medical conditions such as for instance hypertension or in controlling multiple cardiovascular risks (Kalra et al. 2010; Ram 2013; Castellano et al. 2014). Among other things, fixed-dose combination treatment has the advantage to simplify treatment and facilitate the patient compliance to treatment (Sanz et al. 2011). In the area of sleep disorders, combination drug treatment could be useful in managing different aspects of a given sleep disturbance or in those situations in which sleep disorders are associated with other medical conditions (Dikeos and Georgantopoulos 2011; Spiegelhalder et al. 2013). As a matter of fact, a combination of pseudoephedrine and domperidone has already demonstrated promising results in the management of obstructive sleep apnea (Larrain et al. 2010). Although the manufacturing of fixed-dose combination polypills and its development may be very challenging (Guglietta and Guerrero 2009), single formulations that combine different active molecules may become more common in the future also in the area of sleep disorder management.

4.1.4 Innovative Formulations and Delivery Systems

Major advances in the treatment of sleep disorders could also come from development of innovative formulations and new delivery systems applied to old and new molecules. Several new formulation and drug delivery technologies are already available today and actually have already been applied to the development of drugs for treatment of sleep disorders (Settar et al. 2014). These technologies allow to develop innovative formulations and drug delivery systems that improve several pharmacokinetic parameters of a pharmaceutical active molecule such as release, absorption, and metabolism. Undoubtedly, the use of these technologies applied to CNS and sleep disorder drug products will become more common in the future (Brambilla et al. 2014). Technologies such as transdermal drug delivery (Jain et al. 2014; Findling and Dinh 2014), nanotechnologies (Kaur et al. 2014; Krenter 2014), cell-penetrating peptides technologies (Zou et al. 2013), nasal formulation of peptides (Patel et al. 2014; Campell et al. 2012), chronodelivery and pulsatile technologies, (Patil and Shahiwala 2014), and direct delivery to the brain of therapeutic agents such as peptides through the nose (Pardeshi and Belgamwar 2013; Mittal et al. 2014) open up future perspectives on how to develop new drug products based on innovative formulations and drug delivery systems to treat sleep disorders with advantages over current products.

4.2 Indications of Drugs for Sleep Disorders

4.2.1 New Sleep Disorder Indications for Existing Drugs

Re-profiling or Extension of Indications of Marketed Drugs Drugs that were originally developed for a given clinical indication may also be beneficial in other medical conditions for which therefore they could also be developed by conducting new and appropriate studies. This re-profiling or extension of indications of a marketed drug may represent an opportunity for the pharmaceutical industry to further exploit the full potential of a molecule. From a development standpoint, re-profiling of a molecule allows a quicker development because it takes advantage of the pool of the existing data collected during the development for other indications such as CMC, pharmacokinetics, and toxicology data. In addition, multiple indications for the same molecule increases the total revenue and return on investment for the pharmaceutical company and often overcome the unattractive financial return of developing a new drug for only one indication. In the simplest case, that is when the clinical studies required for approval of the new indication are carried out under conditions supported by existing preclinical and clinical data (doses, formulation, duration of drug administration etc.), conducting new and appropriate clinical studies that demonstrate a positive benefit/risk ratio for the new indication may be all it is required to obtain a new approval. In other cases, when the new indication requires higher doses of the drug, different formulations, different duration of treatment, etc., than the original indication, the drug has to be re-profiled more extensively and additional preclinical studies may also need to be conducted such as development of a new formulation, extension of toxicology studies, etc. to make sure that the new conditions under which the drug will be tested for the new indication are safe. However, while drug re-profiling has its advantages, it also presents some limitations such as the multiple indications for which the same molecule may be developed need to share some common biochemical feature upon which the molecule acts.

Some of the new drugs approved for treatment of insomnia available today are actually re-profiling of molecules originally approved for a different indication. This is the case of doxepin a tricyclic antidepressant with H_1 receptor antagonism activity which was originally developed and approved for the treatment of anxiety and depression (Brayfield 2014). The H_1 antagonists are used mainly for the treatment of allergic conditions but are known to have also some sedative activity which however has always been regarded as an annoying side effect of antiallergic therapy (Church and Church 2013). In the case of doxepin however, recently the sedative effect of the molecule was exploited therapeutically and lower than antidepressant and antianxiety doses were developed and approved specifically for the treatment of insomnia (Markov and Doghramji 2010). The case of doxepin can perhaps in the future serve as a precedent to explore the activity and therapeutic potential on sleep disorders of molecules approved for treatment of other conditions.

4.2.2 New Indications of Drugs in Sleep Disorders

As for indications of future sleep disorder drugs, it is likely that some of the trends that are already seen today such as drugs developed for sleep disorders other than insomnia and/or for subsets of patients suffering from a given sleep disorder will become more common.

Drugs Developed for a Broader Variety of Sleep Indications The majority of drug treatments available today for sleep disorders were developed for insomnia and only few drugs are approved for the treatment of other sleep disturbances. This is probably due to that insomnia has been recognized as a distinct disorder for the longest time and that there is large, although incomplete, pool of scientific data available on its pathophysiology, biochemistry, and pharmacology. Moreover commercially insomnia is an attractive market and therefore has captured most of the attention of the pharmaceutical industry.

Hopefully, this will change and it is likely that more drugs to treat sleep disorders other than insomnia will be developed in the coming years. Hints of this trend can already be seen today as for instance in the case of tasimelteon whose development has been directed towards the treatment of circadian rhythm disorders. This represents a new development approach for this class of pharmaceutical agents which in the past, as in the case of Ramelteon, were developed for the treatment of insomnia.

Drugs Developed for Subsets of Patients with Sleep Disorders As our understanding of the different sleep disorders improves, it will become evident that, in some cases, a single disorder may actually represent a heterogeneous group of conditions with some common traits but also characterized by distinct features each of which may be selectively modified by different drugs. This situation may open the door to the development of more selective drugs which will be indicated only for a subset of patient population affected by a specific sleep disorder. Insomnia for instance represents a heterogeneous group of disorders characterized by difficult to fall and stay asleep. However, insomnia patients can actually be broken down into different subgroups with specific and distinct features such as patients that have difficulties to fall asleep or to stay asleep or patients where insomnia is associated with other medical conditions. Each of these distinct features may be the target of different drugs which therefore may be developed and indicated only for that subgroup of patients.

In the future, as we learn more on how genetic, age, race, and perhaps gender affect drug pharmacokinetic and metabolism and the efficacy and safety of a drug, more specific and ad hoc studies conducted in subsets of patients with sleep disorders grouped by age, race, etc. may be required to obtain an indication for that specific subgroup. This fragmentation of indications for sleep disorders on one hand will allow a more scientific and rational approach to the treatment of sleep disorders but on the other hand will also pose some challenging and practical problems for drug developers. For instance, the subset population of the sleep disorder for which a molecule with a given profile could be developed and indicated may not be large enough to justify the cost of development of a new drug. To some extent, this problem can be overcome by strategies such as optimization of development cost and time, higher price of the drug, and regulatory requirements compatible with a fast development time but nevertheless there is a limit to how much one can stretch these strategies. Another approach, to overcome these problems, could be, whenever possible, to develop these drugs not only but also for these sub-indications (i.e., insomnia drug indicated in adult population and pediatric population and elderly population, etc.) by conducting additional clinical studies.

4.2.3 Design and End-Points of Clinical Studies in Sleep Disorders

In order to get an approval for a new drug, clinical studies have to address those points that are considered by regulatory agencies relevant to demonstrate, in addition to safety, a beneficial effect in a given disease. What constitutes a relevant clinical beneficial effect however may be a matter of discussion and a changing concept which may have an impact on the design and end points of clinical studies conducted to obtain an indication for treatment of a given sleep disorder. For instance, in the treatment of insomnia it is important not only to induce and maintain sleep for an adequate duration but also to make sure that the sleep is of "good quality." Moreover, ideally a drug for treatment of insomnia should also have a beneficial effect on all manifestations of the disorder as well as on other medical conditions associated with chronic insomnia (Ram et al. 2010; Center for Disease Control 2011; Bixler 2009; Strine and Chapman 2005). Therefore, in the future we may see that the questions that drug developers have to answer during development may also change with a consequent impact on the design of clinical studies.

Several regulatory agencies publish guidelines to help investigators developing drugs. These publications give some insights from a regulatory standpoint on which parameters should be evaluated in clinical trials, the most appropriate design, the number of studies to be conducted, etc. In the area of sleep disorders, several regulatory guidelines are published which are updated as new concepts emerge in the clinical evaluation of a specific disorder. For instance, in insomnia some recent guidelines were published to reflect the new concepts in this area that affect the way drugs are evaluated to treat this condition (European Medicines Agency 2011). It is likely that in the future more guidelines will be issued as an aid to develop drugs not only in insomnia but also in other sleep disorders as suggested for instance by the increasing attention of the FDA to other sleep disorders such as narcolepsy (FDA 2014).

Conclusions

The drug treatment of sleep disorders has evolved through history and recent times and continuous improvement of the treatment of these conditions have been made. These progresses are the results of a better understanding of the pathophysiology and clinical manifestations of these disorders, advances in pharmaceutical science, and increased public awareness of the prevalence and impact of these conditions on the life of a person and the society. Today, very effective and safe drugs are available to treat several sleep disturbances which in many respects have revolutionized our thinking and treatment of these conditions. It is likely that these advancements will continue in the future with more and specific molecules and drug products developed for treatment of a broader variety of sleep disorders and subpopulation of patients which will be approved on the basis of appropriately designed clinical studies that will demonstrate, in addition to safety, a beneficial effect on old and new relevant clinical parameters.

References

American Academy of Sleep Medicine (2014) International classification of sleep disorders, 3rd edn. American Academy of Sleep Medicine, Darien, IL

Asahi S, Egashira S, Matsuda M et al (2003) Development of an orexin-2 receptor selective agonist [Ala(11), D-Leu(15)] orexin-B. Bioorg Med Chem Lett 13:111–113

Banga AK (2006) Therapeutic peptides and proteins: formulation, processing and delivery system, 2nd edn. CRC Press, Boca Raton, FL

Bixler E (2009) Sleep and society: an epidemiologic perspective. Sleep Med 10:S3–S6

Brambilla D, Luciani P, Leroux JC (2014) Breakthrough discoveries in drug delivery technologies: the next 30 years. J Control Release 14:S0168–S3659

Brayfield A (2014) Martindale: the complete drug reference, 38th edn. Pharmaceutical Press, London

Campell C, Marimoto BH, Nenciv D et al (2012) Drug development of intranasally delivered peptides. Ther Deliv 3:557–568

Castellano JM, Sanz G, Fuster V (2014) Evolution of the polypill concept and ongoing clinical trials. Can J Cardiol 30:520–526

Center for Disease Control (2011) Effect of short sleep duration on daily activity – United States 2005–2008. MMWR 60:239–242

Church MK, Church DS (2013) Pharmacology of antihistamines. Indian J Dermatol 58:219–224

Craik DJ, Fairlie DP, Liras S et al (2013) The future of peptide-based drugs. Chem Biol Drug Des 81:136–147

Dauvilliers Y, Bassetti C, Lammers GJ et al (2013) Pitolisant versus placebo or modafinil in patients with narcolepsy: a double-blind, randomized trial. Lancet Neurol 12:1068–1075

Davies NM, Teng XW (2003) Importance of chirality in drug therapy and pharmacy practice: implication for psychiatry. Adv Pharm 1:242–252

De Lecea L, Kilduff TS, Peyron C et al (1998) The hypocretins: hypothalamus-specific peptides with neuroexcitatory activity. Proc Natl Acad Sci USA 95:322–327

Dey A, Dey A (2013) Phytotherapy against insomnia: extravaganza or an alternative medicine. Pak J Biol Sci 16:148–150

Dikeos D, Georgantopoulos G (2011) Medical comorbidity of sleep disorders. Curr Opin Psychiatry 24:346–354

Dunnick JK, Nyska A (2013) The toxicity and pathology of selected dietary herbal medicine. Toxicol Pathol 41:374–386

Equihua AC, De La Herrán-Arita AK, Drucker-Colin R (2013) Orexin receptor antagonists as therapeutic agents for insomnia. Front Pharmacol 4:Article 163

Ernst E (2006) When natural is not harmless. Int J Clin Pract 60:380

European Medicine Agency (1993) Investigation of chiral active substances. Directive 75/318/ EEC

European Medicine Agency (2010) Reflection paper on considerations given to designation of a single stereo isomeric form (enantiomer) as new active substance in relation to a reference active substance which is a racemic mixture of enantiomers. EMA/651649/2010

European Medicine Agency (2011) Committee for medicinal products for human use. Guideline on medicinal products for the treatment of insomnia. EMA/CHMP/16274/2009/Rev. 1

FDA (2014) Narcolepsy. Patient-focused drug development initiative – summary report

FDA (1992) Development of new stereoisomeric drugs. www.fda.gov/drugs/guidancecomplianceregulatoryinformation/guidances/ucm122883.htm

FDA (2007) Press release: FDA approves first generic versions of Ambien (Zolpidem tartrate) for the treatment of insomnia. www.fda.gov/NewsEvents/Newsroom/PressAnnouncement/2007/ucm108897.htm

FDA (2013) FDA-approved drugs to treat sleep disorders. J Psychosoc Nurs Ment Health Serv 51:9–10

Findling RL, Dinh S (2014) Transdermal therapy for attention-deficit hyperactivity disorders with the methylphenidate patch (MTS). CNS Drugs 28:217–228

Fisher F (1878) Epileptoide schlafzustände. Arch für Psychiatl 8:200–203

Gélineau J (1880) De la narcolepsie. Gazette des Hôpitaux 53:626–628

Gélineau JBE (1881) De la narcolepsie. Surgères, Charente-Inférieure: Imprimerie de Surgères 64

Griffin CE 3rd, Kave AM, Bueno FR et al (2013) Benzodiazepine pharmacology and central nervous system-mediated effects. Ochsner J 13:214–223

Guglietta A, Guerrero M (2009) Issues to consider in the pharmaceutical development of a cardiovascular polypill. Nat Clin Pract Cardiovasc Med 6:112–199

Hirai N, Nishino S (2011) Recent advances in the treatment of narcolepsy. Curr Treat Options Neurol 13:437–457

Horne JA (ed) (2007) Sleepfaring: a journey through the science of sleep, 1st edn. Oxford University Press, Oxford

Institute of Medicine (2006) Sleep disorders and sleep deprivation: an unmet public health problem. The National Academy Press, Washington, DC

Jain A, Jain P, Kurmi J et al (2014) Novel strategies for effective transdermal drug delivery: a review. Crit Rev Ther Drug Carrier Syst 31:219–272

Johns MW (1977) Self-poisoning with barbiturates in England and Wales during 1959–74. Br Med J 1:1128–1130

Kalra S, Kalra B, Agrawal N (2010) Combination therapy in hypertension: an update. Diabetes Metab Syndr 2:44

Kaur IP, Kakkar V, Deol PK et al (2014) Issues and concerns in nanotech product development and its commercialization. J Control Release 14:S0168–S3659

Krenter J (2014) Drug delivery to the central nervous system by polymeric nanoparticle: what do we know? Adv Drug Deliv Rev 71C:2–14

Larrain A, Kapur VK, Gooley TA et al (2010) Pharmacological treatment of obstructive sleep apnea with a combination of pseudoephedrine and domperidone. J Clin Sleep Med 6:117–123

López-Muñoz F, Ucha-Udabe R, Alamo C (2005) The history of barbiturates a century after their clinical introduction. Neuropsychiatry Dis Treat 1:329–343

Malakoutikhan M, Guixer B, Arranz-Gibert P et al (2014) “A la carte” peptide shuttle: tools to increase their passage across the blood–brain-barrier. ChemMedChem 9:1594–1601

Markov D, Doghramji K (2010) Doxepin for insomnia. Curr Psychiatry 9:67–77

Mayer G (2014) Narcolespy. Nervenarzt 85:28–34

Mittal D, Ali A, Md S et al (2014) Insights into direct nose to brain delivery: current status and future perspectives. Drug Deliv 21:75–86

Meolie AL, Rosen C, Kristo D et al (2005) Oral nonprescription treatment for insomnia: an evaluation of products with limited evidence. J Clin Sleep Med 1:173–187

Niewoehner J, Bohrmann B, Collin L et al (2014) Increased penetration and potency of a therapeutic antibody using a monovalent molecular shuttle. Neuron 81:49–60

Nishino S, Okuro M (2010) Emerging treatments for narcolepsy and its related disorders. Expert Opin Emerg Drugs 15:139–158

Nishino S, Sakurai E, Navsimalova S et al (2009) Decreased CSF histamine in narcolepsy with and without low CSF hypocretin-1 in comparison to healthy control. Sleep 32:175–180

Nutt D (2006) GABAA receptors: subtypes, regional distribution and function. J Clin Sleep Med 2: S7–S11

Nutt D, Sathl SM (2010) Searching for perfect sleep: the continuing evolution of $GABA_A$ receptor modulators as hypnotics. J Psychopharmacol 24:1601–1612

Olesen JB, Hansen PR, Erdal J et al (2010) Antiepileptic drugs and risk of suicide: a nationwide study. Pharmacoepidemiol Drug Saf 19:518–524

Patel A, Cholkar K, Mitra AK (2014) Recent development in protein and peptide parental delivery approaches. Ther Deliv 5:337–365

Patil SS, Shahiwala A (2014) Patented pulsatile drug delivery technologies for chronotherapy. Expert Opin Ther Pat 24(8):845–856

Pardeshi CV, Belgamwar VS (2013) Direct nose to brain drug delivery via integrated nerve pathways bypassing the blood-barrier: an excellent platform for brain targeting. Expert Opin Drug Deliv 10:957–972

Prinzmetal M, Bloomberg W (1935) The use of benzedrine for the treatment of narcolepsy. J Am Med Assoc 105:2051–2054

Ram CV (2013) Fixed-dose triple combination treatment in the management of hypertension. Manag Care 12:45–55

Ram S, Sirawan H, Kumar SK et al (2010) Prevalence and impact of sleep disorders and sleep habits in the United States. Sleep Breath 14:63–70

Sakurai T, Amemiya A, Ishii M et al (1998) Orexins and orexin receptors: a family of hypothalamic neuropeptides and G protein-coupled receptors that regulate feeding behavior. Cell 92:573–585

Salter S, Brownie S (2010) Treating primary insomnia – the efficacy of valerian and hops. Aust Fam Physician 39:433–437

Sanchez-Ortuño MM, Belanger L, Ivers H et al (2009) The use of natural products for sleep: a common practice? Sleep Med 10:982–987

Sanz G, Fuster V, Guzman L et al (2011) The fixed-dose combination drug for secondary cardiovascular prevention project: improving equitable access and adherence to secondary cardiovascular prevention with a fixed-dose combination drug. Study design and objectives. Am Heart J 162:811–817

Schenck CH, Montplaisir JY, Frauscher B et al (1986) Chronic behavioral disorders of human REM sleep: a new category of parasomnia. Sleep 9:293–308

Littleton S (ed) (2005) Gods, goddesses and mythology, vol 2. Marshall Cavendish, Tarrytown, NY

Settar M, Sayed OM, Lane ME (2014) Oral-transmucosal drug delivery – current status and future. Int J Pharm 47:498–506

Shepard JW Jr, Buysse DJ, Al C et al (2005) History of the development of sleep medicine in the United States. J Clin Sleep Med 1:61–82

Shorter E (1997) A history of psychiatry. From the era of the asylum to the age of Prozac. Wiley, New York, NY

Siegel E, Steinmann ME (2012) Structure, functions and modulation of $GABA_A$ receptors. J Biol Chem 287:40224–40231

Sleep Disorders Classification Committee (1979) Classification of sleep and arousal disorders. Sleep 2:1–137

Spiegelhalder K, Regen W, Nanovska S et al (2013) Comorbid sleep disorders in neuropsychiatric disorders across the life cycle. Curr Psychiatry Rep 15:364

Strine TW, Chapman DP (2005) Associations of frequent sleep insufficiency with health-related quality of life and health behaviors. Sleep Med 6:23–27

Schwartz JC (2011) The histamine H3 receptor: from discovery to clinical trials with pitolisant. Br J Pharmacol 163:713–721

Taibi DM, Vitiello MV, Bersness S et al (2009) A randomized clinical trial of valerian fails to improve self-reported polysomnographic and actigraphic sleep in older women with insomnia. Sleep Med 10:319–328

Thorpy ML (1990) Diagnostic classification steering committee: international classification of sleep disorders: diagnostic and coding manual. American Sleep Disorders Association, Rochester, NY

Thorpy MJ (2005) International classification of sleep disorders: diagnostic and coding manual, 2nd edn. American Academy of Sleep Medicine, Westchester, NY

Thorpy MJ (2010) History of sleep and man. In: Pollack CP, Thorpy MJ, Yager J (eds) Encyclopedia of sleep and sleep disorders, 3rd edn. Fact on File, New York, NY

Thorpy MJ (2012) Classification of sleep disorders. Neurotherapeutics 9:687–701

Thorpy MJ (2014) International classification of sleep disorders, 3rd edn. American Academy Sleep Medicine, Darien, CT

Wheathley D (2005) Medicinal plants for insomnia: a review of their pharmacology, efficacy and tolerability. J Psychopharmacol 19:414–421

Westphal C (1877) Eigenthümliche mit Einschläfen verbundene Anfälle. Arch Psychiatry 7:631–635

Wick JC (2013) The history of benzodiazepines. Consult Pharm 28:538–548

Wilson S, Nutt D, Argyropoulos S et al (2010) British Association for Psychopharmacology consensus statement on evidence-based treatment of insomnia, parasomnias, and circadian rhythm disorders. J Psychopharmacol 24:1577–1600

Yeung WF, Chung KF, Poon MMK et al (2012) Chinese herbal medicine for insomnia: a systematic review of randomized controlled trials. Sleep Med Rev 16:497–507

Zou LL, JI M, Wang T (2013) Cell-penetrating peptide-mediated therapeutic molecule delivery into the central nervous system. Curr Neuropharmacol 11:197–208

Index

A
Acetylcholine
anticholinergic drugs, 49
cholinergic wake-promoting system, 48
cholinesterase inhibitors, 48–49
milameline, 49
nicotine, 49
receptors, 48
sleep-promoting cholinergic inhibitors, 49
Adenosine
adenosine-mediated sleep-promoting system, 41
receptors, 42
sleep-promoting adenosine receptor agonists, 42
wake-promoting adenosine receptor antagonists, 42
Amantadine, 77
Armodafinil
chemical structure, 212
multiple sclerosis, 222
narcolepsy, 218
obstructive sleep apnea, 219–221
pharmacokinetics, 214
R enantiomer, 213
safety, 224–225
shift work disorder, 223

B
Benzodiazepine Withdrawal Symptom Questionnaire (BWSQ), 183

C
Cataplexy
GHB, 37
narcolepsy type 1, 215
narcolepsy type 2, 215
sleep-promoting orexinergic inhibitors, 46
SXB, 238–240
tricyclic antidepressants, 253
Cell replacement therapy, 253
Central sleep apnea (CSA) syndrome, 76
Cheyne–Stokes breathing (CSB), 76
Cholinesterase inhibitors
donepezil, 49
physostigmine, 48
Chronic insomnia
dose discrimination, lack of, 168
psychiatric comorbidities, 169
short-term efficacy and long-term safety, 167–168
sleep induction and maintenance, 168–169
Chronic obstructive sleep apnea (OSA), 178–179
Chronic primary insomnia
adverse effects, 165, 167
comorbid psychiatric disorders, 164
double-blind trial, 164
efficacy, 165
long-term trial, 165
open-label extension phase, 165
post hoc analysis, 165–166
psychomotor performance, 164–165
rebound insomnia, 164
WASO, 164–165

A. Guglietta (ed.), *Drug Treatment of Sleep Disorders*, Milestones in Drug Therapy,
DOI 10.1007/978-3-319-11514-6

Chronic primary insomnia (*cont.*)
  zolpidem, 166
Chronobiotics, 78
Circadian rhythm sleep–wake disorder (CRSWD), 73, 78
Clinical Global Impression of Change (CGI) scores, 216–217, 241
Clonazepam, 81
Comorbid insomnia
  cancer pain, 180–181
  clinical trials, 170–173
  GAD, 176–177
  MDD, 174–176
  multiple sclerosis, 180
  OSA, 178–179
  Parkinson's disease, 179–180
  perimenopausal and postmenopausal women, 170, 174
  PTSD, 177
  pulmonary disease, 178–179
  rheumatoid arthritis, 179

D

Digit span test (DST), 134
Dopamine
  dopaminergic wake-promoting system, 55–56
  PLMD, 81
  receptors, 56
  RLS, 80–81
  sleep-promoting dopaminergic inhibitors
    monoamine depletors, 58
    postsynaptic D1 and D2 receptor antagonists, 58
    presynaptic D2 receptor agonists, 58
  wake-promoting dopaminergic activators
    dopamine release promoters, 57
    dopamine reuptake inhibitors, 56–57
    postsynaptic D1 and D2 dopamine receptor agonists, 57
    presynaptic D2 dopamine receptor antagonists, 57
DORAs. *See* Dual orexin receptor antagonists (DORAs)
Dorsomedial hypothalamus (DMH), 32
Doxepin. *See* Low-dose doxepin
Drug indications
  design, 284–285
  pathophysiology, biochemistry, and pharmacology, 283
  re-profiling/extension, marketed drug, 282–283
  subsets of patients, 283–284
DST. *See* Digit span test (DST)
Dual orexin receptor antagonists (DORAs)
  advantages, 199, 201–202
  almorexant, 198–199
  SB-649868, 199–200
  suvorexant, 199–201

E

Eszopiclone
  administration dosage, 181
  clinical efficacy and safety
    chronic insomnia, 167–169
    chronic primary insomnia (*see* Chronic primary insomnia)
    clinical trials, 161–163
    improvements, 169–170
    transient insomnia, 161, 164
  comorbid insomnia
    cancer pain, 180–181
    clinical trials, 170–173
    GAD, 176–177
    MDD, 174–176
    multiple sclerosis, 180
    OSA, 178–179
    Parkinson's disease, 179–180
    perimenopausal and postmenopausal women, 170, 174
    PTSD, 177
    pulmonary disease, 178–179
    rheumatoid arthritis, 179
  drug interaction, 181
  hypnotics, 183–184
  mechanism of action, 160
  multicenter trials, 184–185
  pharmacokinetic parameters, 160–161
  pharmacology, 160
  placebo-controlled trials, 184–185
  randomized trials, 184–185
  tolerability profile
    adverse event profile, 182
    BWSQ, 183
    cognitive/psychomotor function, 182–183
    confusion rates, 182
    $GABA_A$ receptors, 182
    memory impairment, 182
    treatment duration, 183

unpleasant and bitter taste, 182
Excessive daytime sleepiness (EDS). *See* Modafinil
Exploding head syndrome (EHS), 73, 79–80

G

GABA. *See* γ-aminobutyric acid (GABA)
$GABA_A$ receptor modulators
anaesthetic site, 36
barbiturate receptor, 36
benzodiazepine receptor, 36
ethanol site, 37
neurosteroids, 37
Galanin
galaninergic sleep-promoting system, 39–40
sleep-promoting galanin receptor agonists, 40
wake-promoting galanin receptor antagonists, 40
γ-aminobutyric acid (GABA)
noradrenergic locus coeruleus, 29, 34, 51
receptors, 34–35
sleep-promoting activators
anaesthetic site, 36
baclofen, 37
barbiturate receptor, 36
benzodiazepine receptor, 36
ethanol site, 37
gaboxadol, 36
GHB, 37
neurosteroids, 37
tiagabine, 37
wake-promoting inhibitors
benzodiazepine binding site, 38
bicuculline and gabazine, 38
flumazenil, 38
$GABA_B$ and $GABA_C$ receptors, 38
GABA receptor autoantibodies, 38–39
Generalized anxiety disorder (GAD), 176–177
Genome-wide association studies (GWAS), 12
γ-hydroxybutyric acid (GHB), 37
GIRK. *See* G protein-regulated inward rectifier (GIRK)
Glutamate
glutamatergic wake-promoting system, 46–47
receptors, 47
REMS-suppressing glutamatergic inhibitors, 48
sleep-promoting glutamatergic inhibitors, 47
wake-promoting glutamatergic activators, 47
G protein-regulated inward rectifier (GIRK), 193

H

Histamine
histaminergic wake-promoting system, 43
receptors, 43
sleep-promoting histaminergic inhibitors, 44–45
wake-promoting histaminergic activators, 44
Hypersomnolence
idiopathic hypersomnia, 73, 77
insufficient sleep syndrome, 73, 78
Kleine–Levin syndrome, 73, 77
long sleeper, 73, 78
traumatic brain injury, 221–222
type 1 narcolepsy, 72, 77
type 2 narcolepsy, 73, 77
Hypnotic drugs
barbiturates, 124–125
benzodiazepines, 124–125
infections/cancer, 183–184
non-benzodiazepines hypnotics, 125
Hypocretin. *See* Orexin
Hypocretin-2/orexin-B (Hcrt-2), 250–251

I

ICSD-3. *See* International Classification of Sleep Disorders (ICSD-3)
Insomnia disorder
alkaloids, 276
biochemical characterization, 90
chemical molecules, 276
classification, 72
clinical development phase
double-blind studies, 99, 100
efficacy and safety data, 88, 97
parallel group studies, 99
phase II studies, 98
phase I studies, 97–98
phase 0 studies (exploratory IND studies), 97
placebo-controlled studies, 97, 100
polysomnography method, 99
proof-of-concept studies, 98
psychometric method, 99
randomized studies, 99, 100
clinical features, 75

Insomnia disorder (*cont.*)
diagnosis, 75
drug development failure, 86, 87
endogenous molecules replacement approach, 89
eszopiclone (*see* Eszopiclone)
herbal and natural products, 275–276
lorediplon (*see* Lorediplon)
low-dose doxepin (*see* Low-dose doxepin)
molecule-based approach, 89
nonclinical development phase
analytical methods development, 91
efficacy and safety data, 88, 90
formulation development, 95–96
manufacturing process, 90–91
nonclinical safety evaluation (*see* Nonclinical safety evaluation)
pharmacokinetic and metabolism characterization, 95
pharmacology development, 96–97
orexin (*see* Orexin)
pharmacology and pharmacokinetic characterization, 90
post-approval studies, 88
formulation development, 101
indication, 101
miscellaneous studies, 103
regulatory agency, 101
safety monitoring (pharmacovigilance), 102–103
prevalence and impact of, 113–114
target-based approach, 88–89
toxicology tests, 90
treatment, 75
International Classification of Sleep Disorders (ICSD-3)
circadian rhythm sleep–wake disorders, 73, 78
degree of standardization, 74–75
hypersomnolence
idiopathic hypersomnia, 73, 77
insufficient sleep syndrome, 73, 78
Kleine–Levin syndrome, 73, 77
long sleeper, 73, 78
type 1 narcolepsy, 72, 77
type 2 narcolepsy, 73, 77
insomnia disorders, 72, 75
isolated symptoms and normal variants, 74, 81
parasomnias
EHS, 73, 79–80
NREM-related parasomnias, 73, 79
sleep-related hallucinations, 73, 79–80
sleep-related breathing disorders, 72, 76–77
SRMD, 74, 80–81

L
Lithium carbonate, 77
Lorediplon
chemical structure, 125
$GABA_A$ receptor modulation
biological effects, 123
Cys loop receptors, 122
endogenous ligand, 123
glycine receptors, 122
$5\text{-}HT_{3A}$ receptors, 122
hypnotic drugs, 124–125
nicotinic acetylcholine receptors, 122
protein subunits, 122–124
sedation, 123, 124
pharmacokinetic profile, 133–136
preclinical and clinical data, 141
preclinical pharmacology
amnesic effects, 129–130
$GABA_A$ subunits, 126–127
hypnotic and ethanol interactions, 130
Lorediplon EtOH Liability Index, 130
middle-of-the-night awakenings, 131–132
muscular tone inhibition, 127–129
Next-Day Hangover Liability Index, 130
physical dependence, 130, 131
polygraphic record, 131, 132
potency, 131
quality of sleep, 132
sedation, 127–128
sleep fragmentation, 131
sleep–wake state, 131, 133
safety, 139, 141
vs. zolpidem
advantages, 141, 142
cortical and muscle electrodes, 130
DST, 134
next-day residual drug effect, 139
number of awakenings (W), 137, 139
parameters, 131–133
PSQ, 139, 140
PVT, 134
SEI, 137, 139
sleep onset, 135–137
TST, 137, 139
VAS, 134
WASO, 136, 138
Lorediplon EtOH Liability Index, 130
Low-dose doxepin
contraindications, 117
dosing, 118
efficacy, 115–116
mechanism of action, 115
neurotransmitters system

acetylcholine, 112
adenosine, 112
histamine, 113
hypocretin, 113
norepinephrine, 112–113
serotonin, 112–113
pharmacokinetics, 115
tolerability, 117

M

Maintenance of wakefulness test (MWT), 240–241
Major depressive disorder (MDD), 174–176
Mazindol
idiopathic hypersomnia, 77
narcolepsy, 77
Melanin
MCHergic sleep-promoting system, 29, 31, 39
wake-promoting inhibitors, 39
Melatonin
chronobiotic melatoninergic activators, 41
melatoninergic sleep-promoting system, 40
sleep-promoting melatoninergic activators, 40–41
wake-promoting melatoninergic inhibitors, 41
Modafinil
chemical structure, 212
depression, 223
idiopathic hypersomnia, 218–219
mechanism of action, 214
metabolism, 214
multiple sclerosis, 222
myotonic dystrophy, 223
narcolepsy type 1 and type 2
CGI-C, 216–217
double-blind crossover study, 217–218
Epworth sleepiness scale, 216
hypothalamic hypocretin, 215
maintenance of wakefulness test, 216
optimal dosing protocols, 217
placebo periods, 216
REM sleep, 215
U.S. modafinil trials, 217
nonmedical use, 225
obstructive sleep apnea, 219–221
Parkinson's disease, 221
pharmacokinetics, 214
pharmacology, 213–214
posttraumatic hypersomnia, 221–222
safety, 224–225
shift work disorder, 222–223
sleep latency, 212–213

N

Narcolepsy, 12
amphetamines, 276–277
cell replacement therapy, 253
experimental models
day–night cycle, 252
embryonic day 19 E19, 252
end of sleep period (ZT12), 252
Hcrt-2/SAP, 250–251
peptide levels, 252
prepro-Hcrt gene expression, 252
sleep deprivation method, 252
toxin, 250
wake-active period (ZT0), 252
Hcrt system, 250
neural transplantation
animal models, 253
beneficial effects, 253
control rats, 255
grafted dopaminergic neurons, 255
hypocretin-immunoreactive somata, 254
inmunohistochemistry, 255
Parkinson's disease, 253
sodium oxybate (*see* Sodium oxybate (SXB))
NARI. *See* Noradrenaline reuptake inhibitors (NARI)
Neuronal network regulation
circadian network, 32–34
REM-sleep network
cortical activation, 30
dreaming, 29, 32
horizontal saccadic eye movements, 29
impairment, 30
muscle atonia, 29, 30, 32
penile erections, 29, 32
PGO spikes, 29
promoting and inhibiting neurons, 30, 31
sleep/arousal network, 28–29
sleep-promoting systems (*see* Sleep-promoting systems)
wake-promoting systems (*see* Wake-promoting systems)
Next-Day Hangover Liability Index, 130
Nonclinical safety evaluation
addiction, 94
drug–drug interactions, 95
next-day effects, 94–95
physical dependence, 94
risk of falling, 94
tolerance, 93–94
Nonrapid eye movement (NREM)
homeostatic response, 10

Nonrapid eye movement (NREM) (*cont.*)
- parasomnias, 73, 79
- physiological regulation, 13
- sleep architecture, 5–6
- sleep efficiency index, 139
- total sleep time, 139
- wake-promoting neurons, 7

Noradrenaline (NA)
- α_2-adrenoceptor agonists, 53–54
- α_2-adrenoceptor antagonists, 53
- arousal regulation, 50
- autonomic regulation, 50
- cellular targets, 52–53
- monoamine depletors, 54
- motor regulation, 51–52
- μ opioid receptor agonists, 54
- NARI, 53
- noradrenergic wake-promoting system, 50
- photomodulation, 52

Noradrenaline reuptake inhibitors (NARI), 53

O

Obstructive sleep apnea (OSA) syndrome, 76
Orexin
- appetite/human metabolism, 204
- benefit–risk ratio, 204
- benzodiazepines, 203–204
- cataplexy, 203
- cerebral cortex, 45
- clinical study, 202
- DORAs
  - advantages, 199, 201–202
  - almorexant, 198–199
  - SB-649868, 199–200
  - suvorexant, 199–201
- GIRK, 193
- hypnagogic hallucinations, 203
- mesolimbic pathways, 203
- morning/daytime sleepiness, 203
- narcoleptic symptoms, 202
- orexin-deficiency state, 45–46
- orexinergic neurons, 45
- orthostasis, 202–203
- overexpression, 194
- pharmacological treatment, 192
- potential side effects, 194
- receptors, 193–194
- REM sleep behavior disorder, 203
- sedating antidepressants, 204
- sleep and wakefulness, 192–193
- sleep paralysis, 203
- sleep-promoting orexinergic inhibitors, 46
- SORA
  - OX1R, 195–196
  - OX2R, 196–197
- wake-promoting nuclei, 45
- wake-promoting orexinergic activators, 46

P

Paradoxical/active sleep. *See* Rapid eye movement (REM)
Parasomnias
- EHS, 73, 79–80
- NREM-related parasomnias, 73, 79
- sleep-related hallucinations, 73, 79–80

Paraventricular nucleus (PVN), 196–197
Parkinson's disease, 179–180
Periodic limb movement disorder (PLMD), 74, 80
Pharmaceutical products
- benefit/risk ratio, 280
- biochemical mechanisms, 280
- biological products, 281
- combination products, 281
- delivery system, 281–282
- histamine H_3 inverse agonists, 280
- innovative formulations, 281–282

PLMD. *See* Periodic limb movement disorder (PLMD)
Polysomnography method, 99
Ponto-geniculo-occipital (PGO) spikes, 29
Post-Sleep Questionnaire (PSQ), 139, 140
Posttraumatic stress disorder (PTSD), 177
Psychometric method, 99
Psychomotor vigilance test (PVT), 134
Pulmonary disease, 178–179

R

Randomized-withdrawal study of the efficacy and safety of Tasimelteon (RESET) study, 265–266
Rapid eye movement (REM)
- sleep behavior disorder, 277
- sleep network
  - cortical activation, 30
  - dreaming, 29, 32
  - horizontal saccadic eye movements, 29
  - impairment, 30
  - muscle atonia, 29, 30, 32
  - penile erections, 29, 32
  - PGO spikes, 29
  - promoting and inhibiting neurons, 30, 31

Restless legs syndrome (RLS), 12, 74, 80–81

S
Safety and Efficacy of Tasimelteon (SET) study, 265
Saporin (SAP), 250–251
SCN. *See* Suprachiasmatic nucleus (SCN)
Serotonin
  receptors, 59
  REMS-suppressing serotonergic activators, 60
  serotonergic wake-promoting system, 58–59
  serotonin receptor agonists, 59
  serotonin reuptake inhibitors, 59–60
  sleep-promoting serotonergic inhibitors, 60
Single orexin receptor antagonists (SORAs)
  OX1R, 195–196
  OX2R, 196–197
Sleep
  architecture, 5–6
  definition, 4–5
  evolution
    deprivation, 19–20
    long-existing theory, 21
    memory, 20
    phylogenesis, 18
    synaptic homeostasis theory, 21
  mechanisms
    dorsal pathway, 7
    GABAergic nonorexinergic/non-MCH neurons, 8
    glutamatergic projection, 8
    MCH/REM-on neurons, 8
    orexinergic neurons, 8
    REM-on and REM-off neurons, 7
    rostral cholinergic system, 8
    ventral pathway, 7
  physiological functions
    autonomic and cardiovascular functions, 16–17
    endocrine functions, 15–16
    hypothalamus, 13
    respiratory functions, 14–15
    sensory and motor functions, 13–14
    thermoregulation and metabolism, 17–18
  regulation
    chronobiology, 8–9
    circadian and homeostatic sleep control, 10–11
    CLOCK and BMAL1, 11, 12
    CRY and PER proteins, 11
    GWAS, 12
    homeostasis, 9–10
    Homer1 gene, 12
    microarray studies, 12–13
    monozygotic and dizygotic human twins, 12
    negative feedback loops, 12
    Per2 gene, 12
    restless leg syndrome, 12
    upregulates transcription, 12–13
Sleep centers, 277–278
Sleep Efficiency Index (SEI), 137, 139
Sleep-promoting systems
  adenosine
    adenosine-mediated sleep-promoting system, 41
    receptors, 42
    sleep-promoting adenosine receptor agonists, 42
    wake-promoting adenosine receptor antagonists, 42
  GABA (*see* γ-aminobutyric acid (GABA))
  galanin
    galaninergic sleep-promoting system, 39–40
    sleep-promoting galanin receptor agonists, 40
    wake-promoting galanin receptor antagonists, 40
  melanin
    MCHergic sleep-promoting system, 29, 31, 39
    wake-promoting inhibitors, 39
  melatonin
    chronobiotic melatoninergic activators, 41
    melatoninergic sleep-promoting system, 40
    sleep-promoting melatoninergic activators, 40–41
    wake-promoting melatoninergic inhibitors, 41
Sleep-related breathing disorders
  central sleep apnea, 76
  cheyne–stokes breathing, 76
  classification, 72
  obstructive sleep apnea syndrome, 76
  treatment, 76–77
Sleep-related movement disorders (SRMD), 72, 76–77
Sleep-related rhythmic disorder (RMD), 74, 80–81
Sodium oxybate (SXB)
  adverse events, 237–238
  deaths, 243–244
  depressant effects, 244
  efficacy
    cataplexy, 238–240
    CGI scores, 241

Sodium oxybate (SXB) (*cont.*)
characteristics, 238–239
EDS, 240–241
hypnagogic hallucinations, 241–242
sleep architecture, 241
sleep paralysis, 241–242
high sodium load, 242
history, 234
human leukocyte antigen, 232
hypocretin/orexin neurons, 232–233
idiopathic hypersomnia, 77
indications, 235
mechanism of action, 235–236
pharmacology, 233–234
precautions, 236
psychiatric complications, 244
respiratory depression, 242–243
sleep-disordered breathing, 242–243
weight loss, 242
SORAs. *See* Single orexin receptor antagonists (SORAs)
SRMD. *See* Sleep-related movement disorders (SRMD)
Subparaventricular zone (SPZ), 32
Suprachiasmatic nucleus (SCN)
DMH, 32
GABAergic output neurons, 33
melatonin/light stimulation, 33–34
SPZ, 32

T
Tasimelteon
advantages, 267
chemical designation, 263–264
guidelines, 267
non-24 h sleep–wake disorder, 263
pharmacodynamics, 264
pharmacokinetics, 264–265
regulatory issues, 266
RESET study, 265–266
safety, 266
SET study, 265
VEC-162, 266
Tegmentopontine reticular nucleus (TPR), 30
Time to maximum concentration (Tmax), 149–150
Total Sleep Time (TST), 137, 139
Transient insomnia, 161, 164
Tuberomamillary nucleus (TMN), 196–197
Type 1 orexin receptors (OX1R), 195–196
Type 2 orexin receptors (OX2R), 196–197

V
Vanda performed clinical trials with tasimelteon (VEC-162), 266
Visual analogue scale (VAS), 134

W
Wakefulness after sleep onset (WASO), 136, 138, 164–165
Wake-promoting systems
acetylcholine
anticholinergic drugs, 49
cholinergic wake-promoting system, 48
cholinesterase inhibitors, 48–49
milameline, 49
nicotine, 49
receptors, 48
sleep-promoting cholinergic inhibitors, 49
dopamine (*see* Dopamine)
glutamate
glutamatergic wake-promoting system, 46–47
receptors, 47
REMS-suppressing glutamatergic inhibitors, 48
sleep-promoting glutamatergic inhibitors, 47
wake-promoting glutamatergic activators, 47
histamine
histaminergic wake-promoting system, 43
receptors, 43
sleep-promoting histaminergic inhibitors, 44–45
wake-promoting histaminergic activators, 44
noradrenaline
α_2-adrenoceptor agonists, 53–54
α_2-adrenoceptor antagonists, 53
arousal regulation, 50
autonomic regulation, 50
cellular targets, 52–53
monoamine depletors, 54
motor regulation, 51–52
μ opioid receptor agonists, 54
NARI, 53
noradrenergic wake-promoting system, 50
photomodulation, 52
orexin
cerebral cortex, 45
orexin-deficiency state, 45–46

orexinergic neurons, 45
sleep-promoting orexinergic inhibitors, 46
wake-promoting nuclei, 45
wake-promoting orexinergic activators, 46
serotonin
receptors, 59
REMS-suppressing serotonergic activators, 60
serotonergic wake-promoting system, 58–59
serotonin receptor agonists, 59
serotonin reuptake inhibitors, 59–60
sleep-promoting serotonergic inhibitors, 60
Willis–Ekbom disease, 74, 80

X
Xyrem®. *See* Sodium oxybate (SXB)

Z
Zolpidem
chronic primary insomnia, 166
vs. lorediplon
advantages, 141, 142
cortical and muscle electrodes, 130
DST, 134
next-day residual drug effect, 139
number of awakenings (W), 137, 139
parameters, 131–133
PSQ, 139, 140
PVT, 134
SEI, 137, 139
sleep onset, 135–137
TST, 137, 139
VAS, 134
WASO, 136, 138
orexinergic system, 278–279
sublingual formulations
CYP3A4 inhibitors, 150
efficacy/balance profile, 149
low dose, 154
mechanism of action, 149
oral bioavailability, 150
pharmacokinetic parameter, 150–151
safety, 154–155
standard dose, 151–153
Tmax, 149–150